Kirit Shah

ENGLISH FOR SCIENCE AND TECHNOLOGY
A Handbook for Nonnative Speakers

THOMAS N. HUCKIN
The University of Michigan

LESLIE A. OLSEN
The University of Michigan

McGRAW-HILL BOOK COMPANY

New York   St. Louis   San Francisco   Auckland   Bogotá   Hamburg
London   Madrid   Mexico   Montreal   New Delhi
Panama   Paris   São Paulo   Singapore   Sydney   Tokyo   Toronto

**ENGLISH
FOR SCIENCE
AND TECHNOLOGY
A Handbook for
Nonnative Speakers**

5 6 7 8 9 0   DODO   9 8

ISBN 0-07-030821-7

This book was set in Souvenir by Waldman Graphics, Inc.
The editors were Phillip A. Butcher, Jim Dodd and Claudia Tantillo;
the designer was Nicholas Krenitsky;
the production supervisor was Diane Renda.
The drawings were done by J & R Services, Inc.

Library of Congress Cataloging in Publication Data

Huckin, Thomas N.
    English for science and technology.

    "Includes chapters 1–15 and 23 of Principles of communication for science and technology by Leslie A. Olsen and Thomas N. Huckin"—T.p. verso.
    Includes index.
    1. Technical writing. 2. English language—Textbooks for foreigners. I. Olsen, Leslie A. II. Title.
T11.H823 1983   808'.0665021   82-20900
ISBN 0-07-030821-7

FOR CHRISTIANE AND OUR PARENTS

Preface   xv

PART I   INTRODUCTION

1   Why Study Technical Communication?   3

PART II   REVIEW OF BASIC WRITING PRINCIPLES

2   Writing Paragraphs   17

3   Parallelism   35

PART III   ELEMENTS IN PREPARING TO COMMUNICATE

4   Audiences   47

5   Basic Types and Patterns of Argument   59

6   Orienting the Nonspecialist:
    The Foreword or Introduction and Summary   87

PART IV   VISUAL ELEMENTS

7   Visual Elements   119

PART V   MAJOR GENRES OF TECHNICAL REPORTING

8   The Oral Presentation   167

9   Technical Reports   179

10   The Proposal   213

11   The Business Letter   237

12   The Technical Article   275

PART VI   READABILITY

13   Making Your Writing Readable:
     Information Selection   289

14   Making Your Writing Readable:
     Information Ordering   303

15   Making Your Writing Readable:
     Editing for Emphasis   321

16 Connectives   341

17 Countability and the Indefinite Article   353

18 The Definite Article   367

19 Relative Clauses   389

20 Cohesion   407

21 Modal Verbs   429

22 Verbs   441

23 Proofreading   453

24 Vocabulary Building   459

APPENDIXES

A Punctuation   485

B Informal Conversational Expressions   495

C Pronunciation   513

   Index   537

CONTENTS

Preface   xv

PART I INTRODUCTION

1 WHY STUDY TECHNICAL COMMUNICATION?   3
   1.1 The Importance of Technical Communication   3
   1.2 The Frequency of Technical Communication   5
   1.3 The Inadequate Communication Skills of
       Many Technical Professionals   7
   1.4 Skills Needed in Technical Communication   9

PART II REVIEW OF BASIC WRITING PRINCIPLES

2 WRITING PARAGRAPHS   17
   2.1 Write a Good Topic Statement   17
   2.2 Develop a Clear Pattern of Organization   23
       Chronological description   23
       Cause-and-effect analysis   24
       Comparison and contrast   26
       Listing   27
       General-to-particular ordering of details   28
       Other patterns   28

3 PARALLELISM   35
   3.1 Types of Lists   36
   3.2 Misleading Parallelism   40
   3.3 Parallelism in Paragraphs and Larger Units   41

PART III ELEMENTS IN PREPARING TO COMMUNICATE

4 AUDIENCES   47
   4.1 Characteristics of Real-World Audiences   50
   4.2 The Nature of Managerial Audiences   51
   4.3 A Procedure for Audience Analysis   52
       Identify the communication's uses and routes   53
       Identify all possible audiences   54
       Identify the concerns, goals, values, and
         needs of the audience   54
       Make communication appropriate for
         managers   55
       Identify the audience's preferences for and
         objections to the arguments   55

5 BASIC TYPES AND PATTERNS OF ARGUMENT 59
  5.1 Expectations about Claims and Proof 59
  5.2 Three Basic Strategies of Argument 64
  5.3 Basic Types of Argument 67
    The argument of fact 68
    The argument of policy 69
    *The relationship between arguments of fact
      and arguments of policy* 71
  5.4 An Effective General Pattern for Arguments
    Stating Criteria 75

6 ORIENTING THE NONSPECIALIST:
  The Foreword or Introduction and Summary 87
  6.1 A General Form for Stating Problems:
    Forewords and Introductions 87
    Step 1: Introducing a problem 88
    *Steps 2 and 3: Identifying your strategy on
      the problem and the purpose of your
      communication* 94
  6.2 A Short Form for Stating Problems 98
  6.3 Guidelines for Choosing between Full-Form
    and Short-Form Problem Statements 101
  6.4 Introductions and Forewords: A Review 102
    The foreword of a technical report 102
    The introduction of a letter 103
    *The introduction of a scientific or technical
      article* 103
  6.5 The Summary 104
    Framing summaries for particular audiences 106

PART IV VISUAL ELEMENTS

7 VISUAL ELEMENTS 119
  7.1 Making a Visual Aid Truly Visual 120
  7.2 Deciding When to Use a Visual Aid 123
    The visual aid for describing or clarifying 123
    The visual aid for highlighting important points 125
    *The visual aid for conventional or easy
      presentation of data* 125
  7.3 Selecting the Best Type of Visual Aid in a
    Given Situation 128
    Conventions of visual perception 129
    Some types of visual aids and their uses 130

7.4 Designing the Visual Aid 137
Making a visual aid relevant 137
Making a visual aid clear 142
Making a visual aid truthful 146
7.5 Integrating the Visual Aid into the Text 148
Making the visual aid independent 149
7.6 Formatting Conventions That Make
Reading Easier 155
7.7 Formatting Conventions That Make
Writing Clearer 159

PART V   MAJOR GENRES OF TECHNICAL REPORTING

8 THE ORAL PRESENTATION 167
8.1 Giving a Formal Oral Presentation 168
Preparation 168
Practice 173
Delivery 175
8.2 Giving an Informal Oral Presentation 176

9 TECHNICAL REPORTS 179
9.1 Foreword and Summary: Organizing Main
Points for Nonspecialist Readers 179
9.2 Claims Before Proof: Organizing Details for
Specialist Readers 180
The abstract 180
Structuring proof and technical discussions 184
9.3 Technical Reports 186
The short informal report 186
The long informal report 189
The formal report 198
9.4 An Example of Adapting Form to Purpose:
The Feasibility Report 199
9.5 A Concluding Note 206

10 THE PROPOSAL 213
10.1 The Organization of a Formal Proposal 214
Title page 216
Abstract 216
Introduction 216
Background 218
Description of proposed research 218
Description of relevant institutional resources 219
List of references 219

		Personnel	219
		Budget	220
		Appendixes	220
	10.2	The Organization of a Short Informal Proposal	220
	10.3	Editing the Proposal	225
	10.4	Getting the Proposal Approved for Submission	226
11		THE BUSINESS LETTER	237
	11.1	Basic Letter Formats	240
	11.2	Letter of Transmittal	247
	11.3	Letter of Complaint	249
	11.4	Response to a Letter of Complaint	252
	11.5	Letter of Request	252
	11.6	Response to a Letter of Request	254
	11.7	Letter of Application for Employment or Admission and Resume	257
		What makes a good applicant?	257
		Designing a good letter of application	259
		Designing a good resume	264
12		THE JOURNAL ARTICLE	275
	12.1	Introduction	276
	12.2	Materials and Method	278
	12.3	Results	279
	12.4	Discussion	280

PART VI READABILITY

13		MAKING YOUR WRITING READABLE: Information Selection	289
	13.1	Establish Your Topic and Purpose	292
	13.2	Use Key Words Prominently	293
	13.3	Explain Important Concepts When Writing for Nonspecialist Readers	293
	13.4	Use Standard Terminology When Writing for Specialist Readers	297
	13.5	Structure Your Text to Emphasize Important Information	297
	13.6	Construct Well-Designed Paragraphs	298
	13.7	Field Test Your Writing	299
14		MAKING YOUR WRITING READABLE: Information Ordering	303
	14.1	Optimal Ordering of Noun Phrases	304

Put given information before new information 304
Put topical information in subject position 305
Put "light" NPs before "heavy" NPs 306
14.2 Ways of Satisfying These Criteria 309
Passive-active alternation 310
Equative shift 311
Indirect object shift 311
14.3 A Procedure for Producing More Readable
Sentences and Paragraphs 312

15 MAKING YOUR WRITING READABLE:
Editing for Emphasis 321
15.1 Combine Closely Related Sentences 322
15.2 Use Intensifiers and Signal Words 329
15.3 Be Concise 333

16 CONNECTIVES 341
16.1 Conjunctive Adverbs 342
16.2 Subordinating Conjunctions 346
16.3 Common Mistakes to Avoid 347

17 COUNTABILITY AND THE INDEFINITE ARTICLE 353
17.1 Uncountable Nouns 353
17.2 Countable Nouns 354
17.3 Counters 359
17.4 "Two-Way" Nouns 361
17.5 Informal Usages 363

18 THE DEFINITE ARTICLE 367
18.1 Special Adjectives 367
18.2 Special Nouns 368
18.3 Generics 372
18.4 Previous Mention 372
18.5 Modifiers Following the Noun 374
18.6 Shared Knowledge 377
18.7 Implied Uniqueness 378
18.8 A Flow Chart for Articles 378
18.9 Post-tests on Articles 382

19 RELATIVE CLAUSES 389
19.1 Grammatical Rules 390
Positioning the relative clause 390
Heading the relative clause
with a relative pronoun 391
Avoiding a duplicate pronoun 393
19.2 Stylistic Rules 394
Substituting that *for* which 395

Omitting the relative pronoun
and auxiliary verb   *396*
Using the -ing form   *399*
Omitting the relative pronoun   *402*

20 COHESION   407
20.1 Full-Form Noun Phrase Repetition   408
20.2 Short-Form Repetition   409
20.3 Short-Form Repetition Using Relative Clauses   410
20.4 Repetition Using Noun Compounds   411
Constructing noun compounds   *412*
Interpreting noun compounds   *413*
Using noun compounds to promote cohesion   *416*
Noun compounds used
as short-form references   *416*
Noun compounds used as names   *416*
Noun compounds used to ease shifting of
noun phrases   *417*
20.5 Repetition Using Pronouns, Articles,
and Demonstratives   418
It vs. this   *419*
As follows   *420*
20.6 Repetition Using Synonymous Terms   422
20.7 Repetition Using Associated Noun Phrases   424

21 MODAL VERBS   429
21.1 Obligation   429
21.2 Probability   433
Making hypothetical statements   *435*
21.3 Ability   437

22 VERBS   441
22.1 Major Tense Distinctions in Scientific and
Technical English   441
The simple present tense   *441*
The simple past tense   *442*
The present perfect tense   *443*
22.2 The Use of the Progressive (-ing) Form
of the Verb   445
22.3 Subject-Verb Number Agreement   447
22.4 Grammatical Irregularities of Particular Verbs   448
Verbs that do not occur in the passive form   *448*
Verbs taking -ing or unmarked complements   *449*
Verbs with irregular forms of inflection   *449*

23 PROOFREADING   453

24 VOCABULARY BUILDING   459
    24.1 Using Contextual Definitions   459
    24.2 Using Contextual Clues   461
          Grammatical function of the word   462
          Meaningful parts of the word   465
          Other words nearby   470
          Other words in parallel   470
          Rhetorical patterns and visual aids   472
    24.3 Consulting an Authority   476
          When to consult an authority   476
          Whom to consult   476
          Dictionaries   477
    24.4 Getting to Know a Word Well   478
          Usage   478
          Related forms   478
          Pronunciation   479

APPENDIXES

A   PUNCTUATION   485
    A.1   Sentence Punctuation   485
          Showing a close relationship between two
             independent clauses   485
          Separating introductory subordinate elements   487
          Indicating parenthetical information   487
          List   488
    A.2   Word and Phrase Punctuation   490
          Possession   490
          Noun compounds   491
          End-of-line hyphenation   491
          Titles of written works   492
    A.3   Capitalization   493

B   INFORMAL CONVERSATIONAL EXPRESSIONS   495

C   PRONUNCIATION   513
    C.1   On Using This Appendix   513
          Diagnosis   513
          Checklist for listener of reading passage   514
          Practice   515
    C.2   Sound Segments   515
          Vowels   515

	Consonants	518
	Consonant clusters	522
C.3	Word Stress	525
C.4	Noun Compounds	529
C.5	Sentence Intonation	530
	Signaling important words	530
	Signaling the sentence type	531
	Signaling the items of a list	532
	Index	537

English for Science and Technology: A Handbook for Nonnative Speakers is a technical writing textbook for engineering and science students who have a fairly good command of English but are not native speakers. It can also be used as a reference book by professionals working in industry.

The book is divided essentially into two parts. The first half (Chapters 1 through 15) covers the basic principles of technical communication and can be used by native speakers as well as nonnative speakers. The remainder of the book, however, is designed exclusively for nonnative speakers; it focuses on the special problems of grammar, style, vocabulary, and pronunciation that nonnative speakers are known to have when using English.

The most important feature of *English for Science and Technology* is its functional-rhetorical approach, which emphasizes the communicative use of language rather than simply its formal aspects. The book does treat formal aspects where appropriate (for example, in discussing report formats, grammar, punctuation, and visual elements), but it places more emphasis on the psychological and rhetorical principles underlying effective communication. As such, it focuses on language in meaningful contexts, not on sentences or words in isolation.

A second important feature of the book is its attempt to provide complete, explicit explanations of the main teaching points. We feel that science and engineering students prefer systematic, step-by-step instruction and that they like to have reasoned explanations for how things function, including how language functions in communication.

This point gives rise to yet another feature of the book, namely, its focus on the *process* by which a student may produce an effective piece of communication for a given audience. Too often, textbooks treat writing merely as a finished product. They present models of good writing for the student to imitate, but they do not provide adequate guidance by which the student can learn to cope with novel situations on his or her own. Thus, we have provided step-by-step procedures that students may follow as guides to the writing process, including a number of flow charts.

Finally, *English for Science and Technology* is designed to be used not only as a course text but as a long-term reference work as well. It thus should be especially useful for (1) university-level students in technical areas who are using English in their studies and who plan to continue into technical careers and (2) practicing scientists and engineers who need a self-instructional reference book in written and oral English for technical communication. To increase the book's usefulness, we have included many exercises; lists of supplementary readings; reference appendices on punctuation, common idiomatic expressions, and pronunciation; and an index. A teacher's manual is available not only for teachers but also for those using the book for self-instruction.

(Note to teachers of both native-speaker and nonnative-speaker students: a companion book, *Principles of Communication for Science and Technology,* designed specifically for native speakers but covering much of the same material, is also available.)

ACKNOWLEDGMENTS

The preparation of this book has been a long but exciting project and one which would not have been possible without the help and encouragement of our families, friends, and colleagues, many professional acquaintances, and many students. We would like to thank all of those who have contributed in one way or another to the completion of this book, but in particular we would like to thank the following people for their special contributions: Christiane Huckin for her sound judgment on many of the most difficult pedagogical and linguistic issues we faced, as well as for other invaluable behind-the-scenes assistance; our parents, grandparents, and siblings for their unflagging moral support; our teachers and later colleagues Dwight W. Stevenson, Richard E. Young, Louis Trimble, and J. C. Mathes for our early training in the field and their later influence and inspiration; our colleagues in the Technical Communication program in the University of Michigan's Department of Humanities—Thomas M. Sawyer, Lisa Barton, Peter Klaver, and James Zappen—and our other Humanities Department colleagues for their ready counsel and support; and the Department of Humanities office staff—Vicki Gordenier, Barbara Cox, and Margaret Cousins—for cheerfully putting up with us even in times of stress.

In addition, we would like to thank many others who have provided special research funds, research materials, intellectual stimulation, and other forms of assistance: The University of Michigan—particularly, the College of Engineering, the Horace H. Rackham School of Graduate Studies, the Center for Research on Learning and Teaching, and the Division of Research Development and Administration—the National Endowment for the Humanities, Hansford W. Farris, Blanchard Hiatt, Julie Wei, W. Ross Winterowd, Francis Christensen, Frederick J. Newmeyer, John M. Swales, Edward P. J. Corbett, Heles Contreras, Virginia J. Tufte, M. A. K. Halliday, Walter Kintsch, Richard C. Anderson, Chaim Perelman, Paul Anderson, Carolyn Miller, Yoshiaki Shinoda, Larry Selinker, Richard Enos, David Kaufer, Joan Stigliani, John Lepp, Linda Flower, David Kieras, Christine Neuwirth, Fay Edwards, Robert Caddell, Hal Schulte, Harold Hargar, David V. Ragone, Alfred Sussman, Patricia Sullivan, Burte Waite, and Nancy Elder.

We would also like to thank Karl Drobnic, Quentin Johnson, Louis Holschuh, John Lackstrom, Carol Romett, Christopher Gould, William Woods, Douglas Wixson, Ruth Falor, and O. Allen Gianniny, our reviewers, for conscientiously and expertly examining various drafts of our manuscript and giving us numerous helpful suggestions. We wish to thank Phillip Butcher,

Claudia Tantillo, Jo Satloff, Irene Nunes, Nicholas Krenitsky, and Diane Renda of the McGraw-Hill staff for their help in producing the book.

Finally, we would like to acknowledge the contribution of many students in our technical communication classes at the University of Michigan. They have been patient and helpful experimental subjects and sources of examples, suggestions, and new ideas. (They would also, no doubt, prefer to remain anonymous.)

Thomas N. Huckin
Leslie A. Olsen

Part I

INTRODUCTION

1

WHY STUDY TECHNICAL COMMUNICATION?

Why should engineers and scientists study technical writing or speaking? Their primary training and interests lie in *technical* areas; most science and engineering students successfully pursue their *technical* subjects without extra writing courses; and practicing engineers and scientists in government and industry work on *technical* projects. It might seem, then, that writing and communication are superfluous to a technical education.

In fact, this is not the case. Scientists and engineers may be technically brilliant, and creative, but unless they can convince coworkers and supervisors of their worth, the technical skills will be unnoticed, unappreciated, and unused. In a word, if technical people cannot communicate to others what they are doing and why it is important, it is they and their excellent *technical* skills that will be superfluous. From this perspective, communications skills are not just handy; they are critical tools for success, even survival, in "real-world" environments.

1.1

THE IMPORTANCE OF TECHNICAL COMMUNICATION

The importance of technical communication has been noted by many technical people, including Russel R. Raney, one-time Chief Engineer, Farm Implement Division, at International Harvester Company. He points out that most industries are composed of internal units that include (at least) a marketing unit, a manufacturing unit, an engineering or design unit, and a management unit. These units, or internal organizations,

> are quite similar to factions in a political party in that, while they all strive toward the same end, they do not necessarily agree among themselves as to the means to be employed in achieving that end. So it is to be expected that the normal operation of the business will be marked by a certain amount of internal conflict as each organization seeks to promote its own particular interest.[1]

This natural conflict, the many details and changes to be managed, and the incredible complexity of a large organization all combine to create a demand for communication skills unforeseen by most science and engineering students:

> In school you can attain your goal of a degree on your own individual efforts with practically no dependence on anyone else. On the other hand, it should

TABLE 1-1 SUBJECTS MOST NEEDED FOR ENGINEERING CAREERS IN INDUSTRY

RANK	SUBJECT
1.	Management practices
2.	Technical writing
3.	Probability and statistics
4.	Public speaking
5.	Creative thinking
6.	Working with individuals
7.	Working with groups
8.	Speed reading
9.	Talking with people
10.	Business practices (marketing, finance, economics)
11.	Survey of computer uses
12.	Heat transfer
13.	Instrumentation and measurements
14.	Data processing
15.	Systems programming
16.	Economics
17.	Ordinary differential equations
18.	Logic
19.	Economic analysis
20.	Applications programming
21.	Psychology
22.	Reliability
23.	Vector analysis
24.	Electronic systems engineering (circuit design)
25.	Laplace transforms
26.	Solid-state physics
27.	Electromechanical energy transformation
28.	Matrix algebra
29.	Computer systems engineering
30.	Operations research
31.	Law practices (patents, contracts)
32.	Information and control systems
33.	Numerical analysis
34.	Physics of fluids
35.	Thermodynamics
36.	Electromagnetics
37.	Human engineering
38.	Materials engineering

SOURCE: Middendorf, Reference 3. © 1980, American Society for Engineering Education.

be evident from the fore-going description of industrial operation that you will be able to accomplish very little by yourself in an organization. I want to point out two significant results of this situation.

First, it means that for the results of your efforts to be incorporated in any end result you must present them effectively in competition with many conflicting considerations. It will, therefore, inevitably be a part of your responsibility to transmit understandably and persuasively the results of your work to other people, usually both verbally and in writing. Since success in engineering school does not depend much on communication abilities, you may not have developed your skills and may mistakenly resent the absolute dependence you will have on them for accomplishment in an organization.[2]

The importance of communication skills is emphasized in survey after survey. For instance, the American Society for Engineering Education conducted a survey[3] to determine which academic subjects are most needed for engineering careers in industry. Responses were received from 4057 engineers, all having at least several years of on-the-job experience behind them. The results, given in Table 1-1, show that communication skills rank above any other type of skill, capturing five of the nine "most-needed" categories. These include technical writing (2), public speaking (4), working with individuals (6), working with groups (7), and talking with people (9). In contrast, technical skills rank toward the bottom of the list. (It must be noted, however, that this ranking was produced by engineers from a variety of disciplines and that not all of the engineers may have needed all of the engineering subjects in the bottom half of the list. Thus, some of these categories might have received somewhat higher ratings from a more homogeneous group of respondents.)

Another survey of 245 Engineers of Distinction[4] produced an even stronger emphasis on communication skills. For both the distinguished engineers and younger engineers hoping for promotion, writing skills were cited as critical or very important to success, as illustrated in Table 1-2.

1.2

THE FREQUENCY OF TECHNICAL COMMUNICATION

The actual amount of time spent on communication tasks varies from job to job, of course, but an average distribution of communication tasks seems to resemble that presented in Table 1-3 by an industrial consultant, Ed Gilbert. He estimates that a junior engineer spends about 20 percent of the time writing and 20 percent talking, for a total of 40 percent. This escalates to approximately 77 percent for a chief executive, mostly in the verbal medium.[5]

These figures on communication time are supported by a recent survey by the University of Wisconsin's College of Engineering, which discovered that

> 75 percent of engineering undergraduates take jobs in industry, where at least 25 percent of an engineer's time is spent in the reporting process. As the engineer moves up the managerial ladder, this time can increase to as much as 80 percent.[6]

Obviously, if an employee can make such time-consuming activities more efficient, both the employee and the company will be well served. One

TABLE 1-2 RESPONSES OF PROMINENT ENGINEERS TO QUESTIONS ON THE IMPORTANCE OF TECHNICAL WRITING

How important is the writing that you do, and is the ability to write effectively needed in your present position?

Critically important	110
Very important	124
Somewhat important	9
Minimally important	0
No response	2

When you select or approve someone for advancement, you must, of course, consider many factors. If an individual is otherwise qualified, can inability to write effectively delay or prevent advancement?

Ability to write is usually critical—has strong effect on selection	63
Ability to write is usually important—often affects selection	153
Ability to write is sometimes helpful—may have some effect on selection	25
Ability to write is not usually important—has little or no effect on selection	1
No response	3

SOURCE: Davis, Reference 4. © 1977, American Society for Engineering Education.

TABLE 1-3 TYPICAL COMMUNICATION TIME WITHIN ORGANIZATIONS

HIERARCHY	PORTION OF TOTAL WORK TIME SPENT ON COMMUNICATION, %		
	VERBAL MEDIA	INFORMAL WRITING	FORMAL WRITING
Chief executive	65	10	2
Division manager	60	15	2
Department manager	45	20	5
Section head	40	20	5
Project head	30	20	10
Engineer or scientist	25	15	10
Junior engineer	20	10	10
Technician	15	10	2

SOURCE: Gilbert, Reference 5.

easy way to improve efficiency is to develop effective strategies for analyzing communication tasks and audiences as well as strategies for producing reports and oral presentations. This text, and any technical communication course based on it, should help writers develop such strategies.

1.3

THE INADEQUATE COMMUNICATION SKILLS OF MANY TECHNICAL PROFESSIONALS

Another survey (of civil, electrical, and mechanical engineers) notes the extreme importance of communication skills, but in addition stresses the inadequate preparation of most engineers for communication tasks. For instance, the results for civil engineering presented in Table 1-4 rank writing and speaking as the most important area of competence in civil engineering practice, but rank about two thirds of recent graduates inferior in this area. Similarly, the results for electrical and mechanical engineering presented in Tables 1-5 and 1-6 rank writing and speaking as the first and second most important areas, respectively, in engineering practice, but rank about half of the recent

TABLE 1-4 CAPABILITIES OF RECENT CIVIL ENGINEERING GRADUATES

IMPORTANCE TO CIVIL ENGINEERING PRACTICE				CAPABILITY OF RECENT C.E. GRADUATES (1–5 YEARS)		
MOST IMP.	IMPOR- TANT	LESS IMP.	AREA OF COMPETENCE	SUPER- IOR	ADE- QUATE	INFER- IOR
137	86	9	Writing and speaking	7	69	142
106	127	9	Structural analysis and design	25	160	28
72	156	13	Soil mechanics and foundation	9	157	46
62	125	47	Water and wastewater treatment	13	144	27
55	131	38	Fluid mechanics, hydraulics, and hydrology	10	136	42
53	153	32	Computer and numerical methods	42	139	30
50	132	49	Economics and finance	4	96	106
47	147	54	Construction methods and equipment	9	107	95
46	105	70	Law, labor, and management	4	81	120
20	148	67	Surveying and measurement	4	139	65
19	110	92	Transportation, highways, and traffic	5	161	30
14	124	51	Materials	6	153	35
13	85	132	Social sciences and humanities	12	137	53

SOURCE: Kimel and Monsees, Reference 7. © 1979, American Society for Engineering Education.

TABLE 1-5 CAPABILITIES OF MECHANICAL ENGINEERING GRADUATES

IMPORTANCE TO MECHANICAL ENGINEERING PRACTICE				CAPABILITY OF RECENT M.E. GRADUATES (1–5 YEARS)		
MOST IMP.	IMPOR-TANT	LESS IMP.	AREA OF COMPETENCE	SUPER-IOR	ADE-QUATE	INFER-IOR
27	33	2	Engineering materials	3	37	10
26	27	4	Writing and speaking	1	24	34
23	31	4	Thermodynamics and heat transfer	10	40	1
22	28	6	Fluid mechanics	4	41	6
18	16	13	Dynamics and vibration controls	9	31	6
16	30	11	Machine design	7	31	10
14	24	19	Manufacturing methods	1	24	23
14	20	15	Individual project experience	1	22	24
13	36	10	Engineering graphics	2	31	16
13	33	13	Computer and numerical methods	13	35	1
11	35	11	Measurements and instrumentation	8	24	14
6	28	13	Economics, law	0	28	19
4	20	32	Social science and humanities	3	36	4

SOURCE: Kimel and Monsees, Reference 7. © 1979, American Society for Engineering Education.

TABLE 1-6 CAPABILITIES OF RECENT ELECTRICAL ENGINEERING GRADUATES

IMPORTANCE TO ELECTRICAL ENGINEERING PRACTICE				CAPABILITY OF RECENT E.E. GRADUATES (1–5 YEARS)		
MOST IMP.	IMPOR-TANT	LESS IMP.	AREA OF COMPETENCE	SUPER-IOR	ADE-QUATE	INFER-IOR
86	66	1	Writing and speaking	4	59	62
56	68	22	Power engineering and energy systems	7	68	38
53	71	22	Digital and hybrid computer systems	34	77	20
44	72	26	Discrete and integrated electronics	18	79	13
42	90	10	Solid-state devices	22	88	14
33	77	41	Automatic control	11	86	16
37	81	24	Communication and information systems	19	81	22
35	78	26	Network theory	15	93	15
22	100	24	Economics and finance	3	73	52
17	59	55	Advanced automation	7	90	19
13	52	78	Antennas and wave propagation	7	75	19
11	45	80	Artificial intelligence	4	72	21
9	58	88	Social sciences and humanities	6	100	22

SOURCE: Kimel and Monsees, Reference 7. © 1979, American Society for Engineering Education.

graduates inferior in communication skills. This lack of communication skills has also been noted by all the surveys cited above and is pungently summarized by the Industrial Advisory Board at the University of Tulsa:

> Engineers cannot communicate—cannot spell, cannot make a sketch, have difficulty in all phases of communication with others. The student of today needs more ability than ever and a key need is to increase the ability to communicate both in speech and graphics.[8]

1.4

SKILLS NEEDED IN TECHNICAL COMMUNICATION

Given the situation just described—that communication tasks are important to success in the technical professions, that communication tasks consitute a relatively large part of a job, and that many (if not most) graduating scientists and engineers have inadequate communication skills—what kinds of things must such people learn to become effective and efficient communicators? Most respondents to that question from industrial or organizational settings stress that communicators should understand the situation, uses, and audiences for a given communication and write for those audiences and uses; that they should have clear organization and logical reasoning; that they should write clear and concise sentences; and that they should follow the standard conventions for grammar, punctuation, and other mechanics. In other words, writers should produce sentences that readers can easily understand, and they should place those sentences in contexts—paragraphs and larger units—whose function is clear.

Two surveys provide support for these generalizations as applied to technical writing. An early survey reported by MacIntosh[9] itemizes 18 typical problems noted by senior managers in science and industry (Table 1-7). A more recent survey of distinguished engineers by Davis[10] reports the same concerns, but in more general terms (Table 1-8).

A third survey ranks the types of communication skills needed on the job.[11] The results of this survey, presented in Table 1-9, suggest that people in technical fields should study significant amounts of both oral and written work and learn to communicate in a variety of forms, especially shorter forms.

The organization of this text reflects the surveys just discussed. It moves from a review of basic principles of technical communication to a consideration of a report's audiences and uses to a more detailed consideration of structure and logic to a discussion of sentence construction. It treats both written and oral communication and includes a unit on formatting and visual aids. It includes appendices on punctuation and mechanics and a number of sample reports illustrating various types of communication.

TABLE 1-7 COMPLAINTS ABOUT TECHNICAL WRITING FROM 182 SENIOR OFFICIALS IN SCIENCE AND INDUSTRY

COMPLAINT	NO. RESPONDING
Generally foggy language	182
Inadequate general vocabulary	173
Failure to connect information to point at issue	169
Wordiness	164
Lack of stressing important points	163
All sorts of illogical reasoning	163
Too much "engineering gobbledygook"	160
Poor overall organization	153
No clear overview	143
No clear continuity	142
Very little concept of writing for anyone but fellow specialists	136
Poor grammar	133
General lack of flexibility to suit circumstances	127
Deliberate obscurity	92
Poor punctuation	81
No sense of proper tone for circumstances	72
Poor adaptation of form to purpose	64
Almost meaningless introduction	54

SOURCE: MacIntosh, Reference 9.

TABLE 1-8 DISTINGUISHED ENGINEERS' CHOICE OF TOPICS TO BE COVERED IN A TECHNICAL WRITING COURSE

TOPIC*	DEGREE OF IMPORTANCE			
	ESSENTIAL	OK	NOT IMPORTANT	NO RESPONSE
Clarity of expression	239	4	0	2
Analyzing a situation and producing a communication to fit the reader's needs	209	31	0	5
Organizing reports and other communications	189	50	3	3
Grammar and syntax	165	75	0	5
Writing a draft and completing the finished document	151	79	7	8
Mechanics (punctuation, abbreviations, capitalization, use of numbers, etc.)	120	103	12	10
Style and tone of expression	117	115	7	6
Finding and using published information	90	127	22	6

*Brevity and conciseness were the most often mentioned additions to this list. NOTE: The question asked was, Some courses in technical writing are good and others are pretty poor. If a course in technical writing is included in the curriculum, what topics should be covered?
SOURCE: Davis, Reference 10. © 1977, American Society for Engineering Education.

TABLE 1-9 RANK OF JOB IMPORTANCE RESPONSES

RANK	COMMUNICATION SKILL	MEAN*	STD. DEV.
1	One-to-one talks with technically sophisticated personnel	5.9391	1.3109
2	Writing using graphs, charts, and/or other illustrative aids	5.3955	1.7141
3	Project proposals (written)	5.3324	1.8167
4	Participation in a small group or committee made up of only technically sophisticated members (oral)	5.2694	1.7733
5	Instructions for completing a technical process (written)	5.2444	1.8332
6	One-to-one talks with nontechnical personnel	5.0718	1.9065
7	Project progress reports (written)	5.0609	1.6301
8	Project proposal presentations (oral)	5.0587	1.9787
9	Writing technical information for technical audiences	4.9157	1.9396
10	Oral presentations using graphs, charts, and/or other aids	4.9139	1.9623
11	Technical description of a piece of hardware (written)	4.8687	1.8793
12	Memoranda	4.8169	1.7175
13	Short reports (less than 10 pages typewritten, double-spaced)	4.7967	1.8507
14	Writing technical information for nontechnical audiences	4.7458	2.1377
15	Project feasibility studies (written)	4.7124	1.9305
16	Project progress report presentations (oral)	4.6667	1.9277
17	Participation in a small group or committee including nontechnical members (oral)	4.6639	1.9608
18	Business letters	4.6257	1.8856
19	Telephone reports	4.5838	1.9682
20	Project feasibility study presentations (oral)	4.4258	2.0780
21	Form completion	4.2842	1.8847
22	Formal speeches to technically sophisticated audiences	4.0452	2.2108
23	Writing in collaboration with one or more colleagues	4.0313	2.0175
24	Laboratory reports (written)	4.0184	2.1800
25†	Long reports (10 or more pages typewritten, double-spaced)	3.8249	2.2120
26	Formal speeches to nontechnical audiences	3.5869	2.2020
27	Writing requiring library research	3.2062	2.0723
28	Abstracts/summaries of others' writing (written)	2.8864	1.9910
29	Articles submitted to professional journals	2.8074	2.1499
30	Reports submitted to professional societies (written)	2.7224	2.0827

*Rating scale: 1 = least important; 7 = most important.
†Skills 25 through 30 were ranked "below average to unimportant."
SOURCE: Schiff, Reference 11. © 1980, American Society for Engineering Education.

REFERENCES

1   Russel R. Raney, "Why Cultural Education for the Engineer?" Speech to the Iowa-Illinois Section of the American Society of Agricultural Engineers, April 29, 1953, quoted in John G. Young, "Employment Negotiations," *Placement Manual: Fall 1981,* College of Engineering, The University of Michigan, Ann Arbor, 1981, p. 49.

2   Ibid., p. 50.

3   William H. Middendorf, "Academic Programs and Industrial Needs," *Engineering Education,* May 1980, pp. 835–837.

4   Richard M. Davis, "Technical Writing: Who Needs It?" *Engineering Education,* November 1977, pp. 209–211.

5   Ed Gilbert, "Technical Communication and Writing: Management's Perspective," speech to the Conference on Teaching Technical and Professional Communication, co-chaired by Dwight W. Stevenson and J. C. Mathes, Department of Humanities, College of Engineering, The University of Michigan, 1975. Used with the permission of J. C. Mathes.

6   Nicholas D. Sylvester, "Engineering Education Must Improve the Communication Skills of Its Graduates," *Engineering Education,* April 1980, p. 739.

7   William R. Kimel and Melford E. Monsees, "Engineering Graduates: How Good Are They?" *Engineering Education,* November 1979, pp. 210–212.

8   Sylvester, loc. cit.

9   Fred H. MacIntosh, "How Good Is Our Product?" Speech delivered at the annual meeting of the Conference on College Composition and Communication, April 8, 1967, Louisville, Ky.

10   Davis, op. cit., p. 210.

11   Peter M. Schiff, "Speech: Another Facet of Technical Communication," *Engineering Education,* November 1980, p. 181.

ADDITIONAL READING

Committee on Scientific and Technical Communications of the NAS-NAE, *Scientific and Technical Communication: A Pressing National Problem and Recommendations for Its Solution,* National Academy of Sciences, Washington, D.C., 1969.

Richard M. Davis, "Technical Writing: Who Needs It?" *Engineering Education,* November 1977, pp. 209–211.

Exploration of Oral/Informal Technical Communications Behavior, American Institutes for Research, Washington, D.C., 1967.

Eugene Garfield, "Communication, Engineering and Engineers," *IEEE Transactions on Professional Communication* **15**(2):123–136 (1972).

H. K. Jenny, "Heavy Readers Are Heavy Hitters," *IEEE Spectrum* **15**(9):66–68 (1978).

Patrick Kelly, M. Kranzberg, S. R. Carpenter, and F. A. Rossini, *The Flow of Scientific and Technical Information in the Innovation Process: An Analytic Study,* Georgia Institute of Technology, Atlanta, 1977.

William Kimel and Melford E. Monsees, "Engineering Graduates: How Good Are They?" *Engineering Education,* November 1979, pp. 210–212.

William H. Middendorf, "Academic Programs and Industrial Needs," *Engineering Education,* November 1977, pp. 209–211.

E. C. Nelson and E. D. Pollack, *Communication Among Scientists and Engineers,* Heath Lexington, Lexington, Mass., 1970.

Peter M. Schiff, "Speech: Another Facet of Technical Communication," *Engineering Education,* November 1980, pp. 180–181.

Hedvah L. Shuchman, *Information Transfer in Engineering,* The Futures Group, Glastonbury, Connecticut, 1981.

Nicholas D. Sylvester, "Engineering Education Must Improve the Communication Skills of Its Graduates," *Engineering Education,* April 1980, pp. 739–740.

Part II

REVIEW OF BASIC WRITING PRINCIPLES

2

WRITING PARAGRAPHS

Unlike readers in the academic world, most readers in the "real world" read *selectively:* rather than thoroughly digesting a piece of writing, they skim-read most of it, skipping from one main idea to another until they come to something that particularly interests them. Such readers are forced to read this way. They are overwhelmed with reading material and simply don't have the time to read *everything* carefully.

As a writer, therefore, you should do everything possible to ease this kind of reader's burden; in particular, you should make your writing *easy to skim-read.* One of the best ways to do this is to write good paragraphs. What makes for a good paragraph in scientific/technical writing? First of all, a good paragraph has *unity:* it focuses on a single idea or theme. Secondly, a good paragraph has *coherence:* one sentence leads to the next in some kind of logical sequence. Finally, a good paragraph has *adequate content:* it has an appropriate selection and number of details to support the main idea of the paragraph. Readers expect to find these qualities in paragraphs, and you as writer should take care not to frustrate their expectations.

There are two principal tools you can use to invest your paragraphs with the qualities just described: (1) a good topic statement and (2) an appropriate pattern of organization.

2.1
WRITE A GOOD TOPIC STATEMENT

The topic of a paragraph is its main idea or theme, i.e., what the paragraph is about. As with a larger piece of writing, readers of a paragraph want to know right away what the topic is. They also like to have some idea of how this topic will be developed. In other words, readers will use whatever cues they can to quickly generate expectations about the paragraph as a whole. This strategy serves two purposes: (1) it allows readers to guess what's coming and thus digest it more easily and (2) it allows them to avoid reading the paragraph altogether if the subject matter holds no interest for them.

You can help your readers, therefore, by providing a good topic statement right at the beginning of the paragraph. It does not have to be confined to a single sentence: often a topic statement is extended over the first two sentences of a paragraph. It should, however, always contain one or more key words directly related to the topic, and it should be as complete a statement of the main idea as possible, without getting into too much detail and making the sentence(s) forbiddingly long. In addition, if possible, it should suggest how the topic will be developed (by comparison and contrast, by cause-and-effect analysis, etc.).

Here is an example of an effective topic statement.

> *Unlike gasohol-powered cars, the fuel cell alternative is virtually pollution-free.* A methanol fuel cell system works through chemical reactions that leave the air clean. A fuel processor breaks the methanol down into carbon dioxide and hydrogen; the hydrogen is then pumped to the cell itself, where it combines with oxygen to form water. Current is then produced when the electrons traded between molecules in this reaction travel through an external circuit. The net products are carbon dioxide, water, and electricity. By contrast, when gasohol is burned in an internal combustion engine, it produces the same nitrous oxides that gasoline does.[1] [italics added]

This topic statement is a good one because it tells the reader immediately what the theme of the paragraph is (fuel cell cars don't pollute) and because it's consistent with how the rest of the paragraph is developed (as a cause-and-effect description of how the fuel cell process works). Notice how the writer has used the key term *fuel cell* in the most important position in the sentence, the main-clause subject position, thus establishing it as the paragraph topic, i.e., what the paragraph is about.

For an example of what *not* to do, here is a paragraph from a student report on whether to use an argon recovery process or a hydrogen recovery process in a proposed ammonia plant:

NEGATIVE EXAMPLE
Utility costs for the argon process are 75 percent greater than for the proposed hydrogen process. Initial capital cost is $5.4 million, roughly three times the hydrogen process cost. However, annual income from the sale of argon, increased ammonia production, and reduced natural gas requirements elsewhere in the plant is 160 percent higher than that generated by the hydrogen process. Present worth analysis shows that the argon process is the better investment. The present worth of the argon process is $10.25 million. The present worth of the hydrogen process is $4.14 million.

Most readers will quickly conclude, on the basis of the first two sentences of this paragraph, that the argon process is more costly than the hydrogen one and should therefore not be chosen. But this is just the opposite of what the writer wants them to understand! For later on—buried near the bottom of the paragraph—the report states that "the argon process is the better investment." This statement is actually the topic statement of the paragraph. By "burying" it, the writer is running a serious risk of having the readers— especially those busy *important* readers!—completely overlook it.

A few simple changes can easily remedy the situation: (1) promoting the topic statement to initial position in the paragraph, (2) combining the next two sentences and subordinating them to the next one, (3) combining the last two sentences, and (4) adding a few words for emphasis:

> The argon process is clearly a better investment than the hydrogen process. Although it has higher utility costs (by 75%) and a higher initial capital cost (by 300%), it generates annual income—from the sale of argon, from increased ammonia production, and from reduced natural gas requirements elsewhere in the plant—that is 160% greater than that generated by the hydrogen process. Present worth analysis shows that the argon process is valued at $10.25 million whereas the hydrogen process is valued at only $4.14 million.

Notice how much more readable the rewritten version is. The topic statement serves to establish the main point and also to suggest how the rest of the paragraph will be developed (as a comparison-and-contrast pattern). The key term *argon process* (or its pronoun equivalent *it*) is used repeatedly in sentence-subject or clause-subject positions, thus keeping the reader's mind focused on it; all reference to the hydrogen process, by contrast, is deliberately subordinated.

The basic principle behind a well-written topic statement is this: by the time a reader has finished reading the first sentence or two of a paragraph, he or she should be able to predict what the rest of the paragraph is generally about and how it will probably be developed. Suppose, for example, you were reading an article on transport across membranes in a popular-science magazine and began reading this paragraph opening:

> The human body is made up of millions and billions of cells, each of which contains, among other substances, millions and billions of protein molecules. . . .[2]

How do you think the rest of the paragraph will go? What will it be about? Do you expect it to elaborate on other substances, on cells, on the human body generally? Probably not. Instead, there seems to be a narrowing down of focus to the term *protein molecules;* this is probably what the paragraph is about. How will this topic be developed? Well, the pattern of development used in this opening sentence is one of classification-division. Perhaps that pattern will be continued. Or maybe the writer has used some other pattern— a general-to-particular ordering of details, say, or a comparison of protein molecules with other kinds of molecules. Maybe the writer has used two patterns together. Of all the possibilities, though, you'd probably expect the classification-division pattern to be continued. It's a general fact about human nature that once we perceive a pattern in something, we expect it to continue—unless, of course, it's explicitly broken. Let's see what happens with this paragraph:

> . . . An average cell contains hundreds of different kinds of proteins, and all of the cells of the human body contain, among them, as many as 100,000 different kinds of proteins. These proteins can perform millions of different functions, a versatility which is largely responsible for the phenomenon called "life."[2]

The writer has indeed continued with the classification-division pattern set up in the first sentence. Notice, too, how the pattern proceeds from general to particular; the grammatical subjects of the three sentences show this progression quite clearly: *The human body. . . . An average cell. . . . These proteins. . . .*

Though paragraphs sometimes exist in isolation, they are usually linked to other paragraphs, forming larger conceptual units. In such cases, either the topic and pattern set up in one paragraph may be carried on to the next or the break between paragraphs can be used to switch to a new topic and/or a new pattern. In any event, it is usually desirable to maintain some kind of continuity when moving from one paragraph to the next. This is most often done by incorporating a key word or term from one paragraph into the first sentence of the next. In the membrane transport article, for example, the first sentence of the next paragraph begins as follows:

> The proteins derive their versatility from their structure—they are made up of chains of molecules of amino acids, substances of which there are 20 different ones in the human organism. . . .[2]

Notice how the writer has picked up on the key word *versatility* from the last sentence of the first paragraph and used it as a transition to the subject matter of the second paragraph.

And what *is* the subject matter of this new paragraph? If the topic statement (above) is a good one, you should be able to predict with some assurance what it is. First of all, two new words appear prominently in this statement: *structure* and *amino acids.* We might suppose, therefore, that the theme of this paragraph is the structure of amino acids. Furthermore, since the writer has begun discussing amino acids by saying that there are 20 different types of them, we might expect the rest of the paragraph to be a discussion of the structural variety of amino acids, perhaps according to a classification-division pattern, a comparison-contrast pattern, or a general-to-particular ordering of details. Let's see how it actually does continue:

> An average protein molecule consists of about 500 molecules of amino acids of different kinds (seldom all 20), arranged in some particular sequence. A sequence of 500 amino acids composed of all the 20 different ones would have as many as 1×10^{60} (1 followed by 60 zeros) possible arrangements,

each arrangement having particular chemical properties and therefore chemical capabilities. From these few facts alone, we can readily appreciate how important the study of the amino acids is to our understanding of proteins, of the cell, and of life.[2]

As you can see, the paragraph as a whole *does* satisfy the expectations raised by the topic statement. It *is* about the structural variety of amino acids, and it *does* follow a general-to-particular pattern of development.

EXERCISE 2-1   For each of the following paragraphs, circle an appropriate topic statement from among the three possibilities given. Be prepared to defend your choices.

A   i Many researchers believe that the cost of platinum will level off in the not-too-distant future.

    ii Fuel cell cars may someday be designed to operate with a catalyst different from the one being used today.

    iii Further development of fuel cell cars, despite their advantages, awaits more compelling economics.

While the soaring price of oil might have made the fuel cell car economically attractive, the rising value of precious metals has had the opposite effect. Fuel cells designed to run on methanol use platinum as a catalyst, and the price of that metal quadrupled during 1979. The cost of platinum alone could add a few thousand dollars to the price of a fuel cell car. Researchers have been trying to decrease the amount of platinum needed in the cells or to substitute a different, cheaper catalyst.[1]

B   i The quantity of coal left in the earth is impressive.

    ii Coal is a more viable source of energy than petroleum.

    iii The mining of coal entails a number of difficulties.

There are known to be 198 billion tons of coal at a depth of less than 1000 feet and lying in beds at least 3.5 feet deep for bituminous coal and at least 10 feet thick for beds of lower grade coal. An equal quantity of coal of the same accessibility is identified as "undiscovered recoverable reserves." In addition, there are even larger quantities of marginally available coal resources, amounting to 1 trillion, 400 billion tons. At 35 million Btu's per ton, coal can provide a great deal of energy for many years to come.[3]

C   i Many molecular biologists are now convinced that the discovery
of movable genetic elements holds the key to the solution of several long-standing mysteries.

ii It has become evident that in eucaryotes, unlike procaryotes, the
genes coding for protein production do not exist as one continuous
stretch of DNA.

iii Wrapping of the DNA on the enzyme with a positive superhelical
sense ensures that the reaction will produce negative supercoiling
once the wrapped segment is translocated.

It seems to go a long way, for instance, toward explaining how the human
body is able to synthesize a million and more different molecular antibody
species, each tailor-made to grapple with a specific antigen. Movable elements may help answer the age-old question of differentiation: how a fertilized egg divides and ultimately becomes, in the course of embryonic
development, many different kinds of tissue cells. Jumping genes may also
provide a mechanism for satisfying scientists who have been arguing that
point mutations alone were far from enough to account for the story of
evolution.[4]

EXERCISE 2-2   Each of the following sentences has been taken from an
original text where it serves as an effective topic statement for a well-written
paragraph. See if you can guess roughly how each paragraph is developed
beyond the topic statement: what the key words are, what the pattern of
organization is.

A   The first modular home to be tested by government engineers for durability exceeded the criteria for the National Bureau of Standards.[5]

B   At the time of its explosion, Mount St. Helens was probably the most
closely watched volcano in the world.[6]

C   The production of an important heavy chemical, nitric acid (HNO_3),
requires large quantities of ammonia.[7]

D   Until 1922, no one knew how a signal crosses the junction between
one nerve cell and another.[8]

E   The basic property of gyroscopic action is that the gyroscope stays
spinning in exactly the same direction in space over both short and
long periods of time.[9]

2.2

DEVELOP A CLEAR PATTERN OF ORGANIZATION

Once you have written a satisfactory topic statement, you'll want to follow it up with a number of supporting statements. These statements should follow a consistent pattern of organization, one that flows naturally or even predictably from the topic statement. That way, you'll satisfy the reader's expectations and allow her or him to process the paragraph as a unified whole.

Some of the most commonly used patterns of organization in scientific and technical writing are chronological description, cause-and-effect analysis, comparison and contrast, listing, and general-to-particular ordering of details. Each of these patterns has certain characteristic features, and by using these features you can make it easier for the reader to perceive which pattern you're using.

Chronological Description

The use of a time frame to tie sentences together is a well-known pattern of organization which you have no doubt used many times in your writing. It is commonly used, for example, to either describe or prescribe a step-by-step procedure: *First connect the vacuum tube . . . then return the plate . . . finally, close the contact key. . . .* It is used to recount a sequence of past events, as when you want to bring a reader up to date, e.g., in a progress report or in the Review of Literature section of a journal article.

The most characteristic features of chronological description are:

time adverbs and phrases	in 1980, last week, at 10:15, first, second, finally, soon after the project began
verb tense sequencing	Originally we *wanted* to. . . . More recently we *have attempted* to. . . . Now we *are trying* to. . . . In the future we *shall try* to. . . .
grammatical parallelism	*Mount* the grating near the end. . . . *Locate* a rider on the scale. . . . *Adjust* the grating. . . . *Read* the distances on the scale. . . .

Not all of these features are likely to be found in any one type of chronological description. For example, descriptions of standard procedures (e.g., test procedures, experimental procedures, assembly instructions) strongly favor the use of parallelism over the other features. Descriptions of past events, on the other hand, tend to rely more on time adverbs and phrases and on different verb tenses.

The following is a well-written paragraph using chronological ordering as its basic pattern of organization:

> Total U.S. research and development spending *is projected* to reach a current-dollar level of $66.7 billion *in 1981,* an increase of 10 percent over *the 1980 projected level* and nearly double the amount spent on these activities *in 1975.* Even in constant dollars, and despite the reductions in R&D programs contained in the Federal *1980* and *1981* budgets, U.S. R&D spending *in 1981 is expected to follow* the growth trend of *the past five years, when* R&D funding *grew* at an average annual rate of better than 3 percent. That growth *resulted* in large part from increased emphasis on searching for means to resolve energy and environmental problems and a resurgence of defense R&D activity. *Between 1975 and 1978,* the last year for which survey data are available, energy *accounted for* one third of the R&D spending increase while, at the same time, amounting to 10 percent of the national R&D effort.[10] [italics added]

This paragraph contains a great deal of information, and it would probably be quite confusing were it not so well structured. Notice how the topic statement provides a clear overview of the paragraph as a whole: it tells you not only what the main theme is (that U.S. R&D spending is continuing to follow a significant growth trend) but also how this theme will be developed (by chronological order, featuring the years 1981, 1980, and 1975). The remainder of the paragraph is then devoted to fulfilling these expectations. Notice how the writer not only has repeated the key dates but also has taken care to use correct verb tense forms when referring to them.

Cause-and-Effect Analysis

This pattern of organization is used in scientific and technical writing for a number of purposes, including (1) making a logical argument, (2) describing a process, (3) explaining why something happened the way it did, and (4) predicting some future sequence of events. Whenever you are describing causes and effects, it is usually best to describe them in straight chronological order: causes *before* effects. That way, you can minimize the number of "traffic signals" you need. Even when the description is clearly one of causes followed by effects, however, your readers will appreciate it if you occasionally insert such signals. The characteristic signals of cause-and-effect analysis include:

connective words and phrases	therefore, thus, consequently, accordingly, as a result, so
subordinate clauses	since, because (of), due to
causative verbs	causes, results in, gives rise to, affects, requires, produces

conditional constructions   when ozone reacts with nitric oxide,
                                             the ozone is destroyed and NO_2 is
                                             formed.

In addition, when causes and effects are described in chronological order, the features associated with chronological description can be used.

A good example of cause-and-effect patterning is this paragraph from a physics text explaining what surface tension is:

> One of the most important properties of a liquid is that its surface behaves like an elastic covering that is continually trying to decrease its area. *A result of this* tendency for the surface to contract is the formation of liquids into droplets as spherical as possible considering the constraint of the ever-present gravity force. Surface tension *arises because* the elastic attractive forces between molecules inside a liquid are symmetrical; molecules situated near the surface are attracted from the inside but not the outside. The surface molecules experience a net inward force; and *consequently,* moving a surface molecule out of the surface *requires* energy. The energy *E* required to remove all surface molecules out of range of the forces of the remaining liquid is proportional to the surface area; *therefore,*
>
> $$E = \sigma A$$
>
> where σ, the proportionality factor, is called the surface tension,
>
> $$\sigma = E/A$$
>
> and is measured in joules/m^2. [italics added][11]

Notice how, in addition to using signal words such as *consequently* and *therefore*, the writer has linked the sentences together in a steplike sequence. This has been accomplished mainly through the following technique: after introducing and discussing a new term in one sentence, the writer then uses it in the next sentence as part of the framework for introducing and discussing the next new term:

Surface tension . . . molecules situated near the surface. . . .

The surface molecules . . . requires energy.

The energy E required . . . is proportional to . . .

the proportionality factor . . . is measured in joules/m^2.

If this kind of step-by-step linking is done properly, it reduces the need to insert many signal words and markers of subordination. (See Chapter 14 for further discussion of this technique.)

Comparison and Contrast

Often, in technical writing especially, you will find it necessary to compare two or more things that are similar in some ways but different in others. This is particularly common in business and industry, where one is constantly wrestling with cost-benefit trade-offs and other choices that must be made from among various alternatives under various constraints.

In writing a comparison-and-contrast paragraph, *try to avoid jumping back and forth from one alternative to another.* Suppose, for example, that you are comparing items X and Y and are using criteria A, B, C, and D to compare them; suppose further that the first three criteria favor X and the fourth favors Y. In such a case, you should present the comparison in terms of these two criteria groupings: first A through C, then D. This will make it easier for the reader to see that product X wins out over product Y in three of the four criteria. And that's what the reader will most likely want to know— the proverbial bottom line. (If you think the reader will be more interested in the details of the comparison, provide a table.)

Characteristic features of comparison-and-contrast paragraphs include:

connective words and phrases	however, on the other hand, conversely, similarly, likewise, in contrast to
comparative constructions	more than, -er, than, less than, as . . . as, rather than, is different from
modal verbs	program X *will* be easy to implement, whereas program Y *would* entail a number of complications . . .
subordinate clauses	while, whereas, but
parallelism	model X is reliable and efficient, whereas model Y is unreliable and relatively inefficient . . .

One final principle of comparison-and-contrast writing is this: *phrase your words so as to reveal your own preference in the comparison.* In other words, don't just make a simple comparison as if you were a neutral observer. Instead, allow your own interpretation of the facts to color your description. Don't say, "Item X weighs 3.2 pounds, and item Y weighs 2.7 pounds"; say, "Item X weighs 3.2 pounds, whereas item Y weighs only 2.7 pounds." After all, *you* are the one who has made the study, and so *you* know what the facts of the matter are. Using some of the features listed above will enable you to let your reader know what your choice is.

Below is a model comparison-and-contrast paragraph. Note how it uses some of the features listed above.

A one-million-fold increase in speed characterizes the development of machine computation over the past thirty years. The increase results from improvements in computer hardware. In the 1940's ENIAC, an early electronic computer, filled a room with its banks of vacuum tubes and miles of wiring. Today one can hold in the hand a computing device costing about $200 that is twenty times faster than ENIAC, has more components and a larger memory, is thousands of times more reliable, costs 1/10,000 the price, and consumes the power of a light bulb rather than that of a locomotive.[12]

Listing

Scientific and technical writing presents frequent opportunities to put information in the form of lists. If you are describing an experiment, for example, you will probably want to make a list of the equipment used. If you are writing a progress report, you may want to make a list of things already done and another list of things still to do. If you're writing a report recommending the development or purchase of some new product, you may want to list its outstanding features or the reasons you're recommending it.

Lists may be either formatted or unformatted. Formatted lists are set off from the rest of the paragraph by means of indentation and/or numbering or lettering. Unformatted lists do not have such visual cues. In both cases, *all items in a list should be cast in parallel grammatical form*. This principle is especially important in the case of unformatted lists, since it can be quite difficult otherwise for the reader to detect the presence of the list. (Parallelism is important not only in lists but also in comparative constructions, in descriptions of procedures, and in other rhetorical patterns; see Chapter 3 for further discussion.)

A second important principle to follow when constructing lists is this: *if the items in a list are not equally important, they should be arranged in descending order of importance.* A list, by definition, is a set of items all of which have something in common and yet are independent of each other. Thus, in principle, these items can be arranged or rearranged any way at all. You can take advantage of this freedom by arranging them with the most important item in the most prominent position, namely, on top. Here is an example:

In addition to coal and nuclear energy, a wide variety of other power sources are also frequently discussed in the news and in scientific literature; unfortunately, most are not yet ready for practical use. Geothermal energy is one of the more practical of proposed new sources. It is already in use in Italy, Iceland, and northern California but is not yet meeting all expectations for it. Solar energy seems an elegant idea because it is inexhaustible and adds no net heat or carbon dioxide to the global environment. Yet present methods of exploiting it make solar energy hopelessly inadequate as a major power source in the next few decades. Sophisticated windmills to generate electricity are also under study by some. Biomass conversion is also getting under way. Some of these sources of energy, which we now generally regard as esoteric, may well prove themselves and make a substantial contribution over the long run if their costs can be brought within reason.[3]

The different power sources discussed in this paragraph constitute an unformatted list:

Geothermal energy . . .

Solar energy . . .

Sophisticated windmills . . .

Biomass conversion . . .

Notice how the effective use of parallelism enables the reader to easily locate the four items making up the list, even though no visual formatting is used.

General-to-Particular Ordering of Details

A final rhetorical pattern commonly used in scientific and technical paragraphs is the ordering of supporting details from the more general to the more particular. Each sentence in this pattern focuses on a smaller frame of reference than the sentence before it.

Here is an example of such a pattern, taken from a discussion of the kinds of bearings needed in flywheel energy-storage systems:

> Magnetic bearings have been developed for aerospace applications, but only recently has their practicality been demonstrated as the heart of energy storage systems. The breakthrough is partly due to the recent development of stronger permanent magnets, such as those made from rare-earth cobalt compounds. Only ten pounds of such magnets could support two tons of rotor. Although the free suspension of a weight with permanent magnets is an unstable condition, an electromagnet servo loop has been used successfully to stabilize the rotor position.[13]

Notice how the topic of the first sentence (and of the paragraph as a whole), *magnetic bearings,* is developed in the second sentence by the writer's focusing on one component of such bearings, the *permanent magnets.* This subtopic is then discussed in even greater detail in the third sentence *(Only ten pounds of such magnets . . .).* Finally, the writer focuses on a subcomponent of these magnets, the *electromagnet servo loop.*

Other Patterns

The patterns of organization described above are not the only ones commonly used in scientific and technical writing. Others—such as classification-division, exemplification, extended definition, and analogy, to mention but a few—are also used. (An example of paragraph development by classification-division, you may recall, was given earlier in this chapter.)

In addition, two or more patterns can often be used together. For example, you might want to give an *extended definition* of a process by describing it in terms of *causes and effects* or in terms of an *analogy* to some more familiar process. You might want to explain something by providing a *list* of *examples*. Or you might want to *divide* some subject into different classes and then *compare and contrast* these classes. Regardless of what kind of combination you elect to use, the important point is to establish at least one pattern of organization that runs through most if not all of the paragraph and thus provides a structural backbone for that paragraph.

EXERCISE 2-3 Each of the following paragraphs follows a particular pattern of organization. Identify the pattern, and point out as many of its features as you can.

A   Flywheels appear to be the ideal energy storage element in solar electric or wind power systems. They can smooth the load on the generators by providing the energy to generate electricity when the sun is not shining or the wind is not blowing. And they can also provide peak power to an electrical load during periods when demand exceeds supply, such as during motor start-ups. Indeed, once spinning, flywheels can deliver energy rapidly for transient load conditions, which makes them especially useful in industrial and agricultural applications.[13]

Pattern: _____

Features: 1._____
          2._____
          3._____

B   *Scalar and Vector Quantities.* Every quantity requires a statement of at least two things: first, a numerical specification, or a magnitude, and, second, the appropriate unit. For one class of quantities the statement of a numeral and a unit is all that is necessary for a complete specification. For example, the quantity volume is completely specified when we say "25 cubic feet." On the other hand, for another class of quantities one must state, in addition, a direction in order to afford a complete specification. For example, the quantity force is completely specified when we say "10 pounds acting vertically upward." A quantity that involves, other than the statement of a unit, only the idea of magnitude is called a *scalar,* while a quantity that needs for its complete specification a direction as well as magnitude is called a *vector.* Quantities such as displacement, velocity, acceleration, force, weight, and torque are vectors: each involves the idea of direction. Quantities such as speed, mass, and energy are scalars; none of these has associated with it a direction.[14]

Pattern: _____

Features: 1._____
2._____
3._____
4._____

C *The Vapor-Compression Refrigeration Cycle.* A simple vapor-compression refrigeration cycle is shown schematically in Fig. 2-1. The refrigerant enters the compressor as a slightly superheated vapor at a low pressure. It then leaves the compressor and enters the condenser as a vapor at some elevated pressure, where the refrigerant is condensed as a result of heat transfer to cooling water or to the surroundings. The refrigerant then leaves the condenser as a high-pressure liquid. The pressure of the liquid is decreased as it flows through the expansion valve, and, as a result, some of the liquid flashes into vapor. The remaining liquid, now at a low pressure, is vaporized in the evaporator as a result of heat transfer from the refrigerated space. This vapor then enters the compressor.[15]

Pattern: _____

Features: 1._____
2._____
3._____
4._____

D One of the newest branches of number theory is *analytic number theory.* A vast and intricate subject, it is largely a creation of the twentieth century. It has been called the science of approximation, for it is concerned mainly with determining the order (relative size) of the errors made when a calculation is approximate rather than exact. Its techniques have had an important impact on many departments of applied mathematics, including statistical mechanics and the kinetic theory of gases, where exact results are sometimes humanly unattainable.[16]

Pattern: _____

Features: 1._____
2._____

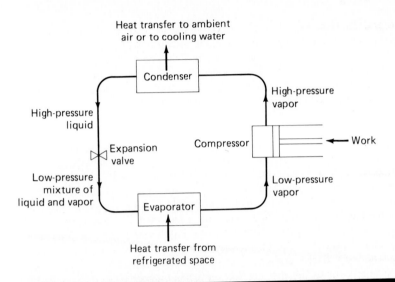

FIGURE 2-1   Schematic diagram of a simple refrigeration cycle.

EXERCISE 2-4   The following passage is taken from a popular-science journal article, where it was divided into seven paragraphs. Using your knowledge of topic statements and patterns of paragraph organization, see if you can determine where the six paragraph divisions should be.

We generally think of volcanoes as sizable mountains that belch lava and smoke from craters at their summits. But the essential element of a volcano is not its aboveground structure but the underground conduit that brings molten rock to the surface. The molten rock is known as magma as long as it is under the ground; after it erupts it is called lava. As the lava flows or explodes from a vent at the top of the conduit, it starts to build a volcanic edifice that may grow into a Vesuvius, a Mauna Loa, or a Mount St. Helens. Few people have witnessed the birth of a volcano, but Dionisio Pulido was present on February 20, 1943, when the Mexican volcano Paricutin made its first appearance in a cornfield on his small farm in the valley of Rancho Tepacua. Pulido, his wife and son, and a neighbor watched as smoke and ash began to rise from a small hole in the middle of the field. The smoke and ash were accompanied by rumbles, hisses, and hot particles of rock that set fire to nearby pine trees. By the next day, ash and rock debris had built a cone 10 meters high; in a week it had grown to 140 meters. By the time Paricutin finally quieted down in 1952, the cone towered 410 meters above the original level of Pulido's cornfield. Paricutin is an unusually large cinder cone, one of four basic types of volcanic edifices. Cinder cones, made up entirely of lava fragments, rarely rise more than 300 meters above their surroundings. They are the simplest type of volcano and are very common in western North America. A second type of volcano is the composite cone, or stratovolcano, built up of alternate layers of lava and ash (lava particles). Mount St. Helens is a composite cone. So are many other spectacular volcanic peaks, including Mount Rainier and Mount Hood in the Cascades and Japan's Mount Fuji. The summits of many composite cones tower 1800 to 2400 meters above their bases. Shield volcanoes are built almost entirely of lava flows. They are broad, gently sloping structures that resemble a warrior's convex shield laid on the ground face up. Many of the world's largest mountains are shield volcanoes; clusters of them form the Hawaiian Islands. Mauna Loa, on the island of Hawaii, is the world's largest active volcano. Its base is more than 4500 meters below the surface of the ocean, and its crest is 4170 meters above the sea level—a total height of more than 8500 meters. Lava domes, the fourth type of volcano, are built up of thick, pasty lava that tends to pile up in a rough knob rather than to flow outward as it emerges from its vent. The main factors that determine whether lava will erupt in a smooth flow or a violent explosion are its chemical composition and the characteristics of the gases that are dissolved in it. Magma is a mixture of several oxides, mainly silicon dioxide. The rock can be principally basalt, andesite, dacite, or rhyolite, which are distinguished one from the other, among other things, by the fact that each has a larger proportion of silicon dioxide than the one before. Basaltic magmas usually erupt in a highly fluid state, and their dissolved gases escape easily, so that the lava flows freely. The more viscous rhyolites and dacites usually erupt explosively. Andesites may erupt in explosions or in flows that are not as smooth as those of basaltic lava.[17]

EXERCISE 2-5   The following passages (extracted from longer reports) were submitted as is by students. As you'll discover, they require very careful reading in order to be understood. Revise them—by rearranging or rewriting sentences, or perhaps by dividing them into two or more paragraphs—so that they can be more quickly and easily understood.

A   ADVANTAGES AND DISADVANTAGES

The balance system is the simplest and cheapest system. However, it has drawn numerous complaints from users in California. The nozzles are heavy, hard to hold in place, and hard to remove due to the necessary airtight seal. The major problem has been that gasoline has been sucked out of the customer's tank back into the storage tank. This occurred when customers attempted to top off their tank and in the process overrode the automatic shutoff. The brands of the nozzles (OPW) were tested by the ARB in actual stations by inserting a small vial in the vapor return hose and checking for liquid in the vial after filling the vehicle. In the ARB tests the nozzles did not fail. However, in the MS test the nozzle failed. A $3\frac{1}{2}$-gallon container fitted with a standard auto filler pipe was used. The container had a mechanism to allow the inside pressure to be increased, and a 22.5% failure rate was reported. In one trial the container was filled to 15 gallons. The vacuum and assisted systems do not permit this overfilling, and no force is required to hold the nozzle. The prices of the three systems vary significantly. The balance system costs $\frac{1}{2}$ the amount of the assisted system, and the vacuum system is the most expensive.

B   TECHNICAL DISCUSSION

The most important part of this modification program is the addition of a larger engine to the Corsair. The eighteen-cylinder Pratt & Whitney R-2800-CB-13 and three-bladed propeller that are the standard fit for an FG-1D are rated at 2500 hp, though there are cases where R-2800s have supplied up to 3300 hp (Greenamyer, 1969) in modified form. One possible alternative to the R-2800, a Curtiss-Wright R-3350-24 Cyclone 18, has been substituted for an R-2800 in an F8F-2 Bearcat. Rated at 2800 hp, this particular example has been pushed to 3800 hp (Shelton, 1971). Another alternative is the twenty-eight-cylinder Pratt & Whitney R-4360-4 Wasp Major. These were raced in the late 1940s on F2G Corsairs with four-bladed propellers. These engines were rated at 3000 hp, though in races they produced up to 4000 hp (Cleland, 1949). After some careful checking, I have chosen to recommend fitting a Pratt & Whitney R-4360-B-20 with a four-bladed propeller, to your Corsair. While rated at 3500 hp, I expect it to produce over 4000 hp after modification. One particular advantage in using the R-4360-B-20 is that it won't increase the aircraft's frontal area, being the same diameter as an R-2800. In contrast, an R-3350 would increase the frontal area by 11% because of its larger diameter. However, in fitting this larger engine, there will be a weight penalty of 1130 pounds. To keep the aircraft's center of gravity from moving too far forward, the firewall will have to be moved back at least a foot.

REFERENCES

1   Joel Gurin, "Chemical Cars," *Science 80,* September/October 1980, p. 96.

2   Julie Wei, "Membranes as Mediators in Amino Acid Transport," *The Research News* **31**(11-12):3 (1980).

3   Adapted from Blanchard Hiatt, "Coal Technology for Energy Goals," *The Research News* **24**(12):3 (1974).

4   Henry Simmons, "DNA Topology: Knots No Sailor Ever Knew," and Ben Patrusky, "Gene Segments on the Move," *Mosaic* **12**(1):9, 41– 42 (1981).

5   Karl G. Pearson, *Industrialized Housing,* Institute of Science and Technology, University of Michigan, Ann Arbor, 1972, p. 39.

6   Henry Lansford, "Vulcan's Chimneys: Subduction-Zone Volcanism," *Mosaic* **12**(2):46 (1981).

7   J. Waser, K. Trueblood, and C. Knobler, *Chem One,* McGraw-Hill, New York, 1976, p. 589.

8   Edward Edelson, "The Neuropeptide Explosion," *Mosaic* **12**(3):15 (1981).

9   Henry Hooper and Peter Gwynne, *Physics and the Physical Perspective,* Harper & Row, New York, 1977, p. 207.

10   *Mosaic* **11**(5):cover 4 (1980).

11   Hooper and Gwynne, op. cit., p. 223–224.

12   Blanchard Hiatt, "Big Computing, Tiny Computers," *The Research News* **30**(4):3 (1979).

13   Alan Millner, "Flywheels for Energy Storage," *Technology Review,* November 1979, p. 33.

14   Joseph Morgan, *Introduction to University Physics,* vol. 1, Allyn and Bacon, Boston, 1963, p. 17.

15   Richard Sonntag and Gordon Van Wylen, *Introduction to Thermodynamics: Classical and Statistical,* Wiley, New York, 1971, pp. 9–10.

16   Julie Wei, "Pure Mathematics: Problems and Prospects in Number Theory," *The Research News* **30**(3):31 (1979).

17   Adapted from Lansford, op. cit., p. 50.

ADDITIONAL READING

David Kieras, "The Role of Major Referents and Sentence Topics in the Construction of Passage Macrostructure," *Discourse Processes* **4**:1–15 (1981).

David Kieras, "Initial Mention as a Signal to Thematic Content in Technical Passages," *Memory and Cognition* **8**:345–353 (1980).

Bonnie J. F. Meyer, *The Organization of Prose and its Effects on Memory,* North-Holland, Amsterdam, 1975.

R. M. Gagne and V. K. Wiegand, "Effects of a Superordinate Context on Learning and Retention of Facts," *Journal of Educational Psychology* **61**:406–409 (1970).

Patricia A. Carpenter and Marcel A. Just, "Reading Comprehension as Eyes See It," *Cognitive Processes in Comprehension,* Lawrence Erlbaum, Hillsdale, N.J., 1977, pp. 109–139.

3

PARALLELISM

All writing—whether of books, of reports, of articles, of sections of articles, of paragraphs, of sentences, etc.—consists of both form and content. Reading is made easier to the extent that the form *reflects* the content: we can then use formal features as cues to guide us as we try to absorb and interpret the content. Selective readers in particular depend on features of form to guide their reading. Such features are easy to spot and thus serve as signposts, allowing the reader to zero in on and quickly process those parts of a text that are of particular interest.

One way in which form reflects content is when we use grammatical parallelism in the writing of lists. If you write out a list, "Things To Do," that looks like this:

> pick up mail
>
> call Hazlitt
>
> work on tailgate project
>
> don't forget to ask RJ for $$

you're using parallelism because you've put each item on the list in the same grammatical form (verb + one or more nouns). If you write the list like this, on the other hand:

> mail needs to be picked up
>
> call Hazlitt
>
> tailgate project
>
> don't forget to ask RJ for $$$

you're not using parallelism.

Of course, when you're just writing to yourself, you can afford to be less careful than when you're writing to someone else. After all, *you* know what you mean, don't you? In writing to other people, however, there will be many times you'll find yourself writing out lists, and in such cases it's important that you make it optimally easy to read them. Here's where parallelism has an important role to play.

Before we describe this role, however, perhaps we should clarify what we mean by the word *list*. What exactly *is* a list? As we refer to it here, a list is any set of two or more independent items, i.e., any two or more items sharing a coordinate (related but nonoverlapping) relationship. This definition includes not only obvious cases such as the one described above but also others that may not be so readily apparent. For example, the following sentence contains a list since two different but related functions (*metabolite transport* and *metabolite signaling*) are mentioned:

> Evidence that the receptor sites on the membrane can serve both for metabolite transport and for metabolite signaling has raised many questions as to the mechanisms of membrane information-transfer.[1]

The following sentence also contains a two-item list:

> A pollution-free inertial-electric system has greater energy efficiency than an internal combustion engine.

Here the items forming a list are the two systems that are being compared.

3.1

TYPES OF LISTS

Lists are especially common in scientific/technical writing. We find lists of experimental apparatus, lists of instructions, lists of task objectives, and so on. Some lists are *fully formatted* as such, with alphanumeric sequencing and vertical alignment used as formal features to make the list stand out. Here is an example, from a company memo:

> REFS:(a)   Government invoice #79-1018A
>
> (b)   Bonded Stores invoice #31285
>
> (c)   DECAS letter 5/20/79
>
> (d)   Bonded Stores invoice #44590

Some lists are *partially formatted,* in the sense that they use either alphanumeric sequencing or vertical alignment but not both, as in this example:

> Factors taken into consideration include the following: (1) size of lot, (2) parking requirements, (3) need for elevator, (4) cost per gross square foot, and (5) expected annual return per gross square foot.

Some lists are completely *unformatted,* having neither alphanumeric sequencing nor vertical alignment:

> Compared to standard bipolar types, VMOS transistors offer higher input impedance, faster switching times, wider operating range, and smaller chip area.

In all cases, however, whether a list is fully formatted, partially formatted, or unformatted, *all of the items making up the list should be cast in the same grammatical form.*

Putting the items of a list in parallel grammatical form allows a busy reader to easily perceive the nature of these items and of the list as a whole. It is especially important to use this principle with unformatted lists. In the VMOS example above, notice how easy it is to glance at the sentence and see what the four criteria of comparison are: . . . *input impedance* . . .*switching times* . . . *operating range* . . . *chip area.* By contrast, notice how difficult it is to perceive the nature of the list, and the items making up the list, in the following badly written example:

NEGATIVE EXAMPLE
The TFC engineers and I found the selection of Hybrid Analog Transmission using FDM (Frequency Division Multiplexer) to be highly reliable, improved security of communications; complete ground isolation, freedom from crosstalk, sparking, short circuit loading; RFI (radio frequency interference), EMI (electric magnetic field), and EMP (electric magnetic polarization) immunity.

What are the major items constituting the list in this example? What role are they playing? You may be able to figure out the answers if you work at it, but many readers—especially managerial readers—are not willing to go to such effort.

As a general rule, you should always endeavor to use parallelism when you are presenting a list in written form. Often this will be a relatively straightforward matter, involving fully-formatted lists made up of simple items. In such cases, all you need do is make minor adjustments so that every item has the same grammatical form. Suppose, for example, you have written a draft of a report about wastewater processing and have included this statement:

NEGATIVE EXAMPLE
The principal processes are:

a)   coagulating and flocculation

b)   removing the solids

c)   nitrogen-removal

d)   disinfection

Surely it would be an easy matter to recast this list in a more grammatically parallel form:

The principal processes are:

a)   coagulation and flocculation

b)   removal of solids

c)   removal of nitrogen

d)   disinfection

Instead of mixing verb forms (-*ing*) and noun forms (-*al* and -*tion*) as in the first version, the second version uses all noun forms. This results in a more sharply defined list; it looks more like the work of a careful, precise technical professional.

But the proper use of parallelism can do far more than just enhance your professional image: it can also help prevent misinterpretations of your writing. To see what can go wrong when parallelism is *not* used, consider the following:

NEGATIVE EXAMPLE
This filter has two important functions: to reject impulse noise signals and passing low frequency command signals without amplitude or phase distortion.

This excerpt could easily be misread (especially by a busy nonspecialist reader) as meaning that two things are rejected: the impulse noise signals and the passing of low frequency command signals. Notice that *reject* and *passing* can be combined to produce such a misreading. Even those readers who can figure out what the writer is trying to say may have to slow down their reading just to make sure what's being rejected.

Such ambiguity is completely unnecessary. By putting the two key verbs in parallel form, the writer can create a much clearer version:

This filter has two important functions: *rejecting* impulse noise signals and *passing* low frequency command signals without amplitude or phase distortion.

Other parallel forms would work equally well: *to reject . . . to pass; the rejecting of . . . the passing of.*

EXERCISE 3-1   Each of the following contains a list that lacks parallelism. See if you can devise an improved version.

A   The building is 140 feet in length, 78 feet wide, and has a height of four stories.

B   Test results indicate that soil #1 is likely to settle, since high plasticity is equivalent to highly compressible.

C   The widest employment of DDT is in the control of insects of public importance (as a mosquito larvicide, a spray for malaria eradication, and to control typhus by dust application).

D   The reason for this is that most small businesses have a lower budget for their managers than do government or industrial managers.

E   In this particular case the most important variables are the following:
  (1)   pressure and temperature of the boiler
  (2)   what type of fuel is required
  (3)   the amount of oxygen
  (4)   fuel temperature

F   As a result of the above problem, this report addresses the following tasks:
  a.   To redesign the mix for the concrete slabs
  b.   An evaluation of the compressive strength with the use of test cylinders for various designs
  c.   Determining both the theoretical and actual material costs per cubic yard

G   We recommend the purchase of the New Orleans heat exchanger because (i) it can be obtained for $25,000, a savings of 80% over the new cost, (ii) production could be increased by 28% if needed, 8% higher than requested, and (iii) it will help recoup losses incurred in our benzene plant.

H   Advantages of this system are:
  1.   Automatically controlled
  2.   Less operating cost

  Disadvantages are:
  1.   May cause slugging of liquid refrigerant to compressors
  2.   Complex automatic controls
  3.   Substantial replumbing of existing system is required

I   This buoy terminal has three components that rotate as a single unit:
  (a)   The Rotating Mooring Bollard allows floating mooring lines to lie on the lee side when weather conditions become bad.
  (b)   Products are carried by a Rotating Cargo Manifold from the terminal-to-tanker hoses to the multiproduct distribution unit in the center of the buoy.
  (c)   A Rotating Balance Arm not only maintains the buoys on an even keel but also provides an accommodation ladder.

J   The purpose of this report is to present evidence that the contracting officer acted within the range of his authority and his final opinion was binding.

3.2
MISLEADING PARALLELISM

Given the fact that readers use parallelism as an aid in perceiving and inter-preting a list, it is important that you *not* use parallelism in situations where lists are *not* involved. Otherwise, readers are apt to be misled and may misanalyze the passage *as* a list, which in turn can lead to comic interpreta-tions or to unnoticed misinterpretations. Here is an example:

NEGATIVE EXAMPLE
Richard Clarke, senior systems programmer, asked me to develop a magnetic tape management system, to reside permanently on the computer, to give better control, and to coordinate the numerous magnetic tapes.

The parallel use of infinitives in this sentence makes it appear that a four-item list is being presented:

Richard Clarke, senior systems programmer, asked me:

- to develop a magnetic tape management system

- to reside permanently on the computer

- to give better control

- to coordinate the numerous magnetic tapes

This would mean, among other things, that it is the writer, not the system, that will "reside permanently on the computer, give better control," etc., surely a strange state of affairs! Of course, most readers would probably figure out what the writer means, but this requires extra time and effort on the reader's part—precisely what good writing avoids.

In cases like this, the writer should break up the misleading parallelism so that only those items that really are part of the list appear in parallel form. In this particular case, only the last three items belong together; the first item should be kept separate. If the last three items were written as relative clauses, say, instead of as infinitives, they would be distinguished from the first item:

Richard Clarke, senior systems programmer, asked me to develop a magnetic tape management system:

- which will reside permanently on the computer

- which will give better control

- which will coordinate the numerous magnetic tapes

This can be simplified by factoring out the relative pronoun and auxiliary verb (*which will*) and making them part of the lead-in:

Richard Clarke, senior systems programmer, asked me to develop a magnetic tape management system which will *reside permanently on the computer, give better control,* and *coordinate the numerous magnetic tapes.*

Compare this version to the original one and see what a difference there is in readability.

EXERCISE 3-2 Each of the following passages contains misleading parallelism. Correct each passage by making appropriate grammatical changes.

A   My present occupation is repairing typewriters, printing machines, and duplicating machines.

B   You learn many reasons why our product failed by reading, observing, talking and listening to our sales people.

C   This technology consists of three methods of scrambling, which are the coupling of light source and optical fiber in low-order mode, splicing, bending and tightening the fiber near its connectors, and installing the scrambler into the existing fiber optic in intervals of 1 km along the route.

3.3
PARALLELISM IN PARAGRAPHS AND LARGER UNITS

Perhaps the most important use of parallelism—one which poor writers usually neglect—is to indicate the presence of listlike structures in paragraphs and groups of paragraphs. Two or more sentences may be related coordinately within a paragraph; two or more paragraphs may be so related. Since these kinds of larger lists are customarily unformatted, parallelism is usually the only formal cue that can be used to guide the reader.

Here is an example of how the failure to use parallelism to indicate paragraph structure can make it difficult to read quickly and selectively, if not accurately:

NEGATIVE EXAMPLE
All-Savers Certificates will not benefit all investors. Investors exceeding a deposit of $7931 ($15,861 joint return) would have an after-tax yield far lower than alternative investments such as money market funds or Treasury bills. Alternative investments would also yield better after-tax yields and no penalty if the certificate was redeemed within the 1-year maturity period.

A common strategy used to skim-read paragraphs is to first read the topic sentence and then read the beginning words of the following sentences to see what kind of support these sentences provide for the topic sentence. In the case above, a person using this technique can easily be misled into perceiving the paragraph as having a general-to-particular structure. The first sentence is clearly the topic sentence. The second sentence supports it by describing a class of investors who would benefit more from alternative investments than from All-Savers Certificates. The third sentence then appears to provide additional details about alternative investments. The diagram below illustrates this structure:

All-Savers Certificates will not benefit all *investors*.

Investors exceeding a deposit of $7931 . . . would have an after-tax yield far lower than *alternative investments*. . . .

Alternative investments would also yield better after-tax yields and no penalty. . . .

Notice how the chain of repeated words links one sentence to the next, each one appearing to be subordinate to the one above it. In actuality, however, the writer means to say that there are not one but *two* classes of investors who will not benefit from All-Savers Certificates: those who exceed a deposit of $7931 and those who redeem their certificates before 1 year has elapsed. The purpose of the paragraph (as we were informed by the writer) is to describe these two classes. The second and third sentences, therefore, constitute a two-item list, with each of the sentences describing one of the two classes. If the writer had cast these two sentences in parallel form, it would be much easier for readers to see this listlike structure:

> All-Savers Certificates will not benefit all investors. Investors exceeding a deposit of $7931 ($15,861 joint return) would have an after-tax yield far lower than alternative investments such as money market funds or Treasury bills. Investors redeeming the certificate within the 1-year maturity period would have to pay a penalty and would also have a lower after-tax yield than with alternative investments.

The diagram below illustrates why this revision has a more transparent listlike structure:

All-Savers Certificates will not benefit all *investors*.

Investors exceeding a deposit of $7931 . . . would have an after-tax yield far lower than alternative investments. . . .

Investors redeeming the certificate within the 1-year maturity period would have to pay a penalty and would also have a lower after-tax yield. . . .

The use of parallelism enables the reader to see very quickly that the second and third sentences constitute a two-item list.

EXERCISE 3-3   Each of the following passages contains information that can be presented in the form of one or more lists. Reconstruct each passage accordingly, using appropriate parallelism.

A   In order to meet the job requirements it became clear that a microcomputer would be required to do this type of work. Microcomputers are very compact and portable. They are easily programmed and can be used for a wide variety of data processing. These computers are normally quite easy to operate and capable of storing large amounts of data both inside and outside of the computer itself. The most attractive part of using a microcomputer is that it is very inexpensive in comparison to other large-scale systems.

B   Reuse of treated water is most applicable where large amounts of water are used and the wastes are not too contaminated. Industrial wastes may be heavily contaminated and therefore may not be very suitable for reuse. The location of the treatment plant and the possible transport of the renovated water are also important factors. The treatment process works most efficiently and economically when dealing with a steady flow of wastewater. A very important point is whether the wastewater will be reused only once or whether it will be reused many times. Multiple recycling results in a buildup of certain dissolved materials, especially inorganic ions, that may make demineralization necessary. Most reuses do not lead to a high degree of recycling.

C   By checking the instrumentation used for this experiment, it was noticed that the collimator is not close enough to the x-ray film, and therefore a lot of neutrons have been lost in this way. But if we minimize this distance, fewer neutrons will be lost by leakage. Also, for better and sharper pictures we need a high resolution which is obtained when neutron beams are as parallel as possible; to achieve a higher resolution we have to either decrease the collimator or increase its length. In this case we cannot increase the collimator length, since it is as long as the depth of the reactor pool, which is a constant. Therefore the only alternative is to decrease the diameter of the collimator for a better resolution.

D   The new design meets all of the important criteria. The new design uses the same sulfur dioxide scrubbing process used in the present scrubbing system. Therefore, the new design gives the same sulfur dioxide removal rate as the present system. The new design includes a regeneration loop for the absorption reactant, thereby cutting the absorption reactant consumption considerably. The new design also terminates with a solid waste product. Thus, a very manageable waste product is produced.

REFERENCES

1   Julie Wei, "Membranes as Mediators in Amino Acid Transport," *The Research News* **31**(11–12):27 (1980).

ADDITIONAL READING

Francis Christensen, *Notes Toward a New Rhetoric: Nine Essays for Teachers*, Harper & Row, New York, 1967.

Part III

ELEMENTS IN PREPARING TO COMMUNICATE

4

AUDIENCES

Before we begin our discussion of audiences, we would ask you to read the partial technical reports presented in Figures 4-1 and 4-2. Each was written by an engineer. Try very hard to spend *no more than 1 minute on each report*. When you have finished reading the two, take another minute or two to decide which you feel is better and why.

Most people at the beginning of a technical writing course prefer the report in Figure 4-1; most at the end prefer the report in Figure 4-2. Why does this happen? If we look more carefully at the two reports, we notice that the one in Figure 4-1 looks very businesslike: it has a nicely indented list and lots of impressive data. Surely it provides just what the senator's public relations officer has asked for—data on the Y-Ships.

However, if we take a second look at this report and consider its audience and use, we get a somewhat different picture. The public relations officer, or someone else on the senator's staff, will be using this report to write a speech for the Y-Ship launching. Since one usually finds the most important information first, if the speech writer is not paying careful attention, he or she may produce an unusual speech:

> Good afternoon, ladies and gentlemen. We are gathered today to launch a Y-Ship of overall length 572′, beam 82′0″, depth to main deck 45′6″, designed draft 28′4″. . .

Obviously, such a speech would not fit the occasion. The information needed for an appropriate speech is at the bottom of the report segment: that this Y-Ship is the first of a new class, that it will link the east and west coasts with the Orient, that these ships are the first merchant vessels in the world to be built almost entirely of low-alloy, high-strength steel, and that they are unusually light and have greatly increased cargo space.

Given the purpose for this report, why do you suppose the writer puts the technical information first and the more general information second? Perhaps because the writer is much more interested in the technical information? This writer has just spent months designing the Y-Ships, and each of the numbers in that list represents many decisions the writer takes pride in. The problem here is that the writer is writing for the writer, not for the reader.

In contrast, if we look more carefully at the report to Lt. Cousins in Figure 4-2, we notice that it is designed for the reader. It puts the information that Lt. Cousins needs at the beginning and saves the details for later; it tells

FIGURE 4-1 Beginning of informal report by a naval architect. (Used with the permission of J. C. Mathes.)

To: XXXX, Public Relations   December 6, 1979

From: YYYY, Naval Architect

Y-SHIPS: VESSEL CHARACTERISTICS

I understand Senator Q's Office has requested data on the Y-Ships for use in connection with his participation in launching the PRESIDENT Y-SHIP at Pascagoula, Mississippi.

Following are physical characteristics of these ships:

        Length overall   572'-0"
        Beam   82'-0"
        Depth to Main Deck   45'-6"
        Designed Draft   28'-4"
        Maximum Draft   30'-7"

        Displacement at 30'-7" Draft   21,230 Tons
        Cargo Deadweight   10,000 Tons
        General Cargo Capacity   770,000 Cu. Feet
        Refrigerated Cargo Capacity   48,000 Cu. Feet
        Liquid Cargo   40,000 Cu. Feet or 1,000 Tons
        Container Capacity (8' × 8' × 20')   144

        Passengers Carried   12 in Eight Staterooms
        Crew   45

Propulsion: Steam turbine developing 24,000 Horsepower driving a single 22'-6" diameter five-bladed propeller.

Cruising Speed: 23.0 Knots
Cruising Radius: 11,600 Nautical Miles

PRESIDENT Y-SHIP is the first of a new class of vessels known as the Y-Ships (Design C4-S-69a). The design was developed by ABC Company, Naval Architects, New York, to meet design and performance characteristics and service requirements established by ST Lines for vessels operating on Trade Route Number 17. This route links both the East and West Coasts of the United States with the Orient and involves a round trip in excess of 30,000 miles. The 23-knot continuous sea speed capability is optimum for the long ocean legs in this service.

The Y-Ships are notable in several respects. They are the first merchant vessels in the world to be built almost entirely of low-alloy, high-strength steel. The result is a weight saving of approximately 330 long tons as compared with a similar ship built of conventional shipbuilding steel. This is reflected in an equivalent increase in cargo capacity.

FIGURE 4-2 Beginning of informal report from a sales engineer. (Used with the permission of J. C. Mathes.)

January 29, 1969

CGC Boutwell
c/o U.S. Coast Guard Base
427 Commercial Street
Boston, Massachusetts

Attention: Lt. (j.g.) G. L. Cousins

Subject: Clutch to disengage turbine starting pump

Gentlemen:

Thank you for the courtesy extended our representative, Ed Driscoll, on his recent visit aboard the Boutwell, at which time he discussed with you your requirement of the clutch to be used to disengage the turbine starting pump from the main generator engine.

We understand you wish to mount the clutch on a shaft which would be turning 720 RPM, and that the duty of the turbine pump is 117 HP. Based on this, the torque requirement is 853 pound feet, and a clutch having 16.3 HP per 100 PRM is indicated. Twin Disc Mode #CL-310 is rated 873 pound feet, and 16.6 HP per 100 PRM, and 135 HP normal duty. The #CL-310 therefore would seem to fill the bill quite nicely.

We refer you to Twin Disc Bulletin #326-B, enclosed. On page 10 you will find the description, capacities, etc., of this clutch. You will also note, it is available in 2-$\frac{1}{4}$" and 2-$\frac{7}{16}$" bore. Accessories are described on page 11. Two possible spider drive arrangements are suggested, as described in Figures 2 and 3 on the back cover of the bulletin. The spiders and their dimensions are indicated on pages 18 and 19. We are pleased to quote as follows:

1. XA5752 Model CL-310 Twin Disc Clutch in standard bore of 2-$\frac{1}{4}$"
   or 2-$\frac{7}{16}$" $131.20
2. Part #3507 Throwout Yoke 2.25
3. Part #3039 Hand Lever 1.49
4. Part #1144-E Operating Shaft 2.18

where to find the details but does not clutter up the beginning of the report with them. Notice, however, what *is* at the beginning: a polite thank you for a recent visit to the ship, a statement about the report's topic, the specifications that need to be met, the acceptability of the proposed clutch (it will meet the specifications), and the costs of the clutch. These are the kinds of information Lt. Cousins needs to do his job: he must be able to see that the proposed clutch will do the job, be installable, and be affordable. Thus, Figure 4-2 presents a report written for the audience. It orders information in a helpful way, pushing generalizations to the top of the report and details to the bottom. It doesn't waste the time of a busy reader.

How do we learn to write for our audience? We must first learn the characteristics of the audience and then figure out what the audience needs to know. Let us consider each of these in turn.

4.1

CHARACTERISTICS OF REAL-WORLD AUDIENCES

If we look at the audiences for whom we write in school, we notice they have the following characteristics. They usually consist of one reader (the teacher) whom the writer knows. This reader is careful, trying hard to figure out what the writer has to say (usually the teacher is trying to decide how much a student has learned in order to assign a fair grade). In addition, the teacher reads the entire paper, is usually interested in the subject, and knows more about it than the writer. If something isn't clear, the teacher can often supply missing information or see where an unclear argument is going. Finally, since the purpose of school writing is to "show what you know," the teacher often puts a great deal of emphasis on the details that show in-depth knowledge.

In contrast to this ideal situation, the writer in a real-world setting has an infinitely more difficult audience. First, this real-world audience often contains many readers having backgrounds very different from each other's and from the writer's: some readers may even come to a report years after it was written. Thus, it's no longer easy to "psych out the professor," that is, to put into a paper just what you know a single reader wants. Second, your readers may have different—even conflicting—desires; often as a writer you can't satisfy everyone. Third, your readers often have little knowledge or interest in your subject and usually read only parts of the report, frequently skipping from section to section. Sometimes, however, your readers are experts in your field and have great interest in the report; such readers might even read your report very carefully, perhaps trying to find holes in your argument. Fourth, although you have spent your entire academic life learning how to pack details into a paper, your real-world audience is usually more interested in generalizations. In fact, some readers don't want details at all, since details just confuse their grasp of the subject. Finally, most readers, especially middle- and high-level managers—the most important readers—read under the worst possible conditions: they have many pressing things on their mind, they are constantly interrupted, and they must read *quickly* in order to get through their mail each day.

Since managerial audiences are so important, let us consider them in more detail.

4.2

THE NATURE OF MANAGERIAL AUDIENCES

Managers are the most important audiences for technical communications because they make decisions that affect projects and careers. Thus, if scientists, engineers, and technicians hope to influence managers favorably, they had better understand how managers work and what will catch their attention.

The nature of managerial work has been studied by a number of researchers, and one of these, Henry Minzberg, has summarized some important characteristics of managers that emerge from the research literature:

1   *Folklore:* The manager is a reflective, systematic planner.
    Fact: Study after study has shown that managers work at an unrelenting pace; that their activities are characterized by brevity, variety, and discontinuity; and that they are strongly oriented to action and dislike reflective activities.

2   *Folklore:* The effective manager has no regular duties to perform.
    Fact: In addition to handling exceptions, managerial work involves performing a number of regular duties, including ritual and ceremony, negotiations, and processing of soft information that links the organization with its environment. [Managers have a wide variety of duties, interests, and needs.]

3   *Folklore:* The senior manager needs aggregated information, which a formal management information system best provides.
    Fact: Managers strongly favor the verbal media—namely, telephone calls and meetings.

4   *Folklore:* Management is, or at least is quickly becoming, a science and a profession.
    Fact: The managers' programs—to schedule time, process information, make decisions, and so on—remain locked deep inside their brains. Thus, to describe these programs, we rely on words like *judgment* and *intuition,* seldom stopping to realize that they are merely labels for our ignorance.[1]

Minzberg also notes that managers fill many roles at one time. By implication, these roles and the constraints noted above largely determine how a manager reacts to communication. According to Minzberg, the fundamental managerial roles are *interpersonal roles:* figurehead, leader of the unit, liaison to external units. However, since they have many contacts from their interpersonal roles and their power, managers also have important *informational roles:* they are the "nerve centers of information." As such, they monitor and disseminate information and serve as the spokesperson for their units:

The processing of information is a key part of the manager's job. In my study, the chief executives spent 40% of their contact time on activities devoted exclusively to the transmission of information; 70% of their incoming mail was purely informational (as opposed to requests for action).[2]

Finally, managers have *decisional roles:* they are initiators of change—supervising up to 50 new projects at a time—disturbance handlers, resource allocators, and negotiators.[3] In short, as Leonard R. Sayles has put it, a manager

> is like a symphony orchestra conductor, endeavouring to maintain a melodious performance in which the contributions of the various instruments are coordinated and sequenced, patterned and paced, while the orchestra members are having various personal difficulties, stage hands are moving music stands, alternating excessive heat and cold are creating audience and instrument problems, and the sponsor of the concert is insisting on irrational changes in the program.[4]

The effects of this environment on communication are severe; managers obviously have little time or attention to spare for careful reading or listening. For instance, Minzberg notes that

1 Half the activities engaged in by the five chief executives of my study lasted less than nine minutes, and only 10% exceeded one hour.[5]

2 These five chief executives treated mail processing as a burden to be dispensed with. One came in Saturday morning to process 142 pieces of mail in just over three hours, to "get rid of all the stuff." This same manager looked at the first piece of "hard" mail he had received all week, a standard cost report, and put it aside with the comment, "I never look at this."[6]

Given this almost impossible set of handicaps, what can the writer do to make a memo or report more easily read and more fully understood? The writer must certainly consider the situation and background of the audience and the uses to which the communication will be put. In addition, the writer has a number of built-in aids for making texts easier to read and understand, including the conventions that will be treated in the rest of this book.

4.3

A PROCEDURE FOR AUDIENCE ANALYSIS

Identifying the audiences, their characteristics, and their needs is one of the most important jobs communicators have; it determines what kind of information must be provided and how it should be phrased. However, identifying the audiences can be difficult since audiences vary with the situation and with the type of communication involved. For instance, technical reports and memos written within an organization usually have a wide range of audiences, perhaps including laboratory technicians, technical specialists, managers, lawyers, and specialists in marketing and finance. In contrast, technical articles written for specialized journals have a narrow range of audience, usually just the specialized readers of the journal. Thus, to help technical communicators identify their audiences effectively, the five-step procedure outlined below has been provided. If technical communicators consciously follow these five steps, they should produce a good assessment of audiences and more effective communication.

Identify the Communication's Uses and Routes

To identify effectively all of their audiences, communicators should think carefully about the uses their communication will have and the routes it may travel, since these will often be more diverse than expected. For instance, consider the situation of a computer engineer working for a company that makes small computers to control industrial machines. The engineer had been asked by his immediate supervisor to improve the program by which a user tells a computer, the Model ABX, what to do. The engineer wrote a new program for the ABX and sent it off to his supervisor with the note, "Here's the program you asked me to write last week to improve the control of the ABX." The supervisor forwarded the engineer's note and program to the head of the Programming Department, who then forwarded it for review to a second computer engineer, to the Head of the Hardware Design Department, and to the Head of the Marketing Division. The second computer engineer knew a bit about the project because she had talked to the first engineer about it briefly over lunch one day, but the two department heads knew nothing about it. Since they could not review the new program themselves, they forwarded it to people in their departments who they hoped could do the review: a junior engineer in Hardware Design and a project engineer in Marketing.

All of these reviewers were faced with the task of evaluating the first computer engineer's new program. The second computer engineer had to check the program for programming adequacy. The Hardware engineer needed to see how difficult it would be to implement the new program, since his department would have to redesign some circuitry to accommodate it if the company adopted it for the ABX. The Marketing engineer had to evaluate the sales potential of the new program. How much would it increase the cost of the ABX? Would it be something customers would want and be willing to retrain their employees to use? In such a situation, this engineer normally surveyed customers, calling them up, explaining the features and costs of a new program, and getting their response to it.

Unfortunately, the first engineer had not anticipated these reviews and so had not included an overview or adequate documentation for the new program or a discussion of its improvements on the old program. Thus, the reviewers could not review the new program as it stood. To do their jobs, they had to go to the writer and interview him to get the information that should have been provided in a short report or memo attached to the new program.

The communication situation at first seemed straightforward for the first engineer. He assumed that he had to report only to his immediate supervisor, who had handed him the assignment. In fact, he communicated with—in addition to his supervisor—his department head, the heads of two other departments, two engineers in different departments, and one in his own. In addition, if the report had been adequate, the Marketing engineer would probably have used it as the basis of a communication to a number of customers. Thus, as illustrated by this example, *technical communicators should identify the uses their communication will have and the routes it will travel.*

Identify All Possible Audiences

Technical people work in organizations, and thus their work has implications beyond any particular time or setting. For instance, they may work on a project that goes on for years and passes through many stages, as illustrated below.

$$\begin{array}{c} \text{Problem} \\ \text{Perceived} \end{array} \rightarrow \begin{array}{c} \text{Problem} \\ \text{Evaluated} \end{array} \rightarrow \begin{array}{c} \text{Product} \\ \text{Designed} \end{array} \rightarrow \begin{array}{c} \text{Product} \\ \text{Evaluated} \end{array} \rightarrow \begin{array}{c} \text{Product} \\ \text{Constructed} \end{array} \rightarrow$$

$$\begin{array}{c} \text{Product} \\ \text{Tested} \end{array} \rightarrow \begin{array}{c} \text{Product} \\ \text{Refined} \end{array} \rightarrow \begin{array}{c} \text{Product} \\ \text{Retested} \end{array} \rightarrow \begin{array}{c} \text{Product} \\ \text{Produced} \end{array} \rightarrow \begin{array}{c} \text{Product} \\ \text{Improved} \end{array} \rightarrow \text{etc.}$$

During each of these stages, someone will provide information about the project which may be used immediately by someone else in the communicator's organization, or by someone outside the organization, or later on by someone connected with the project. Thus, as a second step in audience analysis, *technical communicators should try to identify all of the possible audiences, current or future, for a given communication.*

Identify the Concerns, Goals, Values, and Needs of the Audience

Technical people operate in organizations composed of units with different concerns, goals, values, and needs. Engineers in industrial settings, for instance, must cope with, at least, a marketing unit, a manufacturing unit, an engineering or design unit, and a management unit. That these units have different perspectives is illustrated by the following observation:

> Sales-minded persons tend to recommend the development of machines which resemble those being marketed by successful rival firms—partly because the competitive models are known to be in good demand and partly because a

program of imitation takes a minimum of time to execute. The design engineering department, on the other hand, tends to propose the development of machines which are, at the very least, improvements over the competitor's design, or, in many instances, machines which are totally new by comparison to the offerings of the current market. Although the end result is usually more desirable, the development time required is always much greater and less ponderable because of the unknown factors which may be involved. These viewpoints may be summed up by saying that the sales department takes a short-run view while the design engineering department takes the long-run view. Before we yield to our natural bias as engineers and exponents of progress (whatever that is) and condemn the sales department for its short-sightedness, let us bear in mind that the money is made in the short run, and in the long run we are all dead.[7]

Thus, to ensure that they communicate effectively to their audiences, technical communicators should be aware of the differing perspectives of their audiences; that is, *technical communicators should identify the concerns, goals, values, and needs of their audiences.*

Make Communication Appropriate for Managers
Because technical communicators work in organizations, any project on which they work involves the integration of many perspectives and goals. This integration must be achieved at the technical level but is directed at the managerial level. Thus, to make appropriate integration possible and to gain the needed support of management for their own projects, *technical communicators should make their communication accessible and useful to busy managers.*

Identify the Audience's Preferences
for and Objections to the Arguments
The integration of perspectives and goals mentioned above will necessarily involve compromises or even occasional setbacks when conflicts between units occur. Thus, because there may be others within their organization who are competitors for scarce resources and potential critics of their work, *technical communicators should try to identify those arguments and approaches that will be most effective with their audiences and to anticipate any objections that might be raised.*

REFERENCES
1 Reprinted by permission of *Harvard Business Review.* Excerpt from Henry Minzberg, "The Manager's Job: Folklore and Fact," *Harvard Business Review,* July-August 1975, pp. 50–55. Copyright © 1975 by the President and Fellows of Harvard College: all rights reserved. (Minz-

berg has also written *The Nature of Managerial Work,* Harper & Row, New York, 1973.)

2   Minzberg, op. cit., p. 56.

3   Minzberg, op. cit., pp. 54–59.

4   Leonard R. Sayles, *Managerial Behavior,* McGraw-Hill, New York, 1964, p. 162.

5   Minzberg, op. cit., p. 50.

6   Minzberg, op. cit., p. 52.

7   Russel R. Raney, "Why Cultural Education for the Engineer?" Speech to the Iowa-Illinois Section of the American Society of Agricultural Engineers, April 29, 1953, quoted in John G. Young, "Employment Negotiations," *Placement Manual: Fall 1981,* College of Engineering, The University of Michigan, Ann Arbor, 1981, p. 49.

ADDITIONAL READING

James Adams, *Conceptual Blockbusting: A Guide to Better Ideas,* Freeman, San Francisco, 1974.

Thomas J. Allen and S. E. Cohen, "Information Flow in Research and Development Laboratories," *Administrative Science Quarterly* **14**(1):12–19 (1969).

Aristotle, *The Rhetoric of Aristotle,* translated by Lane Cooper, Prentice-Hall, Englewood Cliffs, N.J., 1932, 1960.

Susan Artendi, "Man, Information, and Society: New Patterns of Interaction," *Journal of the American Society for Information Science* **30**(1):15–18 (1979).

C. West Churchman, *The Design of Inquiring Systems: Basic Concepts of Systems and Organization,* Basic Books, New York, 1971.

Mary J. Culnan, "An Analysis of the Information Usage Patterns of Academics and Practitioners in the Computer Field," *Information Processing and Management* **14**(6):395–404 (1978).

Linda Flower, *Problem-Solving Strategies for Writing,* Harcourt Brace Jovanovich, New York, 1981.

William J. J. Gordon, *Synectics: The Development of Creative Capacity,* Harper & Row, New York, 1961; Macmillan, New York, 1968.

R. Johnson and M. Gibbons, "Characteristics of Information Usage in Technological Innovation," *IEEE Transactions in Engineering Management* **EM-22**:27–34 (1975).

J. C. Mathes and Dwight W. Stevenson, *Designing Technical Reports,* Bobbs-Merrill, Indianapolis, 1976.

Henry Minzberg, *The Nature of Managerial Work,* Harper & Row, New York, 1973; also "The Manager's Job: Folklore and Fact," *Harvard Business Review,* July-August 1975, pp. 49–61.

M. A. Overington, "The Scientific Community as Audience: Toward a Rhetorical Analysis of Science," *Philosophy and Rhetoric* **10**(1):1–29 (1977).

James W. Souther, "What Management Wants in the Technical Report," *Journal of Engineering Education* **52**(8):498–503 (1962).

Richard E. Young, Alton Becker, and Kenneth Pike, *Rhetoric: Discovery and Change,* Harcourt Brace & World, New York, 1970.

5

BASIC TYPES AND PATTERNS OF ARGUMENT

Once a communicator has become aware of a problem, she or he must investigate and construct arguments that a given solution is appropriate and preferred over all others. Since these activities often require significant effort and time, this chapter provides tools for making them more efficient.

An argument is a claim that something should be believed or done, plus proof or good reasons for believing or doing it. This definition suggests that to make effective arguments, you must be able to

1   Make appropriate claims

2   Find and recognize proof or good reasons

3   Know when you have enough proof

As first steps to gaining and refining these skills, you ought to be familiar with the basic strategies of argument, the basic types of argument, and the accepted patterns for arranging arguments.

5.1
EXPECTATIONS ABOUT CLAIMS AND PROOF

When they first begin writing on the job, many engineers and scientists fail to provide enough support, or "backup data," for the claims they make in their reports. This often makes the reports ineffective or even useless when audiences disagree with the writer's conclusion, see it as threatening, or need a solid foundation of support on which to base their own decisions.

For instance, consider the following Foreword and Summary dashed off by a junior engineer.

Subject: Establishing a computer file for our library: Cost estimate

FOREWORD
Frequent reference to the books in our library is essential for the research workers in the office. However, because of imperfections in the card cataloguing system, precious time is wasted in locating books. To solve this problem, we have decided to use the computers for book searching; this will be faster and far more convenient. I have been requested to study the feasibility of such a project under our current budget. The purpose of this report is to estimate the cost for establishing a computer file for our library.

SUMMARY

For my test run on the computer, I experimented with 50 books from the library and estimated the cost for establishing a computer file for these 50 books. Since there are approximately 5000 books in our library, that cost figure was then multiplied by 100 to give the total cost figure for the whole library. It is estimated that $2000 will be needed to establish the computer file for the library.

Assume that you are the department manager to whom this memo was sent, that your department has limited funds, and that you can spend absolutely no more than $2000 on this project, if you can spend even that. Further, you can't waste your money: if you invest in the project, you must get a completed library system in return. A partially completed project will be useless and divert money from other needed projects.

Under these very realistic conditions, what would you want to see as proof for the claims made in the summary? Wouldn't you want to know that the sample of 50 books was a representative sample? How could the writer convince you? Wouldn't you like some proof that the estimated cost for the project is accurate? How would you expect to find that out? Perhaps by seeing a breakdown of the total cost into its component parts—$1500 for labor, $400 for computer time, $100 for punch cards to enter the data? Even if the writer provided those figures, how could you be convinced that the $1500 charge for labor would be sufficient, that you wouldn't be asked to provide more money later on?

Similarly, consider the next Foreword and Summary, written by a test engineer to report the mileage for a weight-reduced T-car.

FOREWORD

Increasing the fuel economy of our automobiles has been the top priority of our company for the past 2 years. In your memo dated October 6, 1981, you requested that I perform a simulated road test for our T-car with a weight reduction of 1000 lb and compare its highway mileage rating (MPG) to our present model. I have completed the test and have placed the results in our file. The purpose of this report is to present the test results and my recommendation.

SUMMARY

Upon completion of five simulated road tests, I have determined that the weight-reduced T-car will only give an average 2.5-MPG increase in highway mileage while its lightness creates handling and safety problems. Therefore, for safety reasons, I recommend that sources other than solely reducing the weight of the car be investigated.

If you had to make decisions about fuel economy based on this summary, wouldn't you want proof that the five road tests were sufficient, that the mileage increased only 2.5 miles per gallon, and that there were handling and safety problems with the car? If you were a hostile audience—perhaps you believed in the weight-reduced car, say, or even helped design it—wouldn't you need to see the same kinds of proof before you believed the claims made in the Summary?

The following exercises provide practice in identifying claims which need to be proved, and then the rest of this chapter considers ways of proving such claims.

EXERCISE 5-1   This exercise presents Forewords and Summaries from four different reports. Predict what claims or numbers need to be proved to make a convincing argument. The fourth sample for analysis also contains a Details section, which you should evaluate for adequacy of proof.

A   FOREWORD
The wind turbine project is a new undertaking of Energy Systems. The goal of the project is to design a marketable system which makes wind energy accessible to private residents. To aid in the selection and development of a preliminary design, Mr. Zondervan asked that I research the practicality of using a flywheel for energy storage. He also requested that I identify the organizations doing development work with flywheel rotors so that further information could be obtained if a flywheel is incorporated in the design. This report addresses these requests by identifying characteristics of flywheel energy storage which affect its practicality and by listing those institutions involved in flywheel research.

SUMMARY
New flywheel designs show much promise as energy storage devices, making possible energy densities up to 40 Wh/bl and reducing costs to as low as $50 per kWh. These figures compare favorably with other storage devices, such as batteries. In addition, the new designs have solved major problems with bearings and energy conversion, making flywheel energy storage quite practical.

Flywheel development projects are being carried on by the following people at the listed institutions.

1. David Rabenhorst, The Applied Physics Laboratory of Johns Hopkins University

2. J. A. Rinde, Lawrence Livermore Laboratory, Berkeley, Calif.

3. A. E. Raynard, AiResearch Division of Garret Corporation, Torrance, Calif.

4. Frances Younger, Wm. Brobeck and Associates

5. R. P. Nimmer, General Electric Company, Corporate Research and Development, Schenectady, N.Y.

B **FOREWORD**

In the past six months, the Production Department has received complaints from workers on the main floor that their working environment has become too noisy. They complain that intense noise from machines around them is nerve-racking and that they can communicate with one another only by shouting. Furthermore, some workers have even said that their hearing is deteriorating. Consequently, I measured the sound level in various spots on the main floor to locate the source of the problem. The purpose of this report is to recommend methods for reducing the noise level on the main floor.

RESULTS

I collected data and compared my results with those obtained in December, 1979; the total noise level produced by the machinery has increased by 30 percent. I then discovered that the noise is produced by the vibration of large housings and coverplates and by friction between loose and worn-out parts of machinery. In addition, the noisy condition on the main floor is worsened by reverberations from bare concrete ceilings and metal walls, which reflect a high percentage of incident sound waves.

RECOMMENDATIONS

My recommendations for reducing the noisy condition on the main floor are as follows:

1. Add acoustic material on the concrete ceiling and metal walls to reduce reverberation.

2. Add vibration-damping material beneath large housings and coverplates to dissipate vibration energy in the form of heat.

3. Replace worn-out parts of machines and fix loose components.

4. Provide earmuffs to prevent damage of the workers' ears caused by prolonged working in noisy condition.

5. Consider noise specifications when buying new equipment to reduce noise production in the future.

C **FOREWORD**

The present economic trends have caused smaller fraternities to have problems competing financially with larger fraternities, dormitories, and co-operatives. Specifically, Alpha Epsilon Zeta Fraternity must address the possibility of expansion into nearby apartments not only to remain competitive but also to gain the many social advantages of a larger membership. The purpose of this report is to demonstrate the economic feasibility of such an annex and to recommend its optimal size and location.

SUMMARY

Rent prices vary only slightly within walking distance. Therefore, the best location is the closest one, that being Plaza Apartments at 605 E. Manroe St.

Rent per member living in the annex will cost the fraternity substantially more than for those living in the lodge. However, excess revenue generated in the board account will counter that expense. The annex must hold at least six members to be successful, and anyone in addition could even cause a slight reduction in bills.

D　**FOREWORD**

In your letter of September 30, 1981, you requested me to obtain information regarding the competitiveness of our new MGC-200 videodisc player. Consequently, I proceeded to test a number of the comparable products and especially the Pioneer VP-1000, which is the top-selling videodisc player. This report contains the results of my studies and recommendations for improving our product.

SUMMARY

The results of my research show that even though the quality of the MGC-200 is superior to most similar products, it will not be competitive with the top-selling Pioneer VP-1000. However, some modifications will substantially improve the specifications of our product. The recommended modifications listed below will cost less than $2000 and will postpone the planned production for approximately two more weeks.

1)   Minimize the power dissipation in the microcomputer playback control system by reducing the number of integrated components in the system.

2)   Reduce the signal to noise ratio to 40 dB. This will cut the price by about 10% and will not have significant effect on the quality of the product.

3)   Provide an optional remote control system. As explained in the details, this will not increase the cost significantly.

4)   Reduce the weight by using a smaller and lighter frame.

DETAILS

The specifications of both the MGC-200 and the VP-1000 are tabulated below for comparison:

		MGC-200	VP-1000
1)	Cost	$984	$749
2)	Weight	57.3 lb	38.6 lb
3)	Power Consumption	78 W	71 W
4)	Video Bandpass	4 MHz	4 MHz
5)	Audio Bandpass	80 MHz	83 MHz
6)	S/N Ratio	50 dB	40 dB
7)	Disc Type	3M	3M, MCA

As pointed out in the summary, an optional remote control system can easily be included. The RU-1000 remote control can be used either through the IRAB sensor and amplifier as a completely wireless infrared remote control or plugged into a remote control jack and onto the photo coupler as a wired remote unit when noise is a problem.

I would appreciate being informed of your decision.

5.2

THREE BASIC STRATEGIES OF ARGUMENT

Over 2000 years ago, the Greek philosopher and rhetorician Aristotle identified three main strategies or bases for argument:

1   Logic and reason

2   The character and credentials of the communicator

3   Emotion

If you look through many types of technical writing, you will notice that most use (or try to use) arguments based overtly on *logic and reason*. For instance,

> Pierlest Inc. should adopt a proposed pollution control device because (1) it will cut pollution to an acceptable level, (2) it will be easy to install, (3) it has a reputation for being very reliable, and (4) it will be cost effective.

In contrast, relatively few arguments are based overtly on arguments of *emotion,* though some are. For instance, the argument "Pierlest Inc. is violating federal pollution standards and will be closed down unless it greatly reduces pollution" appeals to the emotion of fear, the fear of being closed down.

You will note that most arguments are combinations of two or all three strategies, with one predominating. For instance, when you write a report recommending a particular solution to a problem, you probably give a series of arguments based on logic and reason to support your recommendation. However, you also make an argument at least partly based on your character and credentials. Since you were hired because of your qualifications, your recommendation is often accepted at least partly, and sometimes fully, because your readers believe that you're competent to speak on your topic. (It is, of course, important to *look* competent and to reinforce your credentials.)

You will also note that in different situations, different strategies of argument will be more important. As indicated above, most technical communication appeals heavily to logic and reason because this strategy of argument is usually appropriate and convincing. However, sometimes technical communicators need to base arguments on emotion. For instance, when scientists and engineers confront deeply emotional issues, such as the fear a community might have over the location of a nuclear power plant, overreliance on logical arguments can be ineffective. Scientists have often argued very logically and "objectively" about a nuclear plant's backup systems for controlling dangerous situations and the very slight chance of an accident but never convinced the listeners that a plant should be located in their community. For many, the "fact" that a plant has only a one-in-a-million chance of blowing up does not outweigh the fear-producing knowledge that the reactor at Three Mile Island almost broke through several security systems to become a radioactive hazard to the surrounding communities.

EXERCISE 5-2   Read the following passages and identify the strategies being used in each.

A

Social Control of Research

When I promised to speak here on the topic of social control of research, I thought it would be a much easier job than it has turned out to be. I have some practical experience in the management of research programs in the federal bureaucracy, but more relevant was seven years helping the Congress deal with research and development (R&D) matters. There, I learned the intricacies of grooming and passing authorization bills for large research programs, and negotiating complex conference agreements. On a number of occasions I had translated what society—represented by its elected officials—wanted in legislation that established new programs of research and development, or institutions to control such programs and their results, such as the Office of Technology Assessment. Another important part of that congressional experience was informing and cajoling those with responsibility for the appropriations process, in order to assure funding for those authorized programs. I had even received the satisfaction which came from coordinating the override of a presidential veto, only the 89th such veto override in the history of our Republic.

My confidence that social control of research would be an easy topic was further increased by my knowledge of and involvement with a number of AAAS programs on such topics as scientific freedom and responsibility, the scientific and legal interface, regulation of recombinant DNA research, analysis of the federal research and development budget, and other policy issues.

It was with substantial optimism that I set about to brush up on the more formal knowledge extant on the subject. So I decided to back off from my original intention to provide a thoughtful and catholic review of the topic. Rather, I shall attempt to identify its more important aspects from the twin perspectives of the legislative process and ongoing AAAS activities.[1]

B

Recombinant DNA Research

James D. Watson of Harvard, Nobel Laureate for his discovery of the molecular structure of DNA, [said] . . . that theoretically all forms of higher animal life may be capable of clonal reproduction, and that the details of the research related to that, as well as to its implications, had not so far been communicated to the public to any substantial degree. He went on to ask,

> Does this effective silence imply a conspiracy to keep the public unaware of a potential threat to their basic ways of life? Could it be motivated by fear that the general reaction would be a further damning of all science and thereby decreasing even more the limited money available for pure research? Or does it merely tell us that most scientists do live such an

ivory tower existence that they are capable of rationally thinking only about pure science, dismissing more practical matters as subjects for the lawyers, students, clergy, and politicians to face up to in a real way.[2]

C   Hard Times for Basic Research

Despite the impressive growth of the National Science Foundation's budget from its initial $500,000 in 1952 to its almost $900 million in FY 1978, the fortunes of basic research, viewed in relative terms, have not been great. In the last decade funds for basic research have been of the order of 0.15 percent of the Gross National Product and have steadily declined. As Shapley, Phillips, and Roback indicate in their analysis of the FY 1978 R&D budget, funds for basic research, expressed in 1972 constant dollars, dropped 20.6 percent between 1967 and 1975 and, despite increases in both the Ford and Carter budgets, have not yet returned to their 1967 level. The growth rate in R&D funding of 11.3 percent between FY 1976 and FY 1978 has meant about 5 percent more for basic research in the academic sector, hardly anything to quicken the pulse.

Despite the general understanding that the federal government should carry the major responsibility for the funding of basic research, and, indeed, does provide about 70 percent of the monies available for this purpose, unyielding preoccupation with quick practical payoff on the part of both the Congress and the executive leadership has meant year-to-year uncertainty for an intellectual activity that should be built on long-range continuity. At least one fundamental reason for this seems to lie in our inability to make clear to the larger public the place of serendipity in science. The layman seems not to fully appreciate the fact that one cannot provide on cue *all* the antecedent information needed for innovative technology, and therefore that knowledge must be stockpiled like scarce materials against the day when it per chance may be needed. I am confident that this lacuna in the public's understanding of science can be filled in the long run with the right kind of precollegiate science education. But, meanwhile, we must give serious attention to creating a more vigorous, more comprehensive, and more carefully calculated program of education for members of the Congress and their legislative aides than we have been willing to give time to in the past.[3]

D   Legislative and Regulatory Control

With regard to whether research *should* be subject to regulation, we must immediately refer to the U.S. Constitution. The First Amendment to the Constitution reads:

> Congress shall make no law respecting an establishment of religion, or prohibiting the free exercise thereof; or abridging the freedom of speech, or of the press; or the right of the people peaceably to assemble, and to petition the Government for a redress of grievances.

The intent of the authors of the Constitution was recently summarized by Professor Thomas Emerson, Emeritus Professor at Yale Law School, who wrote:

> The process is essentially the method of science. The theory of freedom of expression, indeed, developed in conjunction with, and as an integral

part of, the growth of the scientific method. Locke, following Hobbes, based his philosophical and political theories on the premises of science. And the proponents of free expression were all men who, in the broad sense at least, put their faith in progress through free and rational inquiry. Hence the process they envisaged operates upon the same principles as those that guided the men of science: the refusal to accept existing authority; the constant search for new knowledge; the insistence upon exposing their facts and opinions to opposition and criticism; the belief that rational discussion produces the better, though not necessarily the final judgment. This process did not ignore prior knowledge or opinion, but it did insist upon the responsibility of the individual to challenge such opinion and upon the obligation of all to make reasoned conclusions based upon the evidence.

Our basic principles of academic freedom are rooted in Supreme Court decisions invoking the First Amendment. In *Swezey v. New Hampshire,* Chief Justice Earl Warren, writing for the Court, stated:

> The essentiality of freedom in the community of American universities is almost self-evident. No one should underestimate the vital role in a democracy that is played by those who guide and train our youth. To impose any strait jacket upon the intellectual leaders in our colleges and universities would imperil the future of our Nation.[4]

5.3
BASIC TYPES OF ARGUMENT

There are two main types of argument: arguments of fact and arguments of policy. An argument of fact is an argument that something *is* or *exists* (or *is not* or *does not exist*); it can also be an argument that something *is* or *is not true, necessary, justified,* etc. The key word for the argument of fact is *is* or *are:* this company *is* (or *is not*) discharging pollutants into public waters; increased domestic production of oil and natural gas *is necessary* if the United States is to maintain its standard of living without being too dependent on Middle East oil; the costs of buying a new house *are* too high for the average consumer.

In contrast, an argument of policy is an argument that something *should* or *should not* be done. The key word for this argument is *should:* this company *should* be stopped from discharging pollutants into public waters; increased domestic production of oil and natural gas *should* be encouraged in the United States so that people can maintain their standard of living (or increased production of gas and oil *should not* be encouraged in the United States because it would be too harmful to the environment); the costs of buying a new house *should* be reduced for the average consumer.

Notice that your support for an argument of policy would differ from your support for the corresponding argument of fact; in some cases you would need to make an argument of fact as a subargument in an argument of policy. For instance, if you were to argue that the costs of buying a new

TABLE 5-1   OUTLINE OF THE ARGUMENT OF FACT

Argument of fact: argument that something *is* or *exists* (or *is not* or *does not exist*)

Question or subargument of existence: does the thing actually exist; has something actually happened?

Example A   Are there Soviet troops in Cuba?

Example B   Is Pierlest Inc. discharging material from its manufacturing processes into public waters?

Question or subargument of definition: if it exists or has happened, what kind of thing or event is it?

Example A   If it is granted that there are Soviet troops in Cuba, are the troops educators and advisors or attack troops or something else?

Example B   If it is granted (or proved) that Pierlest Inc. is discharging material from its manufacturing processes into public waters, are the discharged materials regulated by law, considered dangerous, considered nontoxic even in large amounts, etc.?

Question or subargument of quality: if it exists and has been defined, how is it to be judged?

Example A   If Soviet troops are in Cuba and are educators and advisors, are they justified, necessary, present in appropriate numbers, desirable, etc.?

Example B   If Pierlest Inc. is discharging material into public waters, and if those materials are regulated by law (or dangerous, etc.), are the materials present in legal (or safe) amounts, illegal (or unsafe) amounts, desirable (or undesirable) amounts, avoidable (or unavoidable) amounts, etc.?

house should be reduced for the average consumer (an argument of policy), you would need first to argue that the current costs are too high (an argument of fact).

The Argument of Fact

Arguments of fact can be derived from three sources:

1   Questions or subarguments of existence

2   Questions or subarguments of definition

3   Questions or subarguments of quality

These are outlined and illustrated in Table 5-1. It can be seen from the examples in this table that, as you move from the question or subargument of existence to the subargument of quality, you logically build an argument. For instance, if you want to argue that a certain manufacturer should be stopped from discharging pollutants into public waters, you first need to address the question of existence (is this company discharging material into public waters?). You then need to address the question of definition (are the discharged materials regulated by law? dangerous? etc.), and finally you must address the question of quality (are the discharged materials present in illegal or unsafe amounts?). Only when all of these questions have been addressed can you argue convincingly that the company should be stopped from discharging into public waters. You make this main argument by stringing together a chain of subarguments of existence, definition, and quality:

> Subargument of existence:
> The company is discharging material into public waters.
> ↓
> Subargument of definition:
> The materials being discharged are regulated by law and are dangerous.
> ↓
> Subargument of quality:
> The materials being discharged are present in public waters in illegal and unsafe amounts.

Note that your main argument could fail on any of these subarguments. Your argument might fail to establish the subargument of existence. It could happen, for instance, that while there are pollutants in the public waters, another company put them there and not Pierlest Inc. In such a case, placing restraints on Pierlest would be both unfair and ineffective. Similarly, your argument could fail on the subargument of definition. It could be that Pierlest is discharging material into public waters but that these materials are not regulated by law or are not harmful. In such a case, it would probably be wasteful to spend time, effort, and money trying to regulate the company. Finally, an argument could fail on the subargument of quality. Pierlest might well be discharging dangerous materials regulated by law, but only those amounts allowed under the law. In such a case, if you still wanted to stop those discharges, you would have to argue that the law should be changed; you could not effectively argue that the company was breaking the law.

You could represent the "path" of reasoning we have just seen as a flow chart, such as the one in Figure 5-1.

The Argument of Policy

The argument of policy, the second basic type of argument, can be derived from two sources:

1   Questions or subarguments of worth or goodness

2   Questions or subarguments of expediency, advantage, or use

These are outlined and illustrated in Table 5-2.

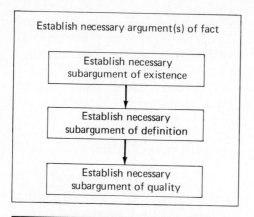

FIGURE 5-1 Partial flow chart for constructing effective arguments: establish necessary arguments of fact.

You should note that arguments based on expediency, advantage, or use are much more frequent in most types of technical writing—indeed, in most areas of our lives—than are arguments based on intrinsic worth, goodness, or merit. People are more likely to believe or do something if it is to their own advantage. Let us consider, for example, the arguments in Table 5-2 and the following situation. Company administrators intend to make false claims about the merits of their product in order to increase sales and profits. You want to argue instead that they should submit honest claims about the product. If you appeal only to the issue of worth (it is good and worthy to be honest), you will probably be unsuccessful. In the value scheme of these administrators, the desire for profit is probably stronger than the desire for honesty. Instead, you are more likely to succeed if you appeal to expediency and advantage: the administrators should make honest claims about the merits of their product because

1   It is advantageous in dealing with customers to have a reputation for honesty

2   Such honesty will protect the company from charges of fraud, expensive law suits, and costly penalties imposed by the government and the courts

This kind of argument better "fits" the supposed value scheme of the administrators. Can you think of a situation in which the appeal to worth or goodness would be as strong as or stronger than the appeal to advantage?

TABLE 5-2 OUTLINE FOR THE ARGUMENT OF POLICY

Argument of Policy: argument that something *should* (or *should not* be done)

Subargument of worth or goodness: is a proposed activity or course of action worthy or good in itself?

Example C   The United States should not protest the presence of Russian troops in Cuba because it is right that they are there (because they have the right to be there).

Example D   Pierlest Inc. should stop polluting public waters because it is right (or they have a social duty) to protect the environment.

Example E   Companies should make honest claims about the merits of their products because it is good and worthy to be honest.

Subargument of expediency, advantage, or use: is the proposed activity or course of action good for the audience in that it is expedient, advantageous, or useful?

Example C   The United States should agree to the presence of Russian troops in Cuba because such agreement would strengthen the presence of U.S. troops on Russian borders.

Example D   Pierlest Inc. should stop polluting public waters because this would improve its public image and thus its sales.

Example E   Companies should make honest claims about the merits of their products because it is advantageous in dealing with customers to have a reputation for honesty and because such honesty will protect the company from charges of fraud, expensive law suits, and costly penalties imposed by both the government and the courts (because it will be advantageous for them to do so).

The Relationship between Arguments of Fact and Arguments of Policy

You have seen that arguments of fact should be made in a logical sequence and that arguments of policy often depend upon prior arguments of fact. You can combine these insights to produce an overall plan or flow chart for ordering arguments of fact and policy, as demonstrated in Figure 5-2.

Notice that if you are only trying to make an argument of fact, you will never "get to" an argument of policy. In contrast, if you are making an argument of policy, you will need to establish all of the necessary parts of the argument of fact: the subargument of existence, the subargument of definition, and the subargument of quality. You do not, however, need to spend a lot of time establishing these subarguments if your audience already agrees to them.

For instance, suppose that you were trying in a speech to present an argument of policy about Pierlest Inc. and its pollution of public waters: "Pierlest Inc. should be required to stop polluting public waters and to follow a cleanup plan advocated by the State Department of Health." If you were talking to an audience that knew nothing about the pollution problem, you would have to carefully prove each part of the argument of fact before you

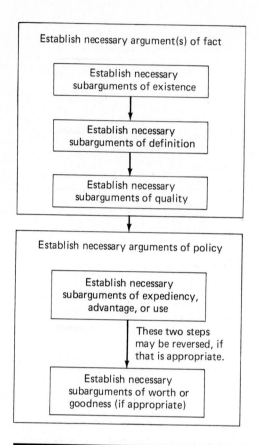

Establish necessary argument(s) of fact

> Establish necessary
> subarguments of existence

> Establish necessary
> subarguments of definition

> Establish necessary
> subarguments of quality

Establish necessary arguments of policy

> Establish necessary
> subarguments of expediency,
> advantage, or use

These two steps
may be reversed, if
that is appropriate.

> Establish necessary
> subarguments of worth or
> goodness (if appropriate)

FIGURE 5-2   Flow chart for constructing effective arguments.

could argue for your policy. However, if everyone already knew that the company had been polluting the waters with a dangerous and illegal substance, you might summarize all of the subarguments of fact in a quick opening statement:

> We all know that Pierlest Inc. has been illegally dumping polyvinyl chloride into the Huron River. We are here to decide what to do about the problem. I would like to argue that Pierlest should be required to follow the cleanup plan proposed by the State Department of Health. The company has a responsibility to the people of the state to clean up the river, and this plan will allow them to do so most quickly and thoroughly and at a minimum cost.

You would then continue (1) to outline the proposed plan, (2) to demonstrate that the plan will allow quick and thorough cleanup of the river, and (3) to

FIGURE 5-3 Introduction to an argument of policy for an audience already accepting the prior argument of fact.

We all know that Pierlest Inc. has been illegally dumping polyvinyl chloride into the Huron River (*argument of fact, already accepted*). We are here to decide what to do about that problem. I would like to argue that the company should be required to follow the cleanup plan proposed by the State Department of Health (*argument of policy, to be proved*). Pierlest has a responsibility to the people of the state to clean up the river (*subargument of worth, perhaps needing to be proved*), and this plan will allow them to do so most quickly and thoroughly (*subargument, advantage to the public, to be proved*) and at a minimum cost (*subargument, advantage to company, to be proved*).

demonstrate that the cost will be minimal. The various subarguments presented are identified in Figure 5-3.

EXERCISE 5-3 This exercise consists of two parts. First, identify the arguments of fact and policy embedded in the short talk below, "The State of Industrial R&D." Notice that much of the proof for individual arguments is missing, since the talk merely gives a short overview, not an exhaustive case analysis. Then, construct several short, general outlines for various arguments of fact and policy. Be sure that you have at least one argument of fact and at least one argument of policy.

The State of Industrial R&D

The role of research and development in industry has changed dramatically in recent years. I can summarize what has happened in a single sentence: Industrial R&D has expanded beyond its traditional function of long-term contributor to corporate growth and has become a real-time contributor to corporate profitability.

In times past, industry's principal technical challenges were the development of new products and the creation of new processes to manufacture these products. As we know, times have changed. In 1977, American industry confronts a host of technological problems that threaten its ability to remain in business—problems such as energy shortages, environmental concerns, skyrocketing materials and manufacturing costs. These are problems that can only be solved *now*.

A new set of R&D priorities has emerged for industry that has made the present as important as the future, maybe more important. And managers of industrial R&D face a perplexing dilemma: How to accommodate these short-term demands without emasculating the long-term programs that are necessary for corporate growth in the years ahead?

This, of course, is a classic predicament of resource allocation. It raises a series of questions that are best answered through a process of research management—a planning/programming approach that integrates industrial research and development efforts with other corporate business activities. Research management is the key to apportioning finite resources across a growing list of R&D demands.

Research management is based on the simple premise that industrial R&D capability is a resource that can help a corporation achieve its business goals. Traditionally, the R&D resource is directed toward the long-term development of technology that will make possible new and/or improved products. At United's Sikorsky Aircraft Division, for example, a long-term R&D program, begun back in the late 1920s, developed the advanced articulated rotor that will lift the next generation of helicopters. This new rotor is a key to the performance, weight-lifting capability, and low vibration of the Blackhawk, tactical utility helicopters we are building for the U.S. Army. The same technology will be carried over to the S-76, a high-performance helicopter designed from the outset for civilian applications.

However, as I mentioned earlier, R&D can also make real-time contributions to corporate profitability. Here are two examples: The costs of raw materials have been climbing 2 percent faster than the overall inflation rate, and these increases represent a significant component of soaring manufacturing costs. Our Pratt & Whitney Aircraft Group is about to bring on line a hot isostatic press developed through company-funded R&D. The press will form jet engine discs by compressing superalloy powder at high temperatures and pressures. The new process needs 70 percent less material to make a turbine disc than conventional forging and machining methods.

Because research and development activities can support diverse corporate goals, it's appropriate to integrate R&D planning into overall corporate planning. This is the starting point of the Research Management process.

The result of coupling technical and business planning is a new perspective on company operations that enables both technical and corporate managers to look ahead at the anticipated benefits of industrial R&D. Specifically, they can judge whether or not the planned technical programs adequately support the company's business objectives. They can ask the kinds of questions that lead to maximum resource utilization and minimum technical risk. Are the goals of the planned R&D programs supportive of corporate strategies? Are important projects being postponed or simply overlooked? Are company activities that need technical support receiving it? If not, what additional resources are necessary to meet the needs? Is there a proper balance between R&D activities designed to develop new products, build new knowledge, and improve existing products? Are proposed R&D programs sized in keeping with their anticipated role in reaching corporate objectives? Are critical programs underfunded? Are less important projects oversupported? Are there any dead-end programs in the plan?

A comprehensive coupling analysis will help everyone involved recognize those R&D activities that are absolutely essential to achieving commercial objectives. Though these projects are on the critical path and carry maximum technical risk in case of failure, they may be small in size and low in perceived priority. At United Technologies, we've found that 10 to 20 percent of the projects in a large R&D program may fall into this category.

Identifying them in advance is the first step in managing technical risk.[5]

TABLE 5-3 OUTLINE FOR THE ARGUMENT OF FACT OR POLICY

I Problem/Introduction/Foreword
 A Direct the audience's attention toward the problem.

II Credentials*
 A If it is useful, give your credentials, i.e., explain why you can speak with authority on the subject, and establish common ground by pointing out shared beliefs, attitudes, and experiences.

III Position/Solution/Summary*
 A Briefly state your position or your proposed solution.
 B Briefly state the major reasons for advocating your position or solution.

IV Background of the Problem*
 A Point out the nature of the problem: (1) its historical background and (2) causes.
 B Explain how it concerns the audience.

V Argument for Position or Solution
 A State the criteria for judgment, i.e., the standards or characteristics any acceptable solution or position must meet. Include explanation where necessary.
 B State your position or solution to the problem, along with any necessary clarification.
 C Demonstrate the soundness of your position or solution by showing how it meets the criteria established in Section IIIA. This step should be accompanied by ample evidence: facts (illustrations, statistics, examples of successful application of the solution) and statements of authority. Be sure to identify the authorities if they aren't widely known.
 D If there are competing positions or solutions, demonstrate the superiority of yours by showing how the others fail to meet the criteria as completely as yours.

VI Conclusion
 A Explain briefly the benefits to be gained by accepting your position or solution or the dangers of rejecting it.
 B Summarize your argument: (1) restate your position or solution (IIIB); (2) restate the reasons your position or solution should be accepted (IIIC).

*Items II, III, and IV may be deleted depending on the situation, the needs of the audience, and the accepted formats for a particular type of technical communication.
SOURCE: Adapted from Richard E. Young, *Rhetoric: Discovery and Change,* Harcourt, Brace & World, New York ,1970, p. 152.

5.4

AN EFFECTIVE GENERAL PATTERN FOR ARGUMENTS

Once you have decided on the basic arguments you are going to make and on their logical order, you need to decide *how* to present the entire speech or report. What goes where? How should things be organized? These are especially difficult questions if you have limited amounts of time and a large amount of information to organize.

One psychologically satisfying and widely useful organization is presented in Table 5-3, which gives an outline for the argument of fact or policy. This organization is so effective that it is still being used over 2500 years after it was first developed in Greece and Rome for speaking in public forums and courts of law. It is particularly useful for technical communicators because almost all technical communication consists of arguments of fact or policy. Thus, by learning this one outline, you can have at your fingertips a general organization for essentially any speech or report.

TABLE 5-4 A SAMPLE ARGUMENT OF FACT:
"ELECTRONIC LOCK DESIGN PROPOSAL"

I Introduction
   A The need for effective yet flexible security systems is more critical today than ever. In many applications the traditional mechanical key or combination lock is no longer adequate. Electronic systems using electronic locks are proving to be the optimal solution, providing advantages unavailable in mechanical systems.
   B Wiley Electronics is about to begin development of an electronic security system of its own. Mr. Silvers has asked me to design a lock to be used as part of such a system.

II

III Summary
   An electronic lock with an 8-digit combination is the optimal lock for Wiley's electronic security system. The proposed lock is effective, simple, and versatile, and it costs only $10.35.

IV Background of Problem
   A Better and more efficient security systems are needed by both industry and the individual.
      1 Most industrial complexes maintain large areas in which stringent security is essential, yet through which large numbers of employees must pass unhindered.
      2 Many individuals also find themselves in situations in which they require a level of security beyond what is normally adequate.
   B Strong interest in better security systems has encouraged the manufacture of many different security systems for the potential customer.
      1 Basically, he/she must choose between ordinary key or combination locks and an electronic lock.
         a A clear trend toward electronic security systems has prevailed over the last few years.
   C Wiley Electronics is about to begin developing a security system to compete in this market.

V Argument
   A The criteria for choosing a system: what does a consumer expect from a security system?
      1 The lock must keep out unauthorized persons and allow those authorized to enter freely.
      2 The lock must be tamperproof.
      3 The lock must have a combination that is easy to alter.
      4 The lock must be easy to interface with a larger controlling system.
      5 The lock must be capable of independent operation.
      6 The lock must have an acceptable cost and market appeal.
   B Solution: the electronic lock meets all of the criteria as well as or better than the mechanical key or combination lock.
   C Argument
      1 The electronic lock is as effective as the other two locks in screening authorized from unauthorized persons.
      2 The electronic lock is more tamperproof.
      3 Its combination is easier to alter.
      4 It is easier to interface with a larger controlling system.
      5 It is as capable of independent operation as the other two locks.
      6 It has an acceptable cost of only $10.35 wholesale.
      7 It will appeal to a wider range of the market than the other two types.
   D Description of the Proposed General Purpose Electronic Lock
      1 Security features of the lock
         a The lock uses an 8-digit combination providing 65,536 different possible combinations for a sufficient level of security for most applications.

(continued)

TABLE 5-4 A SAMPLE ARGUMENT OF FACT:
"ELECTRONIC LOCK DESIGN PROPOSAL" (*Continued*)

     b To provide absolute security, a minor modification will allow the lock to operate with a 16-digit combination, providing over 4 billion unique combinations.

   2 Operation of the lock

     a General operation of the lock

     b Procedure for opening the lock

     c Procedure for setting a new combination

   3 Internal construction and operation of the lock

     a The push-buttons

     b Counter #1

     c The memory

     d The comparator

     e Counter #2

  E Possible alternatives: the "key card" lock

   1 Description: a "key card" lock is the type of lock that accepts a small wallet-size card which is inserted by the operator into a reader. The card is usually impregnated with a magnetic code that the card reader can detect and transmit as a digital signal.

   2 Evaluation: there are two possible means of processing the key card's digital signal, local and remote, both of which can be rejected as inadequate solutions.

VI Conclusion

SOURCE: Adapted from "Electronic Lock Design," *Student Report Handbook,* edited by Leslie A. Olsen, Lisa Barton, and Peter Klaver, Professional Communications Press, Ann Arbor, Mich., forthcoming.

You should note that in any given piece of communication, Sections II (Credentials), III (Position or Solution or Summary), and IV (Background) of the outline may be present or absent, depending on the needs of the audience and the accepted format for the communication. For instance, most technical reports have a heading or title page listing the writer's name and organizational role. These reports do not have a separate section devoted to the credentials of the writer, since those credentials are already established by the role. On the other hand, proposals do have a very important section explicitly presenting the credentials of the proposers. As a second instance, technical reports have a Summary as the second main section of text, whereas a speaker talking to a very hostile audience may decide to hold the Summary or the explicit Statement of Position until after all of the arguments have had a chance to weaken the audience's resistance.

A sample argument to illustrate the general outline is presented in Table 5-4. The numbering of the sample argument corresponds to the numbering in the outline of fact or policy in Table 5-3.

Stating Criteria

A particularly crucial section of the argument of fact or policy is the statement of criteria (Section VA). Criteria are important because they measure the success of *any* solution to a given problem and establish the superiority of *one.* For instance, you can judge the success and superiority of the electronic lock of Table 5-4 only if you know the criteria on which it or any such lock should be judged. By stating the criteria directly, you make it easier for the audience to see them and to agree (or disagree). If the audience agrees with your criteria, and if your position follows from them, then the audience has to agree with your solution as well. (If the audience disagrees with your criteria, it will probably also disagree with your proposed solution or position; if this is the case, you have nothing much to lose by stating the criteria and you may have something to gain.)

Stating criteria has another advantage: it makes it easier for you to create your argument in the first place. For instance, if you want to argue that your electronic lock is acceptable and if the only possible or important criteria are the six stated in Table 5-4, then it is relatively easy to (1) judge your lock by each of the six criteria and then (2) write a section on each judgment. Similarly, if you want to argue that your lock is better than any other lock, all you have to do is to see that your lock meets each criterion as well as or better than any other lock and then to write a section on each comparison. This procedure is illustrated in Section VC of Table 5-4. In some ways, once you have the criteria, the generation of the argument is easy. Note, however, that the hard part is getting the right criteria and making them clear.

Getting the right criteria is sometimes difficult, but there are several general criteria which almost always need to be considered when evaluating a proposed solution:

1 *Effectiveness:* Is the solution effective? Will it solve the problem posed? Why? How do you know?

2 *Technical Feasibility:* Can the solution be implemented? Does it require technology or resources that are unavailable? How do you know?

3 *Desirability:* Would one want to implement the proposed solution? Does it have any undesirable effects? Does it have desirable effects? Why? What are they?

4 *Affordability:* What will the solution cost to implement? To maintain? Is this cost reasonable? Is it affordable? Will it reduce costs in the future? Why?

5 *Preferability:* Is the solution better than or preferred over any other possible solution? Why?

In a particular situation, there may be other criteria that are also appropriate or necessary. As a technical expert, you must know and state these *special criteria.* It may sometimes be difficult to do this, but it should be done, especially for busy readers of important documents who rarely have time to track out arguments and to discover the criteria on which judgments are based.

FIGURE 5-4 A formal statement of criteria.

Introduction

The main conveyer line in Building 9 is used to transport three of Construction Equipment's products. However, each of the three products carried on the main conveyer line has to be switched to a secondary conveyer line which transports them to the shipping area. Currently, three workers are used to carry out this switching process. The products, each of which weighs less than 10 pounds, are lifted off the conveyer line and placed in the appropriate chutes by the workers. This is inefficient and costly, and the labor could be used elsewhere.

The objective of this report is to introduce a successful automatic system to replace the manual system. The new automatic system must meet the following criteria:

    1. It must be compact enough to fit into a 4' × 5' area adjacent to the conveyer line.

    2. It must be economically competitive with the manual system.

    3. It should be of a pneumatic type if possible, because an air supply line is located approximately 10 feet above the conveyer line.

In the following sections I will present the system I have designed. It will automate the current manual process and meet the stated criteria.

How does one actually state criteria? It is usually easiest for readers if criteria are stated formally as illustrated in Figure 5-4. Sometimes criteria can be stated more informally, as illustrated in Figure 5-5, the beginning of a magazine article for a general audience. See if you can identify the criteria being used to evaluate the Intermediate Capacity Transit System, as well as the other modes of mass transportation being compared with it. To give you a head start, we'll look at one sentence, "Subways, of course, are ideal— fast, unobtrusive, impervious to bad weather." The three qualities mentioned are really criteria in disguise: the sentence implies that any ideal mass transportation system should be fast, unobtrusive, and impervious to bad weather.

Sometimes criteria are so obvious that they need not be stated. For instance, look again at Figure 5-5, and especially at the fourth paragraph. Several criteria are implied in the first sentence there: "To make a quiet, unobtrusive yet affordable elevated railway was the goal of UTDC." However, not all of the criteria are stated. This article is talking about mass transit systems, and obviously a quiet, unobtrusive yet affordable elevated railway would be unacceptable if it could carry only 10 people per hour. The article might well have told the reader how many people per hour the system should be able to carry, but any reader would know that it should carry some large number. Further, given the first sentence of the second paragraph, "Yet streetcars and buses are limited to 10,000 passengers per hour," the reader might well assume that the new system should carry more or many more people than the 10,000 mentioned. Are there any other unstated criteria? Are there any criteria missing from the introduction which you think should be considered?

FIGURE 5-5 An informal statement of criteria in a magazine
article. [Charles Maurer, *Popular Science,* **217**(6), 75–76 (1980).]

Tomorrow's Train Is Running Today in Canada

We are standing by the 1.4-mile test track of Ontario's Urban Transportation Development
Corporation. Around us are empty fields that the UTDC hopes will become the foremost
center of urban mass transit in North America. The elevated train I've come to hear and
see—the ICTS (Intermediate Capacity Transit System)—is the vehicle that could bring that
plan to fruition. It's tomorrow's means of mass transportation today. Subways, of course,
are ideal—fast, unobtrusive, impervious to bad weather. But at $80 million per double-track
mile, precious few cities can afford to build them. Few can even afford to maintain the ones
they have.

    Yet streetcars and buses are limited to 10,000 passengers per hour, and to carry
that many they would have to be so close together you could almost climb out on the
vehicles' roofs and walk—and arrive at your destination faster.

    Lightweight, relatively inexpensive rail systems can fill the gap, but if they're to run
at grade level they need a right of way. And with sidewalk-to-sidewalk skyscrapers, there's
no right of way to be had—except over the streets. That's why our largest cities long ago
installed elevated street railways. Sensible monstrosities, but to judge from their noise and
looks, the same firm engineered them that handled hell.

    To make a quiet, unobtrusive yet affordable elevated railway was the goal of UTDC.
Visionary affairs like monorails were out of the question. "They're fine for fairgrounds,"
explained Dick Giles, a mechanical engineer who serves as UTDC's sales manager, "but
for a raft of reasons they're impractical." With monorails, performing maintenance, doing
switching, designing stations, and providing emergency evacuations are all far more difficult. And monorail trains require a higher, more obtrusive track.

    Magnetic levitation was also out—"too complicated, too many developmental risks."
So the train had to run on wheels. To make it unobtrusive, stations had to be short, so
trains had to be short, too—120 feet maximum (four-30-foot cars). Short trains meant fast
trains—45 mph—spaced at intervals of less than one minute apart.

    Spacing trains so closely at those speeds meant computers had to be in control. It
also meant unusually quick accelerating and braking. This, plus the ability to take steep
grades in stride—essential if the railway is to be shoehorned into existing space—ruled
out conventional motors driving steel wheels: too much slippage. And though conventional
motors can drive rubber wheels quietly and surely, the line must be kept free of ice and
snow, impossible in Canada.

EXERCISE 5-4 This exercise should develop your ability to evaluate the
overall structure of a speech or report and, particularly, to state and evaluate
criteria. In Figures 5-6 and 5-7 you will find the beginnings of two different
technical reports, "Analysis of Criteria: Sample 1" and "Analysis of Criteria:
Sample 2." For each report, write a list of criteria that could appear in the
third section of the report. (The third sections are "Why a New System Is
Needed" for Sample 1 and "Introduction" for Sample 2). Note that the
reports may have no specific criteria stated or they may have some—but not

all—of the criteria stated. You need to create a list that moves from the most important criteria to the least important ones. Be sure to consider the five general criteria discussed above—effectiveness, technical feasibility, desirability, affordability, and preferability.

Next, write an outline of an argument of fact or policy that you would like to make. As far as you can, model your outline on the one presented in Table 5-4. Be sure to state a problem at the beginning (see Chapter 6 for a suggested format), derive a list of criteria that any solution to the problem must meet, and then show that your solution meets the criteria and, if appropriate, that it meets the criteria better than other possible solutions.

FIGURE 5-6 Analysis of criteria: sample 1.

Fasano Manufacturing Co.

Interdepartmental Memo

Date: June 24, 1980

To: V. Voros
           Supervisor - Computer Operations

From: R. Sholander
           Analyst - Computer Operations

Subject: Production and Information Control System:
           Preliminary Report on Features and Implementation

Foreword

Mr. Bauer, Vice-President of Production, wants to centralize control of production and purchasing to optimize both inventory levels and assembly shops' loads. New information about purchasing and production is becoming dispersed throughout the system. This makes it difficult for managers and sales representatives to know the status of a job quickly, resulting in delays in dealing with customers. Further, inventory is often at unnecessarily high levels as stock waits to be used in production that has been delayed, and the machine shops are sometimes idle while waiting for parts because purchase orders were delayed or production is ahead of schedule. The computer operations department was given the task of researching and designing solutions to this problem. This report will present my solution and its implementation by computer.

Summary

The solution I propose is an integrated system combining sales forecasting, engineering, inventory control, purchasing, and job shop scheduling. The Production and Information Control System, called PICS, will provide a central data base

FIGURE 5-6   *(Continued)*

which all pertinent information will be collected. It will also provide the following:

- access only to authorized users

- a Materials Requirements Planning system

- optimal shop capacity through simulation procedures

- a status report on any job at any time in the system

- faster response to market demands

This system is well modularized and can be implemented gradually by our departments. A rough time estimate for completion would be about seven months. The cost would be minimal if PICS were operated as a manual system, but this would still require an overwhelming amount of paperwork. A gradual change to a computer system would cost approximately $150,000, but our paperwork could be handled much more efficiently if it were entered directly on a computer disk. Therefore, I suggest a rapid conversion to a computerized approach.

Why a New System Is Needed

Information concerning the inventory system is becoming increasingly more difficult to interpret and understand. Problems arise because each assembly shop keeps its own records of transactions. One problem occurs in purchasing, where orders from shops must be combined to save money in bulk ordering. This combining is often impossible if orders come in separately. A second problem is that it is hard to find the exact status of a job in production if you are not directly involved in the daily operation of that job.

The computer operations department was given the assignment of researching all possible solutions to these problems with particular emphasis on:

- a mechanized system that would facilitate a change to a computerized approach

- a central data bank where all departments contribute and receive information

- a low-cost system that could be phased in gradually

This report presents the alternative I was assigned to investigate and is divided into four sections:

- system overview

- data base requirements

- necessary changes

- cost of new system

FIGURE 5-7 Analysis of criteria: sample 2.

Babies' General Hospital

Interdepartmental Memo

Date: June 20, 1980

To: Nancy Gabrenya, Director
            Medical Information Department

From: Timothy J. Rowell, Principal Engineer
            Management Engineering Department

Subject: REQUEST PROCESSING IN THE CORRESPONDENCE UNIT:
            Recommendations to Improve Productivity and Processing Time Variation

Foreword

The Correspondence Unit of the Medical Information Department at Babies' General Hospital is responsible for supplying medical information to requestors. Because requestors desire this information as soon as possible, hospital management feels that requests should be processed within two weeks. Since the system performance is presently falling well short of this standard, a study of request processing procedures in the Correspondence Unit was undertaken.

This paper reports the results of a three-month study (February through April of 1980) of request processing procedures at Babies' General Hospital. The paper examines the current operations in the Correspondence Unit, identifies the major operational problems, and recommends system improvements which use the currently existing data base and involve minimum implementation costs.

Summary

The request processing system in the Correspondence Unit is a manual operation consisting of five major functions: mail processing, chart retrieving, record screening, photocopying, and mailing. Processing a request currently takes 4.56 weeks.

The smooth flow of requests throughout this system is marred by a number of operational problems. The average worker productivity measured in this study, 59.1%, is below the suggested 65% level. Low worker productivity has caused the backlog of active requests to rise above 2000 requests, thus adding excess processing days. The processing times of many individual requests are increased further by our system of assigning priorities to the requests. Finally, the processing times for some requests are increased even further if the Correspondence Unit does not have the record in question in its own files but must request it from other departments or clinics.

A number of system improvements are recommended to solve these operational problems:

FIGURE 5-7 *(Continued)*

To monitor worker productivity, a productivity index (P.I.) should be used:

$$P.I. = [(2.7) \times (\# \text{ mail processed})$$

$$+ (6.0) \times (\# \text{ records screened})$$

$$+ (4.5) \times (\# \text{ mailings completed})]$$

$$/ [(60) \times (\# \text{ hours worked})]$$

To insure that the backlog of active requests can be processed within two weeks, the backlog size should be maintained near 1000 requests.

To insure equitable request processing with low variance, chronological priorities should be assigned.

Finally, because of the Correspondence Unit's dependence on other departments and clinics, a two-week processing cycle is not feasible for all requests. Thus, management objectives should be to achieve a decreased processing cycle with low variance for all requests and to attain a two-week cycle for those requests with medical records located in file.

REFERENCES

1   J. Thomas Ratchford in *Proceedings of the 31st National Conference on the Advancement of Research,* Denver Research Institute, Denver, 1978, pp. 59–60.

2   Ratchford, op. cit., p. 61.

3   William Bevan in *Proceedings of the 31st National Conference on the Advancement of Research,* Denver Research Institute, Denver, 1978, p. 32.

4   Ratchford, op. cit., p. 64.

5   Wesley A. Kuhrt in *Proceedings of the 31st National Conference on the Advancement of Research,* Denver Research Institute, Denver, 1978, pp. 15, 17–18.

ADDITIONAL READING

Aristotle, *The Rhetoric of Aristotle,* translated by Lane Cooper, Prentice-Hall, Englewood Cliffs, N.J., 1932, 1960. Another good edition of this work is *The "Art" of Rhetoric,* translated by J. H. Freese, Harvard University Press, Cambridge, Mass., 1926, 1967.

Wayne Booth, *Modern Dogma and the Rhetoric of Assent,* The University of Chicago Press, Chicago, 1974.

S. Michael Halloran, "Technical Writing and the Rhetoric of Science," *Journal of Technical Writing and Communication* **8**(2):77–88 (1978).

James A. Kelso, "Science and the Rhetoric of Reality," *Central States Speech Journal* **31**:17–29 (1980).

Carolyn R. Miller, "A Humanistic Rationale for Technical Writing," *College English* **40**(6):610–617 (1979).

Steven Toulmin, *The Uses of Argument,* Cambridge University Press, London, 1958.

Richard E. Young, Alton Becker, and Kenneth Pike, *Rhetoric: Discovery and Change,* Harcourt, Brace & World, New York, 1970.

6

ORIENTING THE NONSPECIALIST
The Foreword or Introduction and Summary

In the real world, you usually communicate about things that are important to you: the proposal you are trying to get funded, the project you worked on for 6 months, a dangerous condition you have just discovered. Unfortunately, as discussed in Chapter 4, you often communicate to audiences who know little about your subject, who may perceive it as unimportant or irrelevant, and who include managers important to your career and to the success of your projects. This situation poses a problem: how do you begin a communication so that the uninformed audience will see its relevance and give it some attention? This chapter addresses that question by presenting a four-step approach for orienting an audience that can be used to begin reports, letters, papers, and oral presentations.* It results in the Foreword and Summary of a technical report or in the Introduction and Summary of a letter, paper, or oral presentation. In particular, this chapter describes

1   A general (or "full") form for posing problems

2   A compressed (or "short") form for posing problems, often found in technical reports

3   Criteria for choosing between the full and short forms

4   A form for stating solutions

6.1
A GENERAL FORM FOR STATING PROBLEMS: FOREWORDS AND INTRODUCTIONS

We all pay more attention to things that interest or seem important to us. One powerful method for developing attention in an audience has been explored by the psychologist Leon Festinger[1] and his coworkers. They argue that the mind—anyone's mind—sometimes finds itself with a sense of unease

---

*We would like to note the influence of our colleagues Dwight W. Stevenson and J. C. Mathes and our ex-colleague Richard E. Young on the evolution of this chapter and the thinking behind it.

about some conflict or inconsistency between two things it believes to be true. When such a conflict or inconsistency occurs, the mind wants it eliminated to avoid chaos, worry, and psychological discomfort. (This is the equivalent of wanting to take off your coat to reduce the physical discomfort of being too hot.)

You can use this observation to catch the audience's attention at the beginning of your communication, in the Foreword of a technical report or the Introduction of a letter or speech. You can introduce an important problem or conflict for the audience and then allow your communication to solve it. Once you've given your audiences a problem they perceive as relevant, they should be willing to hear what you have to say.

Step 1: Introducing a Problem

Richard Young, Alton Becker, and Kenneth Pike[2] have argued that a problem can be seen most easily if it is formed of two directly conflicting terms joined by a word signaling the conflict. The conflicting terms may be perceptions, facts, values, expectations, or beliefs, as indicated in Figure 6-1. Frequently, a value, expectation, or belief in our mind conflicts with an external perception or fact. This pattern is illustrated at the bottom of Figure 6-1, where a belief (the A term) conflicts with a fact (the B term):

> *Our new stereo system must appeal to the teenage market* (belief, A term). *However, it apparently does not appeal to this important market* (perception, B term).

This same pattern appears in the following Introduction from an argument that doctors "are slowly but surely extending cancer patients' lives."

FIGURE 6-1 Outline of a Foreword or Introduction by problem definition, step 1.

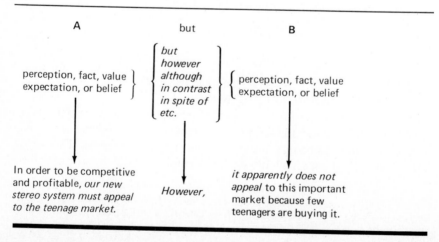

The A term of the problem statement poses the belief or assumption that an individual scientist will find the cure for cancer; the B term presents the conflicting observation that "science rarely works that way." Note that the author starts with an assumption that she believes the audience will readily share: an individual scientist will find the cure for cancer. Note also that she is dealing with a topic—finding a cure for cancer—in which the audience is probably already interested.

IS THE GREAT CANCER CURE ALREADY WITH US?

Since the National Cancer Act of 1971, there has been a continual hubbub about conquering cancer. The public has been led to believe that some morning it will wake up and newspaper headlines will glare: Cancer Cure Found! *The cure, of course, will have come about because some smart or lucky scientist anguished away days and nights until he found the key.*

But science rarely works that way. Breakthroughs come in bits and pieces. And it's exactly in this slow, undramatic way that cancer scientists have greatly increased the number of long-term or even permanent cancer remissions since cancer therapy was first tried during the 19th century. During the 1930's only one out of five cancer patients could expect a remission for five years or perhaps longer. Then one out of three could expect such a remission. Now the rate is approaching one out of two patients. And some of these patients will go on to live out their natural lifespan.[3]

This conflict between a value, expectation, or belief and a perception or fact appears again as the writer of the following Foreword sets up a problem for the managers of The Belle River Power Plant. Note that the author indicates the magnitude of the problem here—replacing coal mills is serious.

The Belle River Power Plant Unit #2 has had over forty years of on-line operation. *Although Unit #2 is still functioning, its increasing demand for both major and minor repairs indicates that its two existing coal mills will need to be replaced soon.* Thus, you requested that we solicit and evaluate bids for the design and fabrication of the two new coal mills. We understand that if you concur, our recommendation will be used as a basis for awarding a purchase order for the design and fabrication of the two new mills. Since the guidelines published by the Public Service Commission require that at least three bidders be considered for such a project, we performed a Dun and Bradstreet analysis of assets and project capabilities on every bidder and then solicited and evaluated bids from the best qualified bidders:

1) American Coal Mill Corporation

2) Babcock and Willis Incorporated

3) Western Electric Incorporated

The following report summarizes our recommendation and supplies the supporting material generated in developing our recommendations.

This author has definitely chosen to phrase the problem statement to catch the audience's attention: managers are concerned that their plants run smoothly and not break down.

Sometimes a good problem statement merely needs to state the A-but-B situation and then move on, as illustrated in the two examples above. However, sometimes a good problem statement needs to state the A-but-B situation and then go on to give details about the A or the B or both. This is illustrated in the following example:

A NEW PUZZLE IN PHYSICS: THE "COSMION"

The latest cherished principle of subatomic physics to be threatened by experiment is the charge independence of the strong interaction. Observations that violate it are reported in the December 30 *Physical Review Letters* as a result of work done partly at the Nuclear Research Center at Demokritos near Athens.

The strong interaction is the force that holds atomic nuclei together. It acts equally upon the electrically charged proton and the neutral neutron. *This independence of charge was crucial in convincing physicists that in the binding of the nucleus they were dealing with a new kind of force* and not a manifestation of electromagnetism as early thoughts on the subject had tended to suppose. *Now the charge independence doesn't look as complete as it once did* thanks to the present work. The work was done by T. E. Kalogeropoulos of Syracuse University and the Demokritos center and nine colleagues from the two institutions.

In the experiments, hydrogen and deuterium nuclei were bombarded with antiprotons to see what happened as the antiprotons met the protons in the nuclei and annihilated each other. Out of such annihilations certain debris is expected, including pi mesons and photons. The basic experimental finding is that there are too many photons, surprisingly too many, almost a whole extra photon per annihilation. The interpretation of this seemingly simple fact leads down a number of important paths in subatomic physics.[4]

At other times a good problem statement needs a line of reasoning leading up to the A or B statement. This is illustrated in the second paragraph of the example immediately above and in the following example from a student paper:

THE EMERGENCE OF GOD-FIGURES IN SCIENCE FICTION

During the twentieth century, man has enchained himself with a world of science, technology, machines and inventions. Arising with this philosophy is the feeling the universe is a machine. Man, who portrays himself as chief engineer, should have every right to probe, study, tinker with and eventually control the mechanism of the world. As scientific man learns more tricks and sheds his old "religious crutch," he is better able to rationally exercise control over nature. *Hence, one would expect science fiction, in keeping with this trend, to depict man as conqueror, leader and master of his future, free of the influence of "gods." But this often is not the case.* Much of science fiction is devoted to the inferiority of man or his inability to execute his free will. Instead of futuristic men finding total fulfillment in expressions of individuality and personal creativity, science fiction often depicts men as passively subservient to a supreme organization as in Orwell's *1984* or Zamiatin's *We*.

The organization with its god-head is idolized and obeyed more fervently than any god ever known to religion. At other times, the controlling figure is a less tangible supernatural entity as in Lindsay's *A Voyage to Arcturus*. No matter what the form, even in science fiction man finds himself at the control of the gods.

If a communicator uses the A-but-B form to identify a problem or conflict for the audience, the communicator must be careful that the A and B terms really conflict or else the whole introduction will not make sense.

EXERCISE 6-1 The following passages were chosen from a variety of scientific and technical sources. See if you can identify the problem stated or implied in each passage, and then try to strengthen or clarify the problem statements. (Although all of these passages are understandable, many would benefit from a sharper problem statement.) Be sure that the magnitude of the problem is clear. Note that several of the passages will be outside your area of specialization; they were selected to help develop your "feel" for the nonspecialist audiences with whom you will often need to communicate.

A   **SUBJECT: ALUMINUM PITTING CORROSION:**
            A PROGRESS REPORT
Foreword: Aluminum is a desirable material for automobile cooling systems because its use provides less corrosion and up to a one-third weight savings and a one-third cost savings over conventional copper-brass cooling systems. However, aluminum exposed to chloride ion solutions undergoes localized corrosion, or pitting.

The reaction of chloride ion with aluminum metal is a well-documented phenomenon, but the interaction between the chloride ion and the passive oxide or hydroxide film which always exists on aluminum surfaces is yet unclear. This interaction is currently being studied by potentiostatic testing to determine the effect of chloride ions on the initiation of pitting corrosion. The purpose of this report is to present the results of these potentiostatic tests.

B   **CREATIVITY**
As engineers, it will be our job to find new and better ways to solve many problems. Since creativity increases one's ability to solve problems, it is logical that a more creative individual will make a better engineer. But most of us do not inherently possess a great deal of creativity. Does this mean that we are doomed to be nothing more than average engineers, hindered by our inability to think creatively? Not at all. Many researchers in this field believe that creative thinking is simply a mental process and that this process can be learned by any intelligent person. In fact, many researchers believe that most people possess a great deal of creative potential which they simply fail to develop.

So if a creative engineer is a good engineer, and if this creative ability can be learned, an important question for many engineers is: how can a person learn to increase her or his ability to work creatively? To learn how, one must first understand what creativity is.

C   SUPERGRAVITY AND THE UNIFICATION OF THE LAWS OF PHYSICS

A catalogue of the most basic constituents of the universe would have to include dozens or even hundreds of particles of matter, which interact with one another through the agency of four kinds of force: strong, electromagnetic, weak and gravitational. There is no obvious reason why nature should be so complicated, and perhaps the most ambitious goal of modern physics is to discover in the diversity of particles and forces a simpler underlying order. In particular, a more satisfying understanding of nature could be achieved if the four forces could somehow be unified. Ideally they would all be shown to have a common origin; they would be viewed as different manifestations of a single, more fundamental force.

In the past 50 years remarkable progress has been made in identifying the elementary particles of matter and in understanding the interactions between them. Of course, many problems remain to be solved; two of the most fundamental ones concern gravitation. First, it is not understood how gravitation is related to the other fundamental forces. Second, there is no workable theory of gravitation that is consistent with the principles of quantum mechanics. Recently a new theory of gravitation called supergravity has led to new ideas on both these problems. It may represent a step toward solving them. [5]

D   LIFE MAY BE HAZARDOUS TO YOUR HEALTH: I

The message we seem to be getting from the media is that our health, if not our very lives, is endangered by the very same technology that supports our existence. The series of warnings is ceaseless: DDT, PCB, polyvinyl chloride, nitrous oxides in the air, unsafe automobiles, radiation, food poisoning, sweeteners, contraceptives, fallout, fluorocarbons, noise pollution, x-rays, cosmetics, . . ., and on and on. One might begin to wonder how we ever managed to survive past the age of 25. Controversies begin and crises of this type often arise because of the uncertainties and complexities of the concept of safety. [6]

E   LIFE MAY BE HAZARDOUS TO YOUR HEALTH: II

As scientists and engineers, one of our daily concerns is with the concepts of risk and safety. It makes no difference what aspect of engineering you eventually become involved with, be it product engineering, manufacturing, quality assurance, sales and marketing, maintenance, or product supervision. In the end, something for which you have been responsible will have the potential for causing harm, big or small. It is essential that we, the bastions of technology, have a thorough understanding of risk, safety, what they involve, and how they are determined.

At one time or another we've all heard the doomsayers who preach that our technology has brought us to the brink of destruction. The solution they offer is full retreat. Before jumping to any such rash conclusions, let's step back and take an historical perspective. To make the comparison fair, we'll stick to modern times and not go back to an era when life was completely miserable.[7]

F   SCIENCE FICTION: A WARNING TO HUMANITY
From the beginning of the human race, through prehistoric times to the present, man has always sought power. He has wanted to control himself and his environment. He has wanted to be able to predict the future. In short, he has wanted to be omnipotent. The examples of this desire, this greed, to become an ideal, perfectly powerful being are many. We have, in the tale of Adam and Eve, a desire to become one's own master. More recently, we have men such as Caesar and Hitler, who wanted to control the world. Even today we see this quest for power, not only in political leaders, but in everyday, common people. People are always trying to make more money, to learn more, in order that they may improve their status.

But this quest for power is not without effect. Things go wrong, people are hurt, and the human condition deteriorates. You may ask "What can be done about this situation?" My answer is that man must be more careful about what he does; he must more closely analyze his situation, what he is planning on doing, and what its effects will be. In this paper, I will present the role that science fiction plays in showing us why we should be more careful about what we do, and what the consequences are if we are not careful.

G   THE AUTOMOBILE: FOR BETTER OR WORSE
Cadillac's chief engineer, Robert Templin, is hardly alone among his colleagues in the industry when he asks, "How much pollution is too much? How much fuel efficiency is enough? Who will pick up the tab for a zero-risk, zero-pollution car?" Never in its history has Detroit been faced with so many problems. Henry Ford himself might have opted for horse-drawn buggies instead of horseless carriages in the same situation.

On the other hand, the search for answers to perplexing problems traditionally has led to inspired and sometimes unorthodox solutions. Necessity is the mother of invention—and for the electronics industry, at least, Detroit's difficulties in meeting regulations have opened doors to electronic applications in cars that otherwise might not have opened so soon, if at all.

This special issue on the automobile reflects both the controversies about Government regulations for cars and the inroads being made by electronics. For instance, author Stork says, on page 51, "EPA is satisfied, from its analyses and studies, that the emissions standards enacted by the Congress can be met without loss in fuel economy, and that even the full statutory emissions standard for nitrogen oxides can eventually be met on that same basis." Yet, on page 47, authors Heinen and Beckman assert, "The simple truth is that there is very direct interaction between emissions and fuel economy. Probably the clearest example of that interaction is the fact that automobiles equipped to meet California's tight emissions-control regulations have consistently demonstrated about 10 percent poorer fuel economy than have comparable cars equipped to meet the less stringent Federal U.S. standards."[8]

H   PASSIVE COOLING SYSTEMS IN IRANIAN ARCHITECTURE
The internal environment of a modern building can be a comfortable one no matter how uncomfortable the external environment is. In general the reason is that energy is freely spent to heat or cool the building. In earlier times, however, when energy was not so readily available and machines such as air conditioners did not exist, the designers of buildings had to rely on other stratagems to maximize the comfort of the internal environment. For example, the traditional architecture of many cultures in climates where the temperatures are uncomfortably hot during the day and uncomfortably cool at night features buildings with thick walls of brick or stone. Such walls are both insulators and reservoirs of heat, so that during the hotter hours of the day the flow of heat from the external environment to the internal one is retarded and during the cooler hours part of the heat stored in the walls warms the internal environment and the rest is lost to the external one. The net result is a flattening of the temperature-variation curve inside the building. In a period when the energy costs of buildings are being intensively reevaluated, such stratagems clearly merit close consideration.

In Iran certain traditional building designs achieve more than a flattening of the temperature curve; they circulate cool air through the building and can even keep water cold and ice frozen from the winter until the height of the long, hot summer of the country's arid central and eastern plains. They do so without any input of energy other than that of the natural environment; hence they can be characterized as passive cooling systems.[9]

EXERCISE 6-2   Survey your experience for a hobby or some other subject you know something about. If you have worked at a job or done an experiment in a laboratory course, then pick a particular problem or experiment you could talk about. For your topic, write an Introduction by problem statement for someone who knows nothing about it. Be sure to define your audience and to keep its interests in mind in setting up the introduction.

Steps 2 and 3: Identifying Your Strategy on the Problem and the Purpose of Your Communication

Once you have established an important problem for the audience, you need to show that your communication addresses this problem. This is done by

1   Indicating the missing information you're providing and the strategy you're taking on the problem (you do this by stating the question or task you're addressing)

2   Announcing the purpose of the communication

For example, let us reconsider the example in Figure 6-1:

Our new stereo system must appeal to the teenage market.

However, it apparently does not appeal to this market.

After having stated the problem directly, we could have tried to solve it by asking questions such as the following:

What is wrong with teenagers?

What is wrong with the stereo system?

Why aren't teenagers buying the stereo system?

Where is it being sold?

Who is buying the stereo system?

If we had asked the first question, we would have assumed that something is wrong with the teenagers and have committed ourselves to this point of view. It may be that something is wrong with them, but it may be that something else is wrong, something we would never see given this perspective. The same complaint could be advanced about the second question. However, the third, fourth, and fifth questions would not have committed us to a limited point of view and thus would probably have produced better information. If, for instance, the cause of poor sales is poor advertising, the last three questions would be much more likely to guide us to this cause.

This example illustrates two important properties of such questions. First, since at the investigation stage we don't want to close off fruitful lines of inquiry, the questions recreating this stage should be open-ended, that is, they should not limit us unnecessarily and should usually begin with such words as *who, what, where, when, why, how,* or *to what degree.* Second, the questions we ask affect the solutions we reach; in fact, the questions define our *strategy* for solving a problem. For instance, all the questions above point to potential causes; they indicate that our strategy for finding a solution is to find the cause of the problem. A different set of questions would have defined a different strategy. If we had asked the following questions, for instance, our strategy would have been to improve the product:

How can we make the stereo more cheaply?

How can we improve the performance of the stereo?

How can we reduce the size and weight of the stereo to make it more portable?

Can we make the stereo look "richer" and more elegant?

Note that if the real problem had been poor advertising, the strategy of improving the product would probably have been ineffective.

How do we define our strategy and our purpose in an Introduction? Let's say that we have explored the question "Why aren't teenagers buying the stereo system?'' and decided that the reason was poor advertising, appearing at the wrong places and at the wrong times. If we were to report on this situation to a vice president and a manager at the stereo company, we might produce an Introduction similar to one of the following:

VERSION 1

In order to be competitive and profitable, our new stereo system must appeal to the teenage market. However, it apparently does not appeal to this important market since few teenagers are buying it. What caused this situation? It appears that a poor sales campaign is the only major cause.

VERSION 2

In order to be competitive and profitable, our new stereo system must appeal to the teenage market. However, it apparently does not appeal to this important market since few teenagers are buying it. I was asked to investigate this situation and to identify the cause. This report will argue that a poor sales campaign is the only major cause.

Notice that each of these has a clear problem statement followed by a statement of question or task and a statement of purpose. Notice also that in version 2 the statement of task is merely an unstated question. "I was asked to investigate this situation and to identify the cause" means "I was asked to find out what caused this situation." The process of writing an Introduction just discussed is summarized in Figure 6-2.

Another example of an Introduction by problem statement follows, written by a student to administrators at his university. He hoped that his report, from which this Introduction was taken, would help decrease or eliminate the use of the test mentioned.

In 1962, the Opinion Attitude and Interest Survey (OAIS) Test, more commonly known as the "raw carrot" test, was first introduced to the University of Michigan as part of the testing program administered during an entering student's orientation. It is a psychological test, testing for motivation, vocational aptitude, vocational interest, and certain factors relating to personality. This exam is supposed to be used as a placement device by the counselors in chemistry and math courses. However, if one surveys the counselors on campus, the test scores are not used in this manner; in fact, they are not used at all. Thus, the largest school at the University—Letters, Sciences, and Arts—stopped "coercing" its students to take this examination. Some schools have followed, but others have not—notably, Nursing, Education, and Engineering.

If the results are not used as the test examiners have projected, then why should the exam be forced on the non-LSA students in Nursing, Education, and Engineering? There is no compelling reason. In fact, the test should not even be encouraged by these colleges for large groups of students; it has internal problems and can be misused and misinterpreted, especially in com-

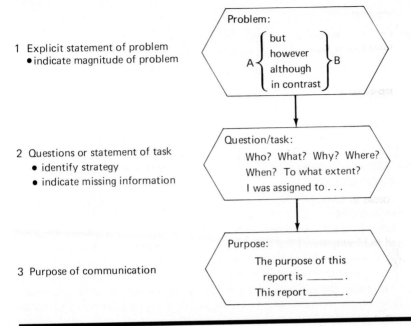

1 Explicit statement of problem
   • indicate magnitude of problem

2 Questions or statement of task
   • identify strategy
   • indicate missing information

3 Purpose of communication

FIGURE 6-2 Outline of Introduction or Foreword,
steps 2 and 3.

bination with other, less volatile sources of information at the counselors' disposal.

Notice that this author has considered his audience carefully in phrasing the problem statement. He has emphasized the fact that the test scores are not used and implied that the university is wasting its resources by continuing to administer the test.

The author of the above report on the OAIS also had three other complaints about the test, which he developed in his report: (1) that it was an invasion of privacy, (2) that it was poorly designed and thus gave poor results, and (3) that it wasted valuable student time during orientation week.

EXERCISE 6-3 Define some audiences for which each of these three complaints could be used in an A-but-B problem statement. You will thus have at least three audiences and three problem statements.

EXERCISE 6-4 Go back to the passages quoted in Exercise 6-1. See if you can identify the question or task and the statement of purpose in each one. Then see if you can add a question or task and purpose for the introduction you wrote in Exercise 6-2.

6.2

A SHORT FORM FOR STATING PROBLEMS

Section 6.1 presented a general form for starting letters, reports, or speeches based on an A-but-B problem statement. This format is commonly used and important. However, if you look at much communication in science and technology, you will find examples which seem to violate the A-but-B format because they seem to be missing either the A term or the B term. Consider, for instance, the following example from a report by a Research Forester to the Superintendent of a State Forest Nursery (notice how the problem statement addresses the concerns of the Superintendent and indicates the magnitude of the problem):

> **FOREWORD**
>
> During my visit to your nursery on September 15, 1979, you and I observed the high mortality of the two-year-old red pine seedlings in bed 19. As you stated, this would lead to a production shortage in 1980. You requested my help in finding the cause of this problem so that it could be corrected in the near future. The purposes of this report are 1) to explain that parasitic nematodes are the probable cause of the high seedling mortality and 2) to recommend a solution to the problem.

What is happening in this example and others like it is that one term in the A-but-B format is unstated, but assumed. It's assumed because it's believed by all members of the audience (the Superintendent and other employees at the nursery) and so obvious that saying it would seem ridiculous. For instance, in the example above, the author merely states the B term: "you and I observed the high mortality of the two-year-old red pine seedlings . . . this would lead to a production shortage in 1980." The author assumes any reader of the report would know the A term: one does not want a high mortality rate of red pine seedlings or a production shortage in 1980. Just stating that they might occur puts the ideas of high mortality and production shortage in clear conflict with a reader's deeply held values and assumptions that they should not occur.

The "short-form" problem statement just described is represented in Figure 6-3, where brackets have been placed around unstated but assumed material. The short form appears frequently in scientific and technical communication, since many of the problems scientists and technical people write about arise from commonly held assumptions. Almost everyone wants and is interested in

Efficiency

Low cost

Competitiveness in cost and performance

FIGURE 6-3 Outline for a short-form problem statement.

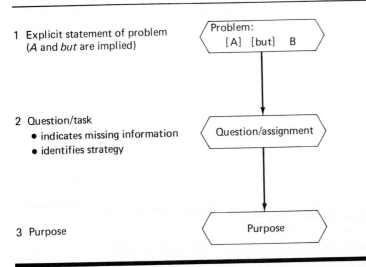

1 Explicit statement of problem
(A and *but* are implied)

Problem:
[A] [but]   B

2 Question/task
  • indicates missing information
  • identifies strategy

Question/assignment

3 Purpose

Purpose

Products of good and even quality

Freedom from pollution violations, labor unrest, and legal action by consumers or the government

Elimination of wasted time, energy, or money

Simplicity rather than complexity in organizational structures and product design

When such values are used in a problem statement, they are usually not stated directly. However, they are there and are often the values needed to catch the interest of managers and others not directly involved with a project. Interestingly, these values also provide the context needed by someone outside a project to understand its real importance and purpose. Look back, for instance, at the example on mortality in red pine seedlings. Could you really understand why someone was worried about parasitic nematodes (the probable cause of the high seedling mortality) without calling on the values and assumptions implied in the Foreword?

We might consider another example of a short-form problem statement from a report written by a civil engineer to the Chief Engineer of a manufacturing company. The readers of this report might well include managers, architects, financial analysts, constructors, and even a vice president. The author rightly assumes that any reader would know that the manufacturing company does not want to damage its building foundations. Again, this is so obvious that an author would not state it, but it is unquestionably there in the mind of the reader.

On January 22, 1980, you requested the analysis of two samples of sand to see if they could be used as a foundation for several high-speed presses. You expressed concern that this sand, which is below the punch press foundations, might settle due to the vibration of the presses, thus damaging your building foundations. Along with the two samples, the following data was forwarded:

Void ratio of white sand (Ottawa Sand) .65

Dry unit weight of black sand (Turkish Emery) 118 lb/ft³

The purpose of this letter is to recommend that neither sand is sufficient for use as a foundation for high-speed punch presses. The vibratory loadings transferred from the machines will cause the foundations to settle.

EXERCISE 6-5 Examine the following Forewords and Introductions from technical articles and reports. Each is missing an overt statement of one of the conflicting terms of its problem statement. See if you can identify or easily reconstruct both the A and B terms of each problem statement.

A **SUBJECT: VACUUM FILTER SLUDGE CAKE:**
ACCURACY OF PRESENT PRODUCTION DATA
Foreword: Slowdowns and necessary repairs have occurred in the incineration area of the Sewage Treatment Plant, costing two million dollars last year alone. These problems may have occurred because of inaccurate data on the amount of sludge cake being produced and delivered to individual furnaces. Thus, Dennis Moore, the Chief Project Engineer, requested that I investigate our data on sludge cake production and suggest improvements in monitoring it if those are necessary. The purpose of this report is to document the unreliability of our present monitoring technique and to recommend a better one.

B **THE SEARCH FOR LIFE ON MARS**
Is there life on Mars? The question is an interesting and legitimate scientific one, quite unrelated to the fact that generations of science-fiction writers have populated Mars with creatures of their imagination. Of all the extraterrestrial bodies in the solar system, Mars is the one most like the earth, and it is by far the most plausible habitat for extraterrestrial life in the solar system. For that reason a major objective of the Viking mission to Mars was to search for evidences of life.[10]

C **SUBJECT: PROPOSED TREATMENT SYSTEM**
FOR CITRUS PROCESSING WASTES
Foreword: In your letter of July 7, 1971, you asked me to suggest a treatment process for the wastewater from your new citrus processing plant. You stated that any treatment process selected should produce an effluent in compliance with federal standards and be economical to operate and maintain. Consequently, I have compared several treatment alternatives based on your data

and criteria. The purpose of this report is to recommend a process for economically and efficiently treating your citrus processing wastes.

D   DRIP IRRIGATION

Some 40 years ago Symcha Blass, an Israeli engineer, observed that a large tree near a leaking faucet exhibited a more vigorous growth than the other trees in the area, which were not reached by the water from the faucet. Blass knew that conventional methods of irrigation waste much of the water that is applied to the crop, and so the example of the leaking faucet led him to the concept of an irrigation system that would apply water in small amounts, literally drop by drop. Eventually he devised and patented a low-pressure system for delivering small amounts of water to the roots of plants at frequent intervals. The technique, as developed by Blass and subsequently refined by him and various manufacturers, consists in laying a plastic tube of small diameter on the surface of the field alongside the plants and delivering water to the plants slowly but frequently from holes or special emitters located at appropriate points along the tube. The concept, which is now called drip irrigation or trickle irrigation, has gained wide acceptance, proving to be particularly valuable in areas that are arid and have high labor costs. An unforeseen benefit is that the system works well with water that is highly saline, as water in arid regions often is.[11]

6.3

GUIDELINES FOR CHOOSING BETWEEN FULL-FORM AND SHORT-FORM PROBLEM STATEMENTS

Sections 6.1 and 6.2 describe two related forms for problem statements, the full form and the short form. The purpose of this section is to summarize guidelines for choosing one form or the other.

As already suggested, you should use the full form of the problem statement when your audience or part of your audience might not know clearly and obviously what each term is. You may choose the short form when one term is so clearly known and believed by all members of your audience that it does not need to be stated to make the problem clear. The possible choices are summarized in Figure 6-4.

For instance, in the competitive business system of the United States, the shared values include those presented in Section 6.2: the desire for low cost; efficiency; good and uniform product quality; freedom from legal prosecution, labor disputes, and pollution violations; simplicity; and elimination of wasted time, energy, or money.

In a scientific or technical field, shared knowledge, beliefs, and expectations are based on the experience common to members of the field. It would be impossible to give an inclusive list of all the shared values that could be unstated in some particular situation. Thus, a communicator must consider various audiences for a communication and produce an Introduction aimed at the least knowledgeable member of the audience.

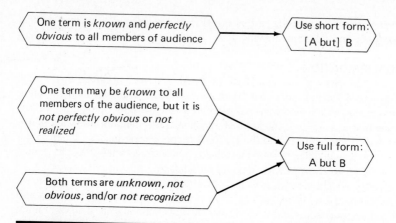

FIGURE 6-4 Guideline for choosing between full- and short-form problem statements.

6.4

INTRODUCTIONS AND FOREWORDS: A REVIEW

Communication about a project can be presented in a variety of formats and to a variety of audiences. The most common of the formats are the technical report, the letter, and the scientific or technical article.

The Foreword of a Technical Report

Technical reports may be aimed at specialists, at nonspecialists, or at both at the same time. They are usually read by a wide variety of audiences—managers, sales people, technicians, financial analysts, specialists in the field, and sometimes even the public. Thus the beginning of such a document must orient audiences of varying backgrounds, experiences, values, and assumptions. To make sure that all readers understand the problem set up at the beginning, the writer of a technical report must write to the least knowledgeable reader. This is the nonspecialist, especially the manager unfamiliar with and uninterested in the details of a particular project.

The technical report orients its variety of readers in its first two sections, the Foreword and the Summary. As with other types of introductions, the purposes of the Foreword are

1   To catch the audience's attention; to place the report in a context so that readers can see how it fits in with other communications and with overall goals; to tell *why* the project was done and *why* it is important

2   To quickly orient the audience to the subject of the report; to define *what* was done; to indicate the missing information the report provides

(these are often covered with an explanation of the writer's technical assignment)

3   To define the purpose and focus of the report

The purposes of the Summary are

4   To quickly present the main results of the project

5   To quickly present the important recommendations and implications of the project

These purposes are summarized in Table 6-1. For illustrations of Forewords, you should reread the many Forewords already presented. The Summary will be treated later in this chapter.

The Introduction of a Letter

Letters have many of the characteristics just mentioned for technical reports. They may be aimed at specialists or nonspecialists or both at the same time; they may be read by a wide variety of audiences; they should be written to the least knowledgeable audience; and, depending on their audiences, they may begin with full-form or short-form problem statements. For examples of letter Introductions, see the letters discussed in Chapter 11, "The Job Letter" and "The Business Letter."

The Introduction of a Scientific or Technical Article

Articles aimed at nonspecialists may begin with an Introduction using either the full-form or the short-form problem statement. In either event, the writer must ensure that both terms of the problem statement are clearly understood.

TABLE 6-1   OUTLINE OF THE FOREWORD AND SUMMARY IN THE TECHNICAL REPORT

FOREWORD		SUMMARY	
PURPOSE	FORM	PURPOSE	FORM
Define context of report to catch audience's attention; tell *why* project was done and *why* it is important	[A but] B	Present main results Present recommendations and implications	Results Recommendations and implications
Indicate missing information; define subject of report; define *what* was done	Assignment or question		
Define *purpose* and focus of report	Purpose and focus		

Several examples of article Introductions aimed at nonspecialists appear in the exercises in this chapter.

6.5

THE SUMMARY

The Foreword and Summary at the beginning of a technical report are the structural features most obviously responsive to the busy managerial or non-expert reader. The Summary provides a compact statement of results, con-clusions, and recommendations to help managerial readers make decisions: it tells what the writer discovered, the implications of the discovery, and the recommendations for action based on the writer's special knowledge—all in terms that the nontechnical manager will find useful.

Some of the questions that a managerial reader will have about a project or study[12] are the following:

1   What is the project's *importance* to the company? Its areas and extent of application?

2   What will it *cost*?

3   Are any *problems* projected because of it?

4   Are there any *implications* to the company from it? Any *more work* to be done? Any *resources* required or involved (people, facilities, equip-ment)? Any *priorities* to be added or changed?

5   Are there any important *dates* or *deadlines* associated with it?

6   Are there any important *recommendations* for future action which the writer could make based on his or her special expertise?

Writers should try to answer these questions in the Summary because they are particularly qualified to do so. They often know more about the answers than anyone else. Further, they are hired to analyze and evaluate data, not just to present it. If they make no effort to analyze their project in the larger context of the organization, they are not really doing their job. Finally, by answering such questions for managers, writers can look excep-tionally competent. Relatively few technical people properly address these issues, and those who do will seem unusually useful to the organization.

What does a Summary that addresses these questions look like? For a sample, you might look at the Summary in Figure 6-5. It comes just after the Foreword, which has posed a problem: there have been frequent delays in an experiment because of the excessive time required to pump down a vacuum system. Since these delays are undesirable, the writer has been asked to reduce them. The Summary reports the writer's solution: buy a gate valve for the vacuum pump. The Summary also tells why this is a good

FIGURE 6-5 Movement from generalizations and claims to data
and support in a short informal report: example 1.

Watson Telephone Laboratories, Inc.
Murray Hill, N.J. 07074

Date: February 5, 1979

To: Dr. J. C. Smith
Manager, Low-Energy Research Group

From: Catherine Doe
Mechanical Systems Engineer
Positron Research Group

Subject: Positron Beam Experiment: Reducing Delays

Foreword: In your letter of January 23, 1979, to Dr. D. Newman, Director of the
Positron Research Group, you expressed concern about the numerous delays in the
positron beam experiment. Dr. Newman attributes most delays to the ten hours
required to pump the vacuum system down from atmospheric pressure to an acceptable operating level and has asked me to investigate means of reducing these
delays. This report recommends the purchase and installation of a gate valve for
the Vac-Ion vacuum pump as the most appropriate method for reducing the delays.

Summary: The purchase of a gate valve for the Vac-Ion pump was found to be the
optimal solution to the delay problem in the positron beam experiment. Addition
of the gate valve will reduce evacuation time to three or four hours, cost no more
than $600, and present no installation problems. The use of a larger fore pump
was seen as the only alternative approach, but this would require extensive modification of the vacuum system, cost approximately $1100, and give no improve-
ment in the evacuation time.

Details: The excessive time required to obtain a suitable vacuum, 10^{-7} torr or less,
is primarily due to contamination of the ion pump. When the vacuum system is
opened to the atmosphere, the entire inner surface of the system and the ion
pump itself become contaminated. The contaminated surface then outgasses considerably upon evacuation, keeping the pressure high for more than ten hours.
The two possible solutions to this problem are to avoid the contamination altogether or to decontaminate more quickly.

The proposed solution would avoid contamination by the installation of a gate
valve on the Vac-Ion vacuum pump. This would allow the ion pump to remain in
operation even while the rest of the system is open to the atmosphere, thus preventing contamination of the pump. This would be a major improvement, since
only the relatively small surface area of the rest of the vacuum system would become contaminated, and evacuation time would be reduced to three or four hours.
A suitable gate valve costs $500 to $600.

The rejected alternative, the purchase of a larger fore pump, would speed the de-
contamination of the vacuum system, but cost too much. Replacing our present
120 ft.^{3}/min. fore pump by an 800 ft.^{3}/min. pump would reduce the evacuation time
to about 4 hours. However, an 800 ft.^{3}/min. fore pump costs approximately $1100,
and its installation would require substantial modification of the vacuum system.

Thus, I recommend the purchase and installation of a gate valve as the best
method for reducing delays in the positron beam experiment.

If you have any questions, I would be happy to answer them.

solution—it will reduce evacuation time to acceptable levels, cost no more than $600, and present no installation problems. Notice that these advantages address the manager's concerns. Before a manager could approve the purchase of the gate valve, she or he would need to know the cost (only $600), the benefits (reduction of delay), and the implications (no installation problems). In addition, the phrase *no installation problems* implies other useful information: that there should be no more problems with the unit and no unusual work or resources required.

A second summary, shown in Figure 6-6, illustrates many of the same points. Again the Foreword has defined a problem: Orange Grove Products needs a method for treating wastewater from its new citrus processing plant, a method which meets criteria of effectiveness and cost. The Summary provides a solution: an aerated lagoon will meet both criteria better than other potential solutions. The Summary also gives general performance and cost data, the kinds of additional information a manager needs to make a decision. (A manager would not approve a solution which was ineffective or too expensive.)

Framing Summaries for Particular Audiences

Once you have defined the audiences for a report, you need to decide how to convey your information so that the audiences can understand and use it. This can be tricky when different audiences need different kinds of information or the same information presented in different ways.

For instance, consider the following situation. You are a geologist working for a township council composed of two business people, one farmer, one school teacher, and one lawyer. The township has so attracted new industry and residents for the past 8 years that water shortages have developed and future growth will be impossible without an expanded water supply. You have been hired to locate, if possible, new wells to increase the township's water supply. After some effort, you have succeeded and must inform the council of your success. Which version of your results would be most appropriate for the township council?

SUMMARY 1

Six test wells were drilled, four of which penetrated suitable sand and gravel.

1   The relationship of geologic units encountered in Test Wells 1 through 6 is interpreted and presented in Plate 1. Two aquifers were identified; an extremely thick aquifer was encountered in Test Wells 2, 5, and 6 ranging in thickness from 78 to 125 feet, and a deep aquifer was encountered in Test Well 1 extending from 150 to 185 feet. There is theorized to be limited hydraulic connection through fine-grained units between the two aquifers. Some degree of connection between the deep aquifer and upper aquifer was established by the rapid rise in water levels in Test Well 1 observed during a heavy rain in September 1979.

2   A conservative estimate of the yield from the upper aquifer is 2.0 mgd. Aquifer data indicate this could be produced by pumping any combination of two wells among Test Wells 2, 5, and 6 (assuming manganese in Test Well 2 proves acceptable). Ultimate production from a French Creek well field could prove to be in the 3.0 to 5.0 mgd range.

SUMMARY 2

We conducted studies and six test drillings to locate an area in which a reliable, high-quality municipal water well field could be developed. We have located such a site.

Results   Four of the six wells drilled were productive and penetrated sand and gravel suitable for development of municipal wells. Each of the four productive wells showed very good water-producing capability, yielding 300 to 500 gallons per minute.

Conclusions   Production wells completed at the test well sites should more than supply the township's current and future needs. Analysis of test data indicates that such production wells could produce over 2 million gallons per day, twice the township's projected requirements. Further, if the water-bearing sand and gravel are as extensive as we project them to be, yields up to 5 million gallons per day could conceivably be developed.

Unless the reader is a geologist, version 1 will probably not make any sense. It contains unfamiliar terms such as *aquifer* and *mgd*. It also contains statements which sound as though they ought to convey information but don't. For instance, Section 2 states, "A conservative estimate of the yield from the upper aquifer is 2.0 mgd." Even if readers guess that *mgd* means *million gallons per day*, they still don't know if 2.0 mgd is a sufficient water supply.

In contrast, version 2 gives the information in a form that nonspecialists can understand. It explicitly states the generalizations the nonspecialist reader needs to know:

1   The geologist has found water.

2   The geologist has found enough water to supply the township's current and future needs.

EXERCISE 6-6   Write a Summary for the reports shown in Figures 6-7 and 6-8. When you have finished, be sure to check your Summary for the important results, conclusions, implications, and recommendations. If you have recently written any other reports, go back now and check their Summaries for appropriateness and completeness. Since Summaries are so important, you should practice writing and rewriting them whenever you have the opportunity.

FIGURE 6-6 Movement from generalizations and claims to data and support in a short informal report: example 2.

To: Ms. J. Jones, Director
                Orange Grove Products
                1135 Halifax Building
                Orlando, Florida 32105

From: Leo Ming Chen, Environmental Engineer
                Camp, Dresser and McKee Inc.
                One Center Plaza
                Boston, Massachusetts 02108

Date: October 29, 1979

Subject: Citrus Processing Waste: Treatment System Proposal

Foreword

    In your letter of July 7, 1979, you asked me to suggest a treatment process for the wastewater from your new citrus processing plant. You stated that any treatment process selected should:

1. exhibit performance effectiveness under average and adverse flow conditions

2. exhibit cost superiority in terms of initial cost and yearly spending

    Consequently, I have compared several treatment alternatives using the data you have supplied and your criteria as a basis for comparison. The purpose of this report is to recommend a process for economically and efficiently treating citrus processing wastes.

Summary

    An aerated lagoon is recommended as the most efficient and economical method for treating citrus processing wastes. Several treatment processes were considered in the selection. These include the activated sludge process, the anaerobic lagoon, and the aerated lagoon. The advantages of the aerated lagoon over the other treatment processes are as follows:

1. The aerated lagoon is the only alternative which could meet the federal pollutions standards under adverse flow conditions. It exhibits significantly better performance under all conditions through more consistent BOD reduction and higher organic loading potential.

2. The aerated lagoon affords significantly lower initial and yearly costs due to its ease of construction, operation, and maintenance. Per lagoon, the estimated initial cost is only $114,000 and the annual operating cost $22,800, approximately half as expensive as the most economical of the other two options.

FIGURE 6-6 *(continued)*

Performance Superiority of Aerated Lagoons

Aerated lagoons consistently produce a better quality effluent than do activated sludge processes or anaerobic lagoons. Aerated lagoons exhibit better BOD reduction and higher organic loading potential under both average and adverse flow conditions than do either of the other treatment schemes.

1. Superior BOD Reduction by Aerated Lagoons

The standard for BOD, as published in the Federal Register of July 1, 1979, states that all discharges into receiving streams shall contain no more than 30 mg/liter of BOD. Table 1 shows aerated lagoons with 95% BOD reduction potential to be capable of producing an effluent in compliance with federal standards under both average and adverse flow conditions. Activated sludge processes and anaerobic lagoons on the other hand can only effectively treat wastewaters of average BOD values.

Table 1: BOD reduction comparison of treatment processes

Secondary Treatment System	BOD Reduction (%)	Effluent BOD of Average Strength Flow (mg/liter)	Effluent BOD of Maximum Strength Flow (mg/liter)
Aerated Lagoon	95	9.5 (OK)	28.2 (OK)
Anaerobic Lagoon	88	22.8 (OK)	79.2 (unacceptable)
Activated Sludge	85	28.5 (OK)	99.0 (unacceptable)

2. Superior Organic Loading Potential of Aerated Lagoons

Seasonal shockloads typical of citrus processing plants are easily handled by aerated lagoons but tend to pose problems for activated sludge processes and anaerobic lagoons. Production within the plant will be a one-shift-a-day operation and may shut down completely on weekends and holidays. The volume of wastewater therefore will fluctuate through the harvesting season, which begins in October and ends in June. A brief performance comparison of how these three systems react to such periods of high loading will show the distinct advantage which aerated lagoons have over the other processes.

1. Aerated Lagoons: Aerated lagoons, because they are not continuous flow processes, can store wastewater until it can be treated. These lagoons can therefore be designed to handle a shock load of unlimited quantity.

2. Anaerobic Lagoons: Anaerobic lagoons also exhibit the same storage capabilities as aerated lagoons. However, because of the nature of anaerobic organisms, obnoxious odors are produced at high organic loadings. Hydrogen sulfide and mercaptans are produced at loading values exceed-

FIGURE 6-6   *(continued)*

ing 240-320 kg BOD/1000 cubic meters, a value very typical of the size of plant in question.

3. Activated Sludge: Activated sludge processes can only tolerate organic loading values not exceeding 260-340 kg BOD/1000 cubic meters. Higher loading rates were followed by bulking primarily caused by Sphaerotilus organisms.

Cost Superiority of Aerated Lagoons

Table 2 shows that aerated lagoons cost less than activated sludge systems and anaerobic lagoons in terms of initial capital as well as operational and maintenance (O&M) costs. The initial capital and O&M costs are usually reflective of the amount of equipment required for each system. Aerated lagoons require only turbine-type aerators and therefore cost infinitely less to build and maintain than activated sludge systems, which need several settling basins and numerous pumps. Anaerobic lagoons, although not needing any equipment, have a higher operating cost than aerated lagoons because of the chemicals required in stabilizing the anaerobic bacteria. The estimated cost of these various systems is based primarily on average flow and BOD loadings. More detailed cost tables appear in the Appendix.

Table 2: Cost Comparison of Treatment Processes

Secondary Treatment System	Cost (× $1000 in 1979 dollars)	
	Capital	Annual
Aerated Lagoon	114.0	22.8
Anaerobic Lagoon	212.0	44.8
Activated Sludge	725.0	39.0

Conclusions and Recommendations

My objective in this report was to recommend a process for economically and efficiently treating the citrus wastes from the new Orange Grove Products plant.

The use of an aerated lagoon is recommended if the ultimate process is to produce an effluent which consistently complies with federal standards. Such a system will also save initial capital as well as annual operations and maintenance costs.

FIGURE 6-7 Sample report 1 for analysis.

DATE: June 6, 1980

TO: Meg A. Lith, Manager
          Research Division
          Scoria Mining Company
          Menhir, Mich. 48003

FROM: Michael A. Newton, Research Assistant
          Phoenix Memorial Laboratory
          47-28 Technology Mall
          Atlanta, Ga. 30081

SUBJECT: Sediment Sample Analysis: Report of Investigation

Foreword: On 14 May you requested an analysis of two sediment samples: TG-1 and TG-2. In response to your request, we have performed a neutron activation analysis on each of the samples to determine the major constituent elements in each. This letter concerns the results of the analysis.

Details: The final analysis is based upon the collected results of two separate procedures: one for short-lived isotopes and one for long-lived isotopes. Although sample preparation and gamma ray spectra analysis for both of the procedures are identical, the process of irradiation is different. The process of analysis is detailed below.

1. Preparation: Sample preparation was performed by obtaining two 1-gram samples from each of the sediments. These samples were then sealed into individual quartz tubes for ease in handling and irradiation.

2. Irradiation: One sample tube from TG-1 and one from TG-2 underwent short half-life activation. The other sample tube from TG-1 and TG-2 underwent long half-life activation.

    2.1 Activation of Short Half-Life Isotopes: Sample activation was performed by exposing the sample tube to a neutron flux for a period of 10 seconds. The sample was allowed to sit for a period of 30 seconds to insure it had decayed to a safe handling level. The sample was then available for analysis.

    2.2 Activation of Long-Lived Isotopes: Sample activation was performed by exposing the sample tube to a neutron flux for a period of 10 hours. The sample was allowed to sit for a period of 5 minutes to insure it had decayed to a safe handling level. The sample was then available for analysis.

3. Analysis: Sample analysis was performed by individually measuring each tube for gamma ray emission. The measurement was performed with a Ge(Li) detector in conjunction with a multichannel analyzer that was

FIGURE 6-7 *(continued)*

controlled by a PDP-8 computer. The resultant gamma ray differential pulse height spectrum was analyzed to determine the following:

<u>3.1</u> The elements present in the sample by referring to the photopeaks present on the spectra.

<u>3.2</u> The concentration of each element present in the sample by referring to the magnitude of the gamma ray emission rate relative to a known sample's emission rate. (The concentration of each element in the known sample had been previously determined.)

Sample composition was then determined by combining the results of the short- and long-lived isotope analysis. Those results were reduced to include only the six major elements of each sediment sample.

<u>Results</u>: The final results of the analysis are shown below for the six major elements in each of the samples.

	TG-1	TG-2
Na-23	31600 ppm	30200 ppm
Mn-56	763 ppm	260 ppm
Fe-59	18800 ppm	46100 ppm
Co-60	4.40 ppm	15.40 ppm
Sc-46	3.63 ppm	12.30 ppm
Pa-233	25.60 ppm	6.98 ppm

Note: ppm denotes concentration in parts per million.

These measurements represent the relative concentration of the elements to within 0.01 part per million.

If any questions arise as to the analysis of the samples, please feel free to contact me.

FIGURE 6-8 Sample report 2 for analysis.

To: J. E. Mauppin
Manager, Engineering Department

From: Christopher Perez
Product Designer, Engineering Department

Subject: Model 704 Angle-Iron Cutters:
Investigation of Recurring Bolt Failure

Dist: B. A. Thomas
Manager, Manufacturing

Date: March 4, 1980

Foreword

Over the past year, Construction Equipment has been plagued by customer complaints dealing with a premature mechanical failure on the new model 704 angle-iron cutter. Apparently a bolt used in the assembly of the cutter is fracturing after minimal usage, causing the machine to be completely inoperable. I was asked by Mr. Mauppin to determine the cause of this recurring failure. The purpose of the report is to present my findings and to suggest corrective measures.

Bolt's Function

The bolt under analysis connects the lever (which raises the cutting edge) to the main body of the cutter. Once this bolt fails, the cutter is inoperable because there is no way to raise the cutting edge.

Forces on the Bolt During Operation

The bolt is subjected to three different forces during the operation of the angle-iron cutter. These forces, listed in decreasing order of importance, are

1. a torsional force, which is produced when the lever is rotated for cutting

2. a shear force, which is also produced when the lever is rotated for cutting

3. a tensile force, which is produced when the nut is tightened down

Since the torsional force is the most significant, the bolt should be designed to withstand the torque produced by the lever. Bolts which have a strong possibility of failing in torsion are usually case hardened to increase their torsional strength.

A series of fractured bolts received from one of Construction Equipment's customers was used to help determine the cause of the recurring failure. Figure 1 shows a detailed sketch of a typical bolt.

Figure 1 Sketch of Failed Bolt

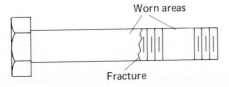

FIGURE 6-8 *(continued)*

As is shown in the sketch, the shank of the bolt revealed the following characteristics:

1. Fracture occurred at the first full thread. This should have been expected because of the stress-concentrating effect of the first full thread.

2. Fracture surface occurred on a plane perpendicular to the bolt.

3. Wear areas occurred where the bolt was in contact with the lever and washers.

Figures 2a and 2b [not included] are photographs (mag. 4.77x) of the mating fracture surfaces of the failed bolt. Examination of the fracture surfaces revealed:

1. a smeared rotary-type appearance in which a great amount of plastic deformation had taken place

2. a final fracture area almost exactly in the center of the bolt

Because of the deformation of the threads in the fracture region, the possibility of a fatigue fracture was ruled out.

SEM Examination of Fracture Surfaces

Further examination of the fracture surfaces at higher magnification with the scanning electron microscope revealed no imperfections in the metal. A small amount of corrosion was found toward the center of the fracture surfaces but because of its location it could not have contributed to the fracture (the corrosion must have occurred after the failure).

Hardness Test

A hardness test was conducted on the shank of the fracture bolt. A value of 74 Rb was obtained, which is quite low for a bolt used in a torsional loading situation.

Cause of Failures

The characteristics of the fractured bolt suggest that failure occurred by a single torsional overload. The source of the torsional loading on the bolt was the lever which raised the cutting edge. Therefore, it is evident that the recurring failures have been due to the inadequate torsional strength of the bolts. The first full threads on the bolts also contributed to the failures because of their stress-concentrating effect.

Recommendation

In order to increase their torsional strength, I suggest that future bolts be case hardened. This will certainly alleviate the failing bolt problem currently faced by Construction Equipment.

REFERENCES

1   Leon Festinger, *A Theory of Cognitive Dissonance,* Stanford University Press, Palo Alto, Calif., 1957.

2   Richard E. Young, Alton Becker, and Kenneth L. Pike, *Rhetoric: Discovery and Change,* Harcourt Brace and World, New York, 1970.

3   Joan Arehart-Treichel, *Science News* **107**:2 (1975), p. 26.

4   *Science News* **107**:2 (1975), p. 20.

5   Daniel Z. Freedman and Peter van Nieuwenhuizen, "Supergravity and the Unification of the Laws of Physics," *Scientific American* **238**:2 (1978), p. 126.

6   Ray Berry, "Caution: Life May Be Hazardous to Your Health," *Michigan Technic* **96**:4 (1978), p. 6.

7   Berry, ibid.

8   Ronald K. Jurgen, "The Automobile for Better or Worse," *IEEE Spectrum,* November 1977 **14**:11, p. 31.

9   Mehdi N. Bahadori, "Passive Cooling Systems in Iranian Architecture," *Scientific American* **238**:2 (1978), p. 144.

10 Norman H. Horowitz, "The Search for Life on Mars," *Scientific American* **237**:5 (1977), p. 52.

11 Kobe Shoji, "Drip Irrigation," *Scientific American* **237**:5 (1977), p. 62.

12 Richard W. Dodge has prepared an extended list of questions managers want answered for various types of projects. The list (shown below) appeared in "What To Report," *Westinghouse Engineer* **22**(4-5):108–111 and was based on information provided by James W. Souther, "What Management Wants in the Technical Report," *Journal of Engineering Education* **52**(8):498–503 (1962).

Problems	Materials and Processes
What is it?	Properties, characteristics, capabilities?
Why undertaken?	Limitations?
Magnitude and importance?	Use requirements and environment?
What is being done? By whom?	Areas and scope of application?
Approaches used?	Cost factors?
Thorough and complete?	Availability and sources?
Suggested solution? Best? Consider others?	What else will do it?
	Problems in using?
What now?	Significance of application to Company?
Who does it?	
Time factors?	

Field Troubles and Special Design Problems

Specific equipment involved?

What trouble developed? Any trouble history?

How much involved?

Responsibility? Others? Westinghouse?

What is needed?

Special requirements and environment?

Who does it? Time factors?

Most practical solution? Recommended action?

Suggested product design changes?

Tests and Experiments

What tested or investigated?

Why? How?

What did it show?

Better ways?

Conclusions? Recommendations?

Implications to Company?

New Projects and Products

Potential?

Risks?

Scope of application?

Commercial implications?

Competition?

Importance to Company?

More work to be done? Any problems?

Required manpower, facilities and equipment?

Relative importance to other projects or products?

Life of project or product line?

Effect on Westinghouse technical position?

Priorities required?

Proposed schedule?

Target date?

ADDITIONAL READING

Elliot Aronson, "The Rationalizing Animal," *Psychology Today,* May 1973, pp. 46–52,119.

Richard S. Burington, "The Problem of Formulating a Problem: General Considerations," *Proceedings of the American Philosophical Society* **105**(5):429–443 (1960).

Scott Consigny, "Rhetoric and Its Situations," *Philosophy and Rhetoric,* Summer 1974, pp. 175–186.

Otto Dieter, "Stasis," *Speech Monographs* **17**(4):345–369 (1950).

Richard W. Dodge, "What To Report," *Westinghouse Engineer* **22**(4–5):108–111 (1962).

J. C. Mathes and Dwight Stevenson, *Designing Technical Reports,* Bobbs-Merrill, Indianapolis, 1976.

Henry Minzberg, "The Manager's Job: Folklore and Fact," *Harvard Business Review,* July–August 1975, pp. 49–61; *The Nature of Managerial Work,* Harper & Row, New York, 1973.

Part IV

Visual Elements

7

VISUAL ELEMENTS

So far, this book has talked about the production of words, either written or spoken. There are times, however, when words alone are not the best way to transfer information or points of view, times when words need to be combined with visual aids, formatting (the use of white space and indenting), or other visual elements. For instance, as suggested later in this chapter, appropriate formatting can make a technical report much easier to read—so much easier that the formatting becomes *necessary* given the limitations on the time and attention of an audience. The same can often be said of other visual elements, such as drawings, figures, charts, or graphs, which can quickly summarize an important point or present it in a different way.*

For instance, consider the following situation. Figures 7-1 and 7-2 present the same information but in different forms, Figure 7-1 in prose text and Figure 7-2 as a visual element. The discussion in Figure 7-1 requires 134 words and forces its reader to learn its points linearly, that is, one at a time as they appear in the discussion. In contrast, the chart in Figure 7-2 presents the information quickly and holistically; that is, it presents the whole in a clearly organized pattern that can be seen at a single glance.

Psychologists have shown that you can increase the strength and memorability of a message simply by repeating it or, even better, by repeating it in a different form. Thus, when a visual presentation is added to a verbal one, the combination can produce *a much stronger and more easily remembered message* than either presentation alone. Further, a visual aid can present *a compact summary* of the main points of a verbal text. (Have you ever heard the expression "a picture is worth a thousand words"?) Finally, a visual element can often summarize in *a more memorable form* than words alone can.

Given these advantages of visual aids, a communicator ought to be able to use them effectively. This involves knowing

1   How to make a visual aid effective

2   When to use the visual aid

3   How to select the best type of visual element in a given situation (e.g., pie chart, bar graph, line graph)

4   How to integrate the visual aid into the text

---

*We would like to note the efforts and influence of our colleague Lisa Barton on the evolution of this chapter. She was a major contributor to the thinking behind Sections 7.1 and 7.3 and coauthored an early lecture on visual aids from which those sections are derived.

FIGURE 7-1 Safety records of surface and underground mining, presented in prose text form. [From Edmund A. Nephew, "The Challenge and Promise of Coal," *Technology Review* **76**:2 (December 1973), pp. 27–28.]

Safety: Room for Improvement

The safety record of surface mining—in terms of both deaths and non-fatal injuries per ton of coal mined—is spectacularly better than deep mining, which provides at best a hostile, hazardous environment for the miner.

The human costs of moving from surface to underground mining take the form of higher injury and death rates and of greater occupational hazards in general. In 1971, 86 per cent of all coal mining fatalities occurred in underground mines. In the same year, only half of the total coal production came from deep mines. Over the seven-year period 1965 to 1972, 1,412 lives were lost in the underground coal mining industry in the production of 2,335 billion tons of coal. This amounts to an average of 0.606 deaths per million tons of coal—a fatality rate more than five times greater than that of the coal surface mining industry. In recent years, falls-of-roof have accounted for 40 per cent of the deaths in underground mines and coal haulage accidents for about 20 per cent. Annual fatalities from dust and gas explosions fluctuate greatly, but over the years there has been a declining trend.

A similar safety disparity between surface and deep mines holds for non-fatal injuries as well.

During the time period from 1968 to 1971, the safety performance in deep mining of the nation's top ten coal producers ranged from 0.28 to 1.52 deaths per million man-hours. The differences are even more marked for the category of non-fatal injuries—2.72 to 72.13 injuries per million man-hours. Since the passage of the Coal Mine Health and Safety Act of 1969, most companies have greatly strengthened their safety programs, and this increased emphasis on safety may bring significant improvements in fatality and injury rates. Indeed, the wide range of safety performances cited above makes it clear that—even without a technological breakthrough—much can be done to narrow the safety gap existing between deep and surface mining.[1]

7.1

MAKING A VISUAL AID TRULY VISUAL

Take about 2 to 5 seconds to look at Table 7-1 and then cover it up. Do not look at any of the following tables or discussions. Now try to write down the main points made by the table. When you have finished this, look at the presentation of the same information in Table 7-2 and see if you can quickly add any more main points to your list. Do this before you continue.

Typically, people who read only Table 7-1 note (1) that job satisfaction declines in each of the two main groups of occupations. These readers will *sometimes* notice (2) that there is a large difference in job satisfaction between the two groups—that is, that most of the first group is relatively satisfied (93 to 75 percent satisfied) whereas most of the second group is much less satisfied (only 52 to 16 percent satisfied). Very few readers of only Table 7-1 will notice (3) that the job satisfaction of skilled printers is higher than that

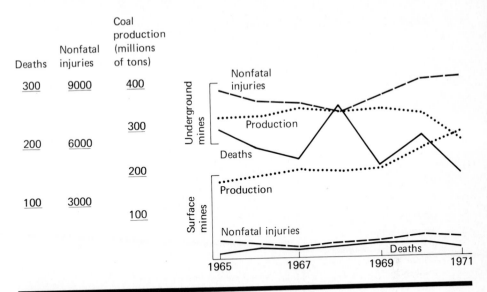

FIGURE 7-2 Safety records of surface and underground mining, presented as a visual element. [From Edmund A. Nephew, "The Challenge and Promise of Coal, *Technology Review* **76**:2 (December 1973), p. 28.]

TABLE 7-1 PROPORTIONS OF OCCUPATIONAL GROUPS WHO WOULD CHOOSE SIMILAR WORK AGAIN

PROFESSIONAL AND WHITE-COLLAR OCCUPATIONS	PERCENT	SKILLED TRADES AND BLUE-COLLAR OCCUPATIONS	PERCENT
Urban university professors	93	Skilled printers	52
Mathematicians	91	Paper workers	42
Physicists	89	Skilled auto workers	41
Biologists	89	Skilled steelworkers	41
Chemists	86	Textile workers	31
Firm lawyers	85	Blue-collar workers	24
School superintendents	85	Unskilled steelworkers	21
Lawyers	83	Unskilled auto workers	16
Journalists (Washington correspondents)	82		
Church university professors	77		
Solo lawyers	75		
White-collar workers (nonprofessional)	43		

SOURCE: Based on a study of 3000 workers in 16 industries, conducted by the Roper organization; on Wilensky's study of Detroit workers and professions; and on a study of Massachusetts school superintendents by Neal Gross, Ward Mason, and W. A. McEachern. From *Psychology Today,* February 1973, p. 39.

TABLE 7-2 ALTERNATE ARRANGEMENT FOR "PROPORTION IN OCCUPATIONAL GROUPS WHO WOULD CHOOSE SIMILAR WORK AGAIN"

PROFESSIONAL WHITE-COLLAR OCCUPATIONS	PERCENT	
Urban university professors	93	....
Mathematicians	91	....
Physicists	89	....
Biologists	89	....
Chemists	86	....
Firm lawyers	85	....
School superintendents	85	....
Lawyers	83	....
Journalists (Washington correspondents)	82	....
Church university professors	77	....
Solo lawyers	75	....
White-collar workers (nonprofessionals)	43	....
SKILLED TRADES AND BLUE-COLLAR OCCUPATIONS		
Skilled printers	52	....
Paper workers	42	....
Skilled auto workers	41	....
Skilled steelworkers	41	....
Textile workers	31	....
Blue-collar workers	24	....
Unskilled steelworkers	21	....
Unskilled auto workers	16	....

of nonprofessional white-collar workers. These last two observations (points 2 and 3) are very hard to "see" in the format used in Table 7-1.

In contrast, most readers of Table 7-2 easily and quickly note all three observations, as well as a few other, more subtle ones, simply because of the format of the table. Notice that Table 7-2 makes it *visually* quite clear that the job satisfaction ratings of the two groups overlap and that the skilled trade and factory workers as a group are less satisfied than the professionals.

This comparison illustrates an important point: to be most effective a visual aid should present information in a truly *visual* form. Table 7-1 presents information in a chart, but it does not use most of the chart's visual possibilities. It nicely arranges the information in two groups and in descending order within the two groups. However, it does not visually indicate the descending order. To really understand the chart, readers have to do a lot of mental work. For instance, they have to notice

1   That for the left column the first four categories have satisfaction levels of 93, 91, 89, and 89 percent

2   That these four levels are all quite close to each other

3   That the next levels go from 86 to 75 percent

4   That these are pretty close to each other but lower than the first group

5   That the lowest category is only 43 percent

6   That this is pretty low, too, etc.

7.2
DECIDING WHEN TO USE A VISUAL AID

Communicators often wonder *when* they should use a visual aid in a communication. Three suggested principles for deciding this are to use a visual aid

1   Where words alone would be either impossible or quite inefficient for describing a concept or an object

2   Where a visual aid is needed to underscore an important point, especially a summary

3   Where a visual element is conventionally or easily used to present data

Let us consider each of these uses in turn.

The Visual Aid for Describing or Clarifying

A communicator often has to describe an object or concept which is hard to describe efficiently in words alone. For instance, consider the following passage and then try to sketch out the image created in your mind by the passage. Do not look at Figure 7-3 before making your sketch.

SKYLAB DESCRIPTION
The Skylab cluster is an assembly of modified and specially built space hardware which together provides over 12,750 ft. of working volume and weighs in at 199,750 lbs. The Workshop is a Saturn IVB stage modified and outfitted for manned habitation; it contains living quarters, food storage and preparation and waste management facilities and the attitude control thrusters, and it carries the two ill-fated solar arrays (one shows clearly in its extended position in this drawing) whose electric input is routed to the power system in the Airlock Module (AM). The latter, serving as a passageway between the Workshop and

the docking facilities, is the focus of many of Skylab's technical systems: atmospheric and thermal controls, power control and distribution, and communications and data handling. The Multiple Docking Adapter (MDA) has two docking ports—the axial port shown in use here and a contingency radial port at the bottom in this diagram; it also contains some space research equipment, including the Earth Resources Experiment Package. The Apollo Telescope Mount (ATM) is a sophisticated solar observatory—the first U.S. manned scientific telescope in space. Here as well are attitude and experiment pointing controls for the entire Skylab cluster and a solar array and associated battery system adequate to power the ATM's equipment. The drawing shows the Apollo Command and Service Module (CSM), used to transport the crew to and from orbit, docked to the MDA.[1]

As suggested earlier, try to sketch out the image created in your mind by the passage before you look at the drawing of the Skylab presented in Figure 7-3. When you are finished, compare your sketch to the drawing. If you are like most readers, your sketch captures some of the features of the Skylab: it probably shows that the *Airlock Module* comes between the *Workshop* and the *docking facilities* and that the *Apollo Command and Service Module* is attached to the *Multiple Docking Adapter*. However, it may not show that the *Multiple Docking Adapter* is the same as the *docking facilities*, and it probably will not show the appropriate sizes and shapes of the various parts of the Skylab system. Clearly, the formal drawing presents lots of information about sizes, shapes, details, and arrangements which are not presented in the words and which *could not* be presented there, given the limited

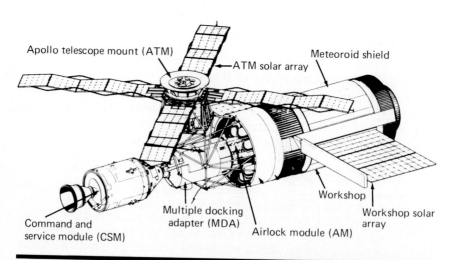

Apollo telescope mount (ATM)

ATM solar array

Meteoroid shield

Command and service module (CSM)

Multiple docking adapter (MDA)

Airlock module (AM)

Workshop

Workshop solar array

FIGURE 7-3 The Skylab cluster. [From William Schneider and William D. Green, Jr., "Saving Skylab," *Technology Review* **76**:3 (January 1974), p. 44.]

number of words in the original description and even the extra space occupied by the visual aid. In such a situation, you obviously need a visual aid.

Sometimes you will need to describe a concept or object which, even if you have an unlimited number of words, seems impossible to describe with words alone. Consider, for example, the passage below describing the pterosaur, a prehistoric flying reptile, and its near relatives.

> Pterosaurs evolved in the early part of the Mesozoic era. Their appearance preceded that of the earliest birds by about 50 million years and followed that of the earliest dinosaurs by about 20 million years. It is hypothesized that the precursors of the pterosaurs (and of the birds and of both orders of the dinosaurs, the ornithischians and the saurischians) were the reptiles called the thecodonts, which for their part evolved from the early, lizard-shaped reptiles known as eosuchians. Among the pterosaurs the rhamphorhynchoids first appeared some 50 million years before the pterodactyloids. The last pterodactyls died out with the dinosaurs 64 million years ago.[2]

If you are having trouble placing all of the prehistoric creatures just described, see if their relationships are clearer in Figure 7-4. In this case, the conceptual system described is too complex for most readers to "see" by words alone but quite easily grasped with the help of the visual aid.

The Visual Aid for Highlighting Important Points

There are many times when you need a visual aid to bring out an important point or to summarize data or a line of argument. For instance, Chapter 6 presents arguments about Introductions and Forewords: that they are based upon psychological principles and that they can be represented by a formula consisting of a problem statement in a certain format, a question or an assignment, and a thesis. These arguments included discussions and several examples and ended with a summary of the Introduction. This summary is shown in Figure 7-5, which usually recalls the information in Chapter 6 even for those readers who have not recently reviewed it.

The Visual Aid for Conventional or Easy Presentation of Data

Some types of data are almost always presented in a visual aid. These include cost summaries, budget calculations, frequency spectra, electrical circuits, architectural plans, and the relative positions of geographical areas. As a professional, you ought to be aware of the typical visual forms in your field and use them when it is appropriate.

For instance, consider the sound spectra in the following extract, a short overview of an interesting research project. In just over 500 words, a problem is introduced, the project's research model and technique are explained, the assumptions set out, and the findings reported. Note in this example that using the visual aid is not only more conventional but also more efficient than using words alone: an equally detailed verbal description of the cry spectra would take much more space. Note also *where* the visual aid is placed

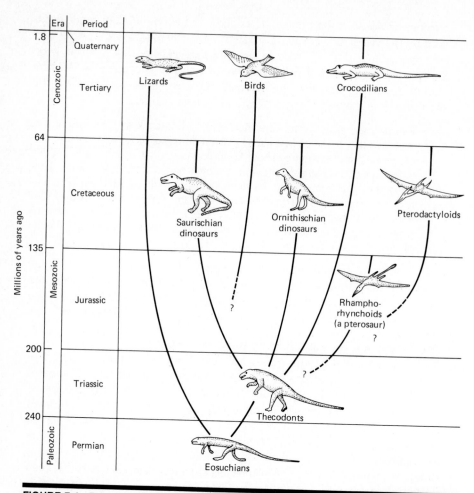

FIGURE 7-4 The evolution of the pterosaurs.[From Wann Langston, Jr., "Pterosaurs," *Scientific American* **244**:2 (February 1981), p. 126.]

in this passage: it gives the detailed backup data needed to substantiate a scientific claim. It would be hard to imagine another place in this brief overview where the extra proof and detail of a visual aid would be needed to prove a central claim.

BABIES' CRIES GIVE CLUES TO DISEASES

The long, loud, crystal-shattering sounds coming from a newborn infant indicate his good health according to current research by Howard L. Golub and Kenneth N. Stevens of M.I.T.'s Research Laboratory of Electronics. Although scientists in this country and abroad have failed for 20 years to relate cry profiles to infant ailments, their recent tests with a computer cry model showed surprisingly accurate results.

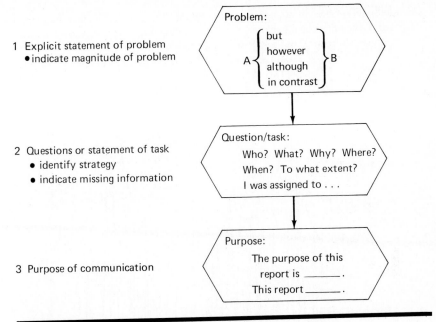

1 Explicit statement of problem
 • indicate magnitude of problem

Problem:

A $\left\{ \begin{array}{l} \text{but} \\ \text{however} \\ \text{although} \\ \text{in contrast} \end{array} \right\}$ B

2 Questions or statement of task
 • identify strategy
 • indicate missing information

Question/task:
Who? What? Why? Where?
When? To what extent?
I was assigned to . . .

3 Purpose of communication

Purpose:
The purpose of this
report is _____ .
This report _____ .

FIGURE 7-5 Outline of an Introduction or Foreword by
problem statement.

Mr. Golub developed the model by using the accepted acoustic theory of adult speech production and modifying it to fit babies. Since adults and children differ in anatomical proportion, ratios like pharynx length to mouth length, nasal tract length to vocal tract length, and central nervous system disparities had to be adjusted accordingly. Then, using the tape-recorded protests of 55 apparently healthy newborns during a blood test (called the P.K.U. heel stick), Mr. Golub subdivided the cries into 88 variables, such as pitch, intensity, fundamental frequencies, and resonances.

He then used a computer to compare the cries of healthy babies with those of 43 babies having known or suspected health abnormalities. Recordings of these latter cries were supplied by Dr. Michael Corwin, resident in pediatrics at the Upstate Medical Center, Syracuse, N.Y. Of these 43 cases Mr. Golub's computer analysis located 19 of 21 infants with severe jaundice and 9 of 10 babies suffering from respiratory difficulties. Fifteen healthy babies from Dr. Corwin's group were also correctly identified from their cries alone.

Mr. Golub's research team found that a specific ailment is associated with a characteristic cry pattern. Assuming that most infant pathologies alter the acoustically relevant structures, it follows that some aspects of the cry will be correspondingly changed.

For example, if the infant has respiratory distress one would expect a shorter cry due to a change in the dynamics of respiratory muscles responsible for oxygen intake, or a change in vocal tract resonances resulting from a constriction of the pharynx—a possible explanation for the tragic sudden infant death syndrome. In such a case one might also expect a higher frequency sound since the vocal tract acts much like a pipe organ acoustically—a narrow or squeezed pipe giving a higher frequency sound.

Interestingly, Golub and Stevens found precisely these effects, as illustrated by the following two charts.

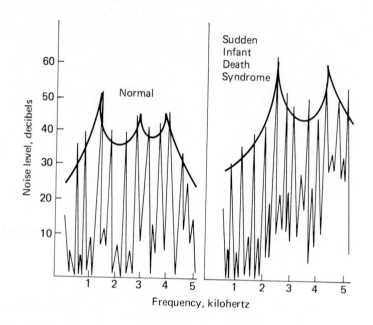

The chart at the left shows the short-time (25 milliseconds) spectrum of the mid-portion of the cry of a normal infant. The chart at the right plots the short-time (25 milliseconds) spectrum of the mid-portion of the cry of an infant that later died of sudden infant death syndrome. Notice the very high first resonance and large amount of noise at the high end of the spectrum, possibly indicating a constriction of the vocal tract near the pharynx.

As a sidelight of his research on jaundice cases, Mr. Golub discovered that some levels of biliruben—a by-product of the breakdown of hemoglobin—previously considered "safe" are likely harmful. Early treatment for such a condition could prevent brain damage. Other abnormalities likely detectable with the model include bacterial meningitis and deafness.[3]

7.3

SELECTING THE BEST TYPE OF VISUAL AID IN A GIVEN SITUATION

When you design a particular visual aid, you are consciously or unconsciously making certain decisions. You are deciding that the particular type of aid you choose (a line graph, bar chart, pie diagram, photograph) is the best *type* to

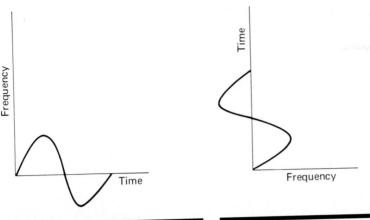

FIGURE 7-6 Preferred location of independent variable on a graph.

FIGURE 7-7 Unconventional location of independent variable on a graph.

make your point and that the arrangement and highlighting of material on the page is, again, the best to make your point. Unfortunately, there is little information available on which to base such decisions. If you are like most writers, you probably choose one type of visual aid over another simply because it is the first thing you think of using.

The purpose of this section is to sketch out some better or more conscious reasons for choosing. The section will first identify some conventions of visual perception in Western cultures and then examine several common types of visual aids to see what they do and do not show well.

Conventions of Visual Perception

Given the way Western societies view the world and read, there are a number of general statements we can make about our expectations of visual information. First, we expect written things to proceed from left to right. Note that in scientific and technical graphs, we place the independent variable on the x axis so that the more important variable moves from left to right. For instance, we plot time on the x axis and frequency on the y axis, as illustrated in Figure 7-6. This pattern is so pervasive that Figure 7-7 looks at best odd and at worst disturbing.

Second, we expect things to proceed from top to bottom, and, third, we expect things in the center to be more important than things on the periphery. Fourth, we expect things in the foreground to be more important than things in the background; fifth, large things to be more important than small things; and sixth, thick things to be more important than thin things. Note that type that is larger, thicker, or bolder than the surrounding type is usually more important: a heading, a title, or an especially important word

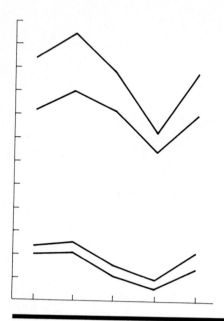

FIGURE 7-8 The shaping of expectation by visual arrangement: a sample for analysis.

in a passage. Seventh, we expect areas containing a lot of activity and information to contain the most important information. Eighth, we expect that things having the same size, shape, location, or color are somehow related to each other. (Notice the way the curves in Figure 7-8 are grouped; all labels have been deleted to free you to "see" this point.) Lastly, ninth, we see things as standing out if they contrast with their surroundings because of line thickness, type face, or color. (You should note that warm or hot colors—red, yellow, and orange—stand out more than cool colors—blue and green.)

Notice how many of the above expectations are verified in Figure 7-9. The dark lines tracing the bus, streetcar, and rail routes clearly stand out from the lighter print and the white background. Conceptually, San Francisco Bay and the Pacific Ocean are merely background for the lines of the transit system; both the bay and the ocean are on the edge. In contrast, the most important part of the visual aid, the transit lines, is in the center in dark, thick lines. You might look for such features in other visual aids and try to explain why the designer made the choices you see.

Some Types of Visual Aids and Their Uses

There are six main types of visual aids with which a scientist or engineer should be familiar: (1) line graphs, (2) bar graphs, (3) pie charts, (4) tables,

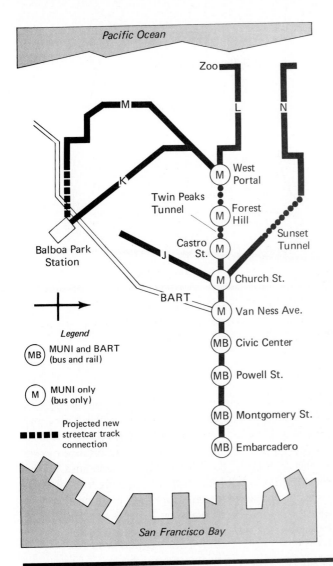

FIGURE 7-9   The five lines of the San Francisco Municipal Transit System.

(5) photographs, and (6) line drawings. Each of these types has particular strengths and weaknesses, and to use any one appropriately, you must decide what point you are trying to make and then select the type of visual aid which makes that kind of point well.

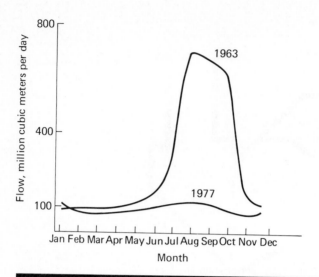

FIGURE 7-10   River flow before (1963) and after (1977) construction of the Aswan High Dam on the Nile River. [From Julie Wei, "Aswan and After: The Taming and Transformation of the River Nile," *Research News* **31**:7 (The University of Michigan, July 1980), p. 21.]

LINE GRAPHS   Line graphs *show well* continuity and direction as opposed to individual or discrete points, direction as opposed to volume, and the importance of a nodal point, if there is one. These characteristics are illustrated in Figure 7-10. Line graphs *do not show well* the importance of one particular point which falls off a node, the relationship of many lines, or the intersection of three or more lines. If it is important to be able to trace each line on a graph, you should probably not put more than three or four on a single graph, especially if they intersect frequently, or you may produce a graph as hard to follow as the one in Figure 7-11.

BAR GRAPHS   Bar graphs *show relatively well* the discreteness or separateness of points as opposed to their continuity, volume as opposed to direction, the relationships among more than three or four items at a time, the contrast between large and small numbers, and the similarities and differences between similar numbers (you can see both that 17 percent and 18 percent are about the same and that 18 percent is a bit larger than 17 percent). These characteristics are evident in the variant of the bar graph presented in Figure 7-12 and in Figure 7-13. Bar graphs *do not show well* the absolute values of the items measured, though you can indicate absolute values with labels, as shown in Figure 7-13.

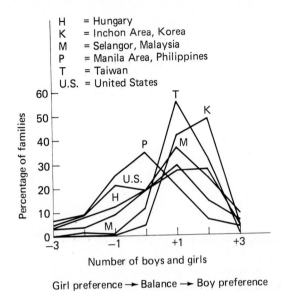

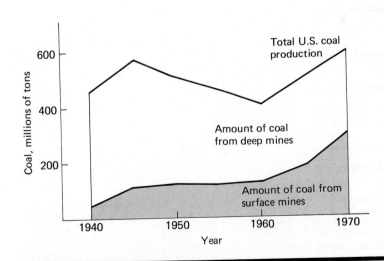

FIGURE 7-11   Preference of families for girls versus boys in six
countries.

FIGURE 7-12   Growth of surface mining in U.S. coal production.
[From Edmund A. Nephew, "The Challenge and Promise of Coal,"
Technology Review **76**: (December 1973), p. 22.]

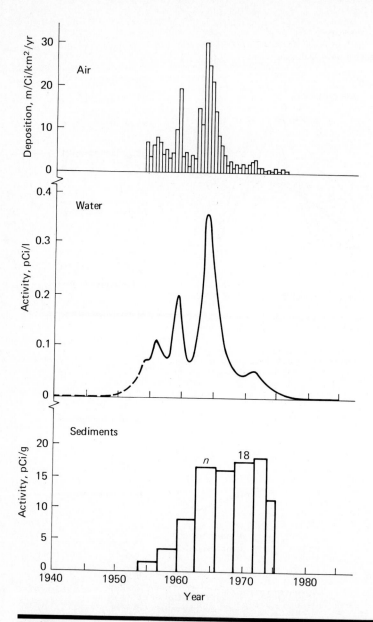

FIGURE 7-13 Comparison of cesium-137 fallout in the Lake Huron air-water surface, water, and sediment. [From Julie Wei, "Towards Cleaner Water," *Research News* **29** 8/9 (The University of Michigan, August-September 1978), p. 22.]

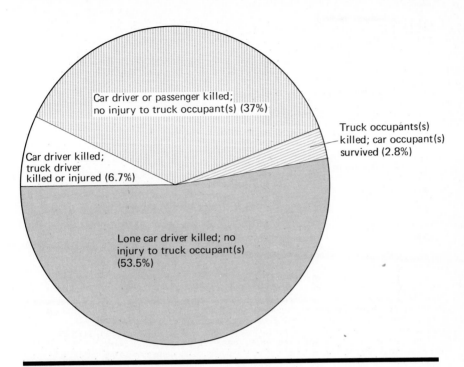

Car driver or passenger killed; no injury to truck occupant(s) (37%)

Truck occupants(s) killed; car occupant(s) survived (2.8%)

Car driver killed; truck driver killed or injured (6.7%)

Lone car driver killed; no injury to truck occupant(s) (53.5%)

FIGURE 7-14 Distribution of fatalities in 181 fatal car-truck crashes. [From Daniel J. Minahan and James O'Day, "Fatal Car-into-Truck/Trailer Underride Collisions," *HSRI Research Review* **83**: (The University of Michigan Highway Safety Research Institute, November-December 1977), p. 7.]

PIE DIAGRAMS Pie diagrams *show relatively well* the relationship among three or four items which total 100 percent, the contrast between large and small percentages, and the similarity between relatively similar percentages (they show well that 27 percent and 29 percent are about equal). Pie diagrams *do not show well* the small difference between two similar percentages (you can't usually see the differences between 27 percent and 29 percent). They also *do not show well* absolute values (unless you label the parts of the pie) or the relationships among more than five or six parts; with too many parts, it is hard to see the relationships of part to part and part to whole. These strengths and weaknesses are illustrated in Figure 7-14.

TABLES Tables are convenient for presenting lots of data and for giving absolute values where precision is very important. However, since they present items one at a time in columns, they emphasize the discrete rather than the continuous and make it very difficult to show trends or direction in the data.

TABLE 7-3 ANNUAL ENERGY SAVINGS FROM SOLAR ENERGY

	SOLAR SPACE AND DOMESTIC HOT WATER SYSTEMS		SOLAR TOTAL-ENERGY SYSTEMS		SOLAR BASE-LOAD ELECTRIC POWER PLANTS	
YEAR	NUMBER OF DWELLINGS (MILLIONS)	ANNUAL ENERGY SAVINGS (10^{15} B.T.U.)	FLOOR AREA OF BUILDINGS (10^4 M^2)	ANNUAL ENERGY SAVINGS (10^{15} B.T.U.)	INSTALLED CAPACITY (10^6 KW)	ANNUAL ENERGY SAVINGS (10^{15} B.T.U.)
1980	0.3	0.04	—	—	—	—
1985	3	0.4	52	0.24	—	—
1990	8.4	1.2	200	0.92	18.7	1.4
1995	14	1.9	400	1.9	69	5.3
2000	20	2.7	610	2.8	137	10.4
2005	27	3.6	850	3.8	208	15.2
2010	34	4.7	1,090	5	284	22
2015	41	5.9	1,350	6.2	363	30
2020	49	7	1,620	7.6	445	37.6

Assuming the most favorable development, annual energy savings of 52.2×10^{15} B.t.u.—over 1,160 million tons of oil—may be realized by the use of solar energy by the year 2020. The estimates for dwellings are based on the use of approximately 10^8 B.t.u. annually and a heating plant efficiency of 70 per cent; those for total-energy systems on consumption of 6.67 million kwh/yr in a plant of 10,000 m^2 floor area, with an efficiency of 50 per cent; those for power generation on the construction of solar plants with total capacity of 12.5×10^6 kwh each year beginning in 1995, the plants having 40 per cent efficiency.
SOURCE: From *Technology Review* **76**:2 (December 1973), p. 39.

Tables are not predominantly visual: the reader's mind must translate each number into a relationship with each other number, as already described in the job satisfaction example at the beginning of this chapter. Thus, for maximum *visual* impact, tables should probably be a last choice as a visual aid and used only when it is important to provide a great deal of information with precision in a very small space. As an illustration, consider Table 7-3.

PHOTOGRAPHS Photographs *are useful* when you do not have the time, the money, or the expertise to produce a complicated line drawing; when you are trying to produce immediate visual recognition of an item; when you are emphasizing the item's external appearance (as opposed to its internal structure or a cross section); and when you are not concerned with eliminating the abundant detail a photograph provides. While photographs can be air-brushed to eliminate some undesired detail, they still are *not preferred* when you need to focus on some one aspect by eliminating a lot of detail and when you have the time and resources to produce a good line drawing.

LINE DRAWINGS   The term *line drawing* includes several types of drawings which focus on external appearance, physical shape, function, or relationship. These include "simplified photos," maps, anatomical drawings, parts charts, and drawings of models (such as atomic or molecular models) or objects from any field of science or engineering. Also included are flow charts, organizational charts, schematic charts, block diagrams, architectural plans, and blueprints.

While there are many types of line drawings, all of them share certain functions. They allow you to show things which you can't normally see in a photograph because of size, location, or excessive detail. They also allow you to easily highlight a particular shape, part, or function.

For instance, consider the photograph and drawings of a eukaryotic cell and its compartments shown in Figure 7-15. Notice that the photograph, an electron micrograph, has so much detail that at first it is hard to identify the parts of the cell. Further, because of their size, it is hard even with the drawings for reference to clearly see either the cisternae and ribosomes identified in the lower right corner or the boundary of the nucleus. Obviously, the drawings included with the photograph are much clearer, allow highlighting and focusing impossible in the photograph, and can present important details missing in the photograph.

7.4
DESIGNING THE VISUAL AID

Once you have decided *where* a visual aid is needed and *what type* it should be, you must design it so that it is as relevant, clear, and truthful as possible. This will usually be at least a two-stage process: designing a rough copy and then producing the finished copy. If you work for a company which has an art or illustration department, you may be able to get a technical illustrator to produce the finished copy for you and to counsel you in the design stage. However, even if you have such help, you should be the real designer of the visual aid: you have the best knowledge of the subject and best know the purpose of the aid and the context in which it is being used.

Making a Visual Aid Relevant

Since you place a visual aid in a text to make a point, you should be sure that it makes the point you intend. For instance, suppose that you are discussing expected energy saving from the use of solar energy in the future. You have posed three possible sources of the savings—residences, total energy systems such as industrial parks and shopping centers, and solar-based electric power plants—and have broken down the specific savings as illustrated in Table 7-4.

Now that you have your data, you want to construct a visual aid to

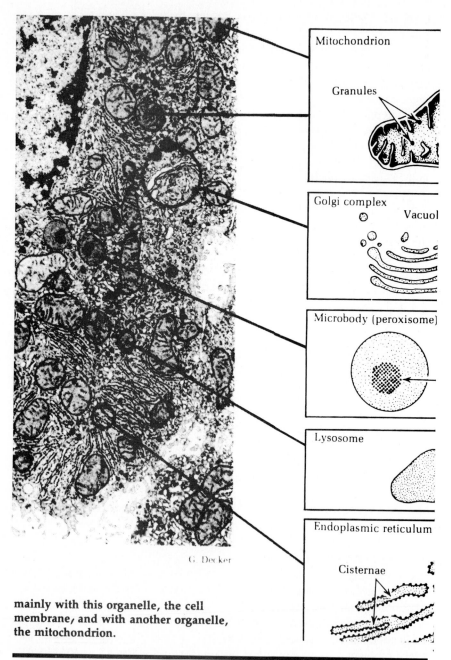

G. Decker

mainly with this organelle, the cell
membrane, and with another organelle,
the mitochondrion.

Mitochondrion

Granules

Golgi complex

Vacuol

Microbody (peroxisome)

Lysosome

Endoplasmic reticulum

Cisternae

FIGURE 7-15   A eukaryotic cell and its compartments. [From
Albert L. Lehninger, *Biochemistry* (New York, Worth Publishers,
1975), pp. 32–33.]

TABLE 7-4   EXPECTED ANNUAL SAVINGS FROM SOLAR ENERGY

| | | ANNUAL SAVINGS (10^{15} BTU) | |
YEAR	RESIDENCES	TOTAL ENERGY SYSTEMS	SOLAR-BASED ELECTRIC POWER PLANTS
1985	0.4	0.24	—
1990	1.2	0.92	1.4
1995	1.9	1.9	5.3

show the growth in savings and the relative contributions of each source. You construct five possible versions of a visual aid, presented in Figures 7-16 through 7-20, and now have to choose the one most appropriate to your point. On what basis do you choose? What are the differences among the five visual aids?

First let us consider the bar graphs. Among the bar graphs, Figure 7-16 presents the most information in the smallest space and the clearest vision of total growth; however, in comparison to the other charts, it obscures the comparisons between items in the same year and between the same item in different years. Figure 7-17 obscures the total growth but makes the comparisons already mentioned much clearer, especially between the same item in different years. On the other hand, Figure 7-18 clarifies the comparison between items in the same year but obscures comparisons between years. The line graphs in Figures 7-19 and 7-20 have the same strengths and weaknesses as their respective bar graph counterparts, but in addition they also bring out more strongly the idea of direction and rate of change.

So how do you choose one (or two) from among the group? You pick the one which best matches the focus you wish to take in your report or talk. If you are not much concerned about total growth but want to focus on the contribution of each area for savings, then you would probably choose Figure

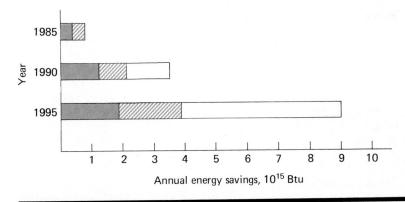

FIGURE 7-16   Annual energy savings from solar energy: version 1.

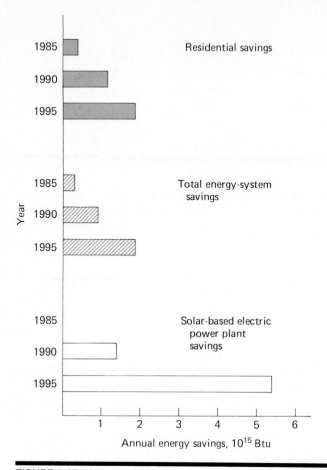

FIGURE 7-17 Annual energy savings from solar energy: version 2.

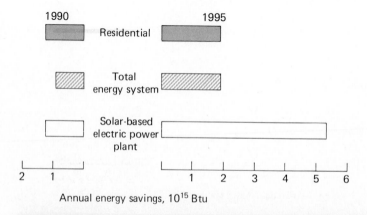

FIGURE 7-18 Annual energy savings from solar energy: version 3.

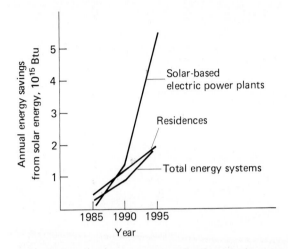

FIGURE 7-19 Annual energy savings from solar energy: version 4.

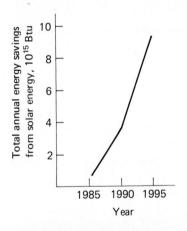

FIGURE 7-20 Annual energy savings from solar energy: version 5.

7-17. If you are interested in the growth of the contribution of each area, you would probably choose Figure 7-19. If you are primarily interested in the increase in total savings, you would probably choose Figure 7-16 or 7-20.

Since relevance and clarity are so important, let us consider one more example illustrating them, the visual aid in Figure 7-21. This visual appeared in a report outlining resources and long-term investment possibilities for a particular small city. The aid was supposed to show that significant production on a long-term basis could be achieved from shale wells and medina wells (two types of natural gas wells) and that such wells would make good long-term financial investments. Unfortunately, a reader could easily notice the

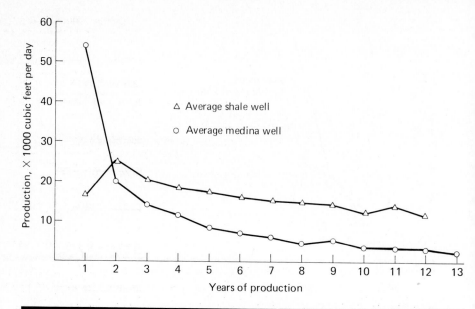

FIGURE 7-21 Production potential for average shale and medina wells: version 1.

sharp initial drop in the medina well curve and see the flat parts of both curves as low relative to the initial high production of the medina well. Such a comparison *visually* suggests that production is low after 2 years and that neither well would be a good long-term investment, although the medina well might be a promising short-term investment. However, this is the direct opposite of the intended point.

It turns out that these types of wells are financially sound investments if they produce on the average above 4000 cubic feet per day. If we add this information to the visual as a reference line, it dramatically changes the visual's impact. The modified visual, Figure 7-22, clearly (and visually) reveals the wells' adequate productivity and long-term investment potential.

Making a Visual Aid Clear

Making a visual aid clear involves two separate activities: making it conceptually clear and making it technically clear. Making it conceptually clear means having a clearly defined and relevant point and a good form for the point. Conceptual clarity is discussed above. Technical clarity is a simpler matter and will be treated here. It involves having an informative title, appropriate headings and labels, and enough white space so that an audience has the best possible chance of finding the "right" meaning for the visual aid. For

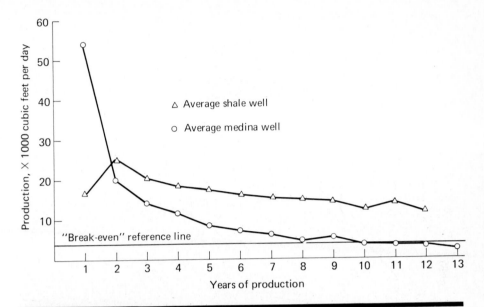

FIGURE 7-22 Production potential for average shale and medina
wells: version 2. [Adapted from Moody and Associates, Inc.,
*Natural Gas Development Feasibility Study: Report for the School
District of Erie, Pennsylvania* (Meadville, Pennsylvania, October
1980), p. 6.]

example, suppose Figure 7-21 had been entitled "Productivity Decline of
Medina and Shale Wells." Given the lack of a break-even reference line, this
title would have focused the reader's attention even more strongly on the
downward slope of the medina production curve and on the comparatively
low long-term production rates. Presumably, this would have made the visual
even less effective in communicating its real point.

To really see the benefit of proper labeling and sufficient white space,
look at the series of graphs presented in Figure 7-23. Graph *a* is an extremely
bad example of a visual aid since it has *none* of the labeling information
usually presented. Graphs *b* and *c* present more information, but still not
enough to really get the message across. (Notice that graph *c* lacks enough
information even though it provides everything except the title and two critical
labels.) Graph *d* provides an adequate title and labels, but the grid in the
background is so obtrusive that a reader can hardly see the important lines
and labels. Finally, graph *e* provides adequate information and enough white
space to let it be seen; from these, a careful and hardworking reader can
probably figure out the message. (You should note that version *d* is typical
of most student reports, which are done quickly and checked mainly for
accuracy rather than readability.)

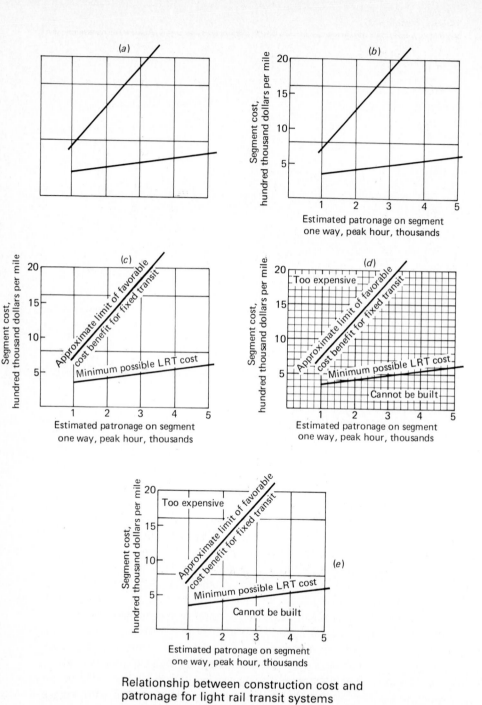

Relationship between construction cost and
patronage for light rail transit systems

FIGURE 7-23 The necessity of labels, headings, and titles in visual aids.

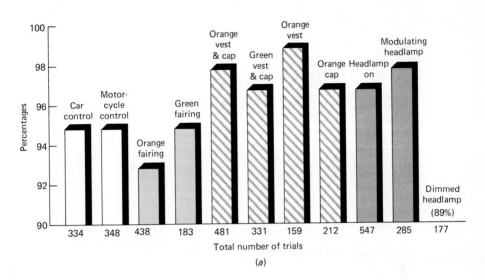

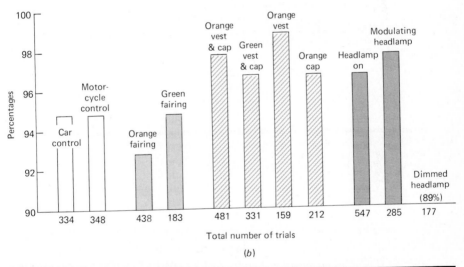

FIGURE 7-24 Elimination of unnecessary detail in visual aids.

Another use of white space to unclutter a visual aid is illustrated in Figure 7-24. Figure 7-24a has no need of three-dimensional bars; in fact, they add distracting detail. Don't you find Figure 7-24b, with its two-dimensional bars, easier to read?

You should note that color can be as misused as the graphic flourishes noted above. If you use a color other than black or white, you should be able to justify the inclusion of each additional color and of each word printed in a color other than black.

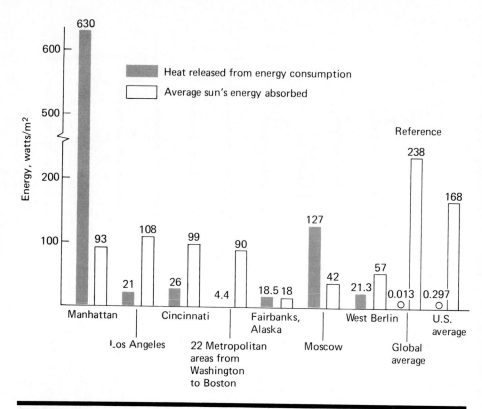

FIGURE 7-25 Use of white space to show structure in visual aid data. [From David C. White, "The Energy-Environment-Economic Triangle," *Technology Review* **76**:2 (December 1973), p. 18.]

Another use of white space showing structure in a visual aid's data is Illustrated in Figure 7-25. In this figure, the consumption-absorption units for various cities and areas are grouped together and extra white space is inserted between the city and area units. This helps to define the structure of the data for the reader. Note this same use of white space in Figure 7-27.

Making a Visual Aid Truthful

Making a visual aid truthful is important for at least two reasons. First, a visual aid which falsifies information or misleads your audience may hurt your reputation since it may lead people to see you as dishonest. (However, if you *are* honest, you probably want to be *seen* as honest.) Second, a false or misleading visual aid may hurt your argument. If members of your audience are upset or dissatisfied with your argument, they may try to discredit

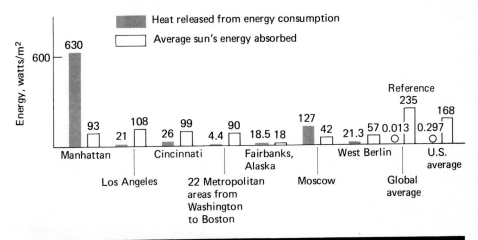

FIGURE 7-26 Distortion of significant differences by inappropriate use of scale.

it, and one of the easiest ways to do that is to point out a false or misleading spot. How can you believe the arguments of someone who is demonstrably false or misleading?

There are several ways to create a false or misleading visual aid. You can inadvertently make a poor choice in the type of visual used for presenting your information and thus obscure your point or make a point quite different from what you intended. This is probably the most frequent cause of misleading visual aids.

In a more serious falsification, you can distort the data. This is sometimes done by obscuring significant differences through inappropriate scale, as illustrated by Figure 7-26. Compare Figure 7-26 with Figure 7-25 and you will notice that Figure 7-25 preserves the zero point for reference and indicates the missing part of the scale by a cut in the y axis and in the Manhattan bar; however, it does not obscure the significant differences in energy consumption and absorption among all of the other cities and areas, as Figure 7-26 does.

Data are sometimes distorted by exaggerating insignificant differences with a suppressed zero, as illustrated by Figure 7-24. Compare the y axis of Figure 7-27 with the y axes in Figure 7-24. Notice that Figure 7-24 suppresses the zero point and greatly exaggerates the differences among the bars by the limited scale range on the y axes. It suggests that the audience responded very differently to the various safety devices. In contrast, Figure 7-27 creates a very different effect: without the suppressed zero, it suggests that the audience responded similarly to the various devices.

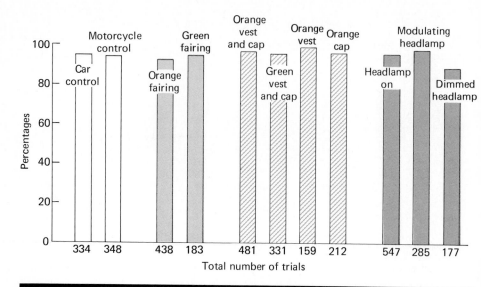

FIGURE 7-27 Preserving truthfulness by appropriate use of scale. [Adapted from Paul L. Olson, Rich Halstead-Nussloch, and Michael Sivak, "Means of Making Motorcycles More Conspicuous," *HSRI Research Review* **10**:2 (The University of Michigan Highway Safety Research Institute, September-October 1979), p 9.]

7.5

INTEGRATING THE VISUAL AID INTO THE TEXT

Once you have decided to use a visual aid in a particular spot in the text, you must incorporate it so that it seems to belong there. This is easier said than done.

The visual aid needs to be tied to the text and explained since it appears in the text and must make sense to readers. In addition, if the communicator does not explain the importance of the visual aid—its main point, limitations, assumptions, and implications—then readers will have to provide these pieces of information for themselves. As a general rule, when readers are put in this position, they will—at least sometimes—see points or implications different from those the communicator wants them to see or perhaps even completely miss the communicator's point.

The easiest way to integrate a visual aid with the text is to explain its main points and any special implications a reader should note. You might refer back to the passage on cry spectra in Section 7.2 for a good example of such an explanation. Another good example occurs in Figure 7-28, which comes from an anthropology text describing the evolution of human beings. Note that the explanation of the visual aid gives the thesis the aid is illus-

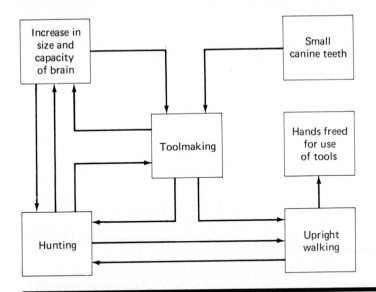

FIGURE 7-28   Tying a visual aid to the text with an explanation.
(From Richard E. Leakey and Roger Lewin, *Origins: What New Discoveries Reveal about the Emergence of Our Species and Its Possible Future,* E. P. Dutton, New York, 1977, p. 73.)

trating: *many characteristics have evolved as a result of the positive feedback mechanism.* It also provides an analysis of the aid and its parts. Other good examples of visual aids that are well integrated and explained occur in the sample reports in the Appendix.

Making the Visual Aid Independent
In addition to being tied to the text, a visual aid also needs to be independent since it will sometimes be read by itself, without any commentary. This occurs whenever a reader leafs through a book or article just looking at the graphs and tables, or when a listener does not pay close enough attention to a speaker and suddenly finds only the visual aid left from which to find the speaker's point. It also occurs when a speaker needs to make an oral presentation, remembers a particularly fine visual from some report, and has it photocopied for the presentation. In this situation, the visual aid is totally independent of the text, and if it has not been properly titled and labeled, it may be difficult to understand.

A communicator can make a visual aid independent by following the advice in Section 7.4, "Designing the Visual Aid": make a relevant, important point and use enough headings, labels, and white space. If these can be done well, the communicator has done his or her job.

EXERCISE 7-1

A Design a visual aid to illustrate the data in each of the following four passages.[4]

1   All of the 841 fatal car-truck collisions that occurred in Michigan and Texas during the periods studied are arranged in three groups: (1) car into rear of truck or tractor-trailer, (2) car into side of truck or tractor-trailer, and (3) all other car-truck collisions (including head-on collisions and truck-into car collisions).

    In the Michigan data the total of 490 collisions consists of 48 rear impacts, 46 side impacts, and 396 other types of car-truck collisions. In the Texas data the total of 351 collisions consists of 46 rear impacts, 41 side impacts, and 264 other types of car-truck collisions. The combined total for both states is 94 rear impacts, 87 side impacts, and 660 other types of car-truck collisions. Thus the total number of cases in which underride could have occurred (181) represents 21.5 percent of the total 841 cases. ["Underride" occurs when the car "rides under" the truck.]

2   The frequency of underrides is high in the 181 cases of a car's striking the rear or side of a truck, tractor, or tractor-trailer. In the Michigan data, underride occurred in 41 of the 48 rear impacts and in 30 of the 46 side impacts. In the Texas data, underride occurred in all of the 46 rear impacts and in 35 of the 41 side impacts. In the combined totals, underride occurred in 92.5 percent of the rear impacts and 74.7 percent of the side impacts. The total underrides constitute 18.1 percent of the 841 fatal car-truck accidents.

3   *What crashed:* The 181 car-truck crashes involved 126 tractor-trailer combinations and 55 straight trucks (including two tractors). The tractor-trailers were predominantly vans and flatbeds; the straight trucks were mainly stake or dump trucks. The passenger cars were categorized into three groups by size—full-size cars, intermediates, and compacts—and distributed as follows:

a.   In Michigan, 22 percent of the cars involved were compacts, 20 percent were intermediates, and 58 percent were full-size cars.

b.   In Texas, 36 percent were compacts, 11 percent were intermediates, and 53 percent were full-size cars.

c.   For the total sample (combining the Michigan and Texas samples), 29 percent were compacts, 16 percent were intermediates, and 55 percent were full-size cars.

4   *When:* Car-truck fatal crashes are predominantly nighttime phenomena. Inclement weather is a minor contributor to these accidents, but low visibility at night is a major factor. This will be discussed under "how

TABLE 7-5 FREQUENCY OF IMPACTS AT VARIOUS POINTS OF THE TRUCKS IN THE 47 FATAL ACCIDENTS

POINT OF TRUCK	DIRECTION OF IMPACT	NUMBER OF IMPACTS
Left front wheel area	From left side	1
Left middle	From left side	1
Left rear wheel area	From left side	4
Right front wheel area	From right side	1
Right middle	From right side	5
Right rear wheel area	From right side	2
Left rear	From rear	10
Middle rear	From rear	12
Right rear	From rear	5

and why'' below, but for the moment please note that 67.4 percent of the 181 fatal crashes occurred at nighttime and 32.6 percent occurred during the day.

B   Where necessary, rewrite the passages in part A of this exercise to introduce and present the visual aids you designed.

C   Evaluate the following two passages[4] to see if any visual aids (other than the tables) are desirable. If necessary, rewrite the passage to introduce and present the aids.

1   Points of impact on the truck or tractor-trailer are of interest in considering the potential effectiveness of underride devices. While more than half of the rear impacts were relatively direct (into the rear center of the truck or tractor-trailer), 42 of the 94 occurred at the rear corner of the truck or trailer. Some side impacts occurred in wheel areas of the truck or trailer, and some occurred in places where underride was possible. Table 7-5 shows the number of impacts to various areas of the trucks involved in the collisions.

2   A variety of circumstances or combinations of circumstances precipitated the 181 fatal car-truck accidents studied here. The more common ones are tabulated in Tables 7-6 through 7-9. In rear-collision situations, the truck or tractor-trailer most often was either moving forward ahead of the car or turning left. In side-collision situations, there were three notable circumstances: the tractor/trailer was astride the road jackknifed, astride the road entering or exiting a driveway or ramp, or astride the road making a left turn or attempting a U-turn. About an equal number of times the car driver or the truck driver failed to obey a traffic signal—usually with disastrous consequences for the car occupants.

TABLE 7-6 CIRCUMSTANCES LEADING TO FATAL CAR-INTO-TRUCK UNDERRIDE ACCIDENTS IN MICHIGAN AND TEXAS

MOVEMENT OF INVOLVED TRUCK	PLACE CAR STRUCK TRUCK	NUMBER OF INCIDENTS
Moving forward, ahead of car	Rear	7
Turning left	Rear	8
Slowing or stopping for signal	Rear	3
Crossing centerline or lane line	Side	2
Stopped or parked	Side	1
Jackknifed in path of car	Side	1
Entering or exiting driveway in path of car	Rear	1
	Side	1
Making U-turn in path of car	Rear	2
	Side	4
Running stop sign or failing to yield	Rear	1
	Side	5

TABLE 7-7 CIRCUMSTANCES LEADING TO FATAL CAR-INTO-TRACTOR-TRAILER UNDERRIDE ACCIDENTS IN MICHIGAN AND TEXAS

MOVEMENT OF INVOLVED TRACTOR-TRAILER	PLACE CAR STRUCK TRACTOR-TRAILER	NUMBER OF INCIDENTS
Moving forward, ahead of car	Rear	22
Turning left	Rear	19
	Side	1
Slowing or stopping for sign	Rear	11
Stopped or parked	Rear	5
	Side	1
Jackknifed in path of car	Side	11
Running stop sign or failing to yield	Side	7
Crossing centerline or lane line	Side	1

TABLE 7-8 CAR MOVEMENT IN FATAL CAR-INTO-TRUCK ACCIDENTS

MOVEMENT OF INVOLVED CAR	PLACE CAR STRUCK TRUCK	NUMBER OF INCIDENTS
Running stop sign or failing to yield	Side	5
Crossing centerline or lane line	Side	6

TABLE 7-9   CAR MOVEMENT IN FATAL CAR-INTO-TRACTOR-TRAILER ACCIDENTS

MOVEMENT OF INVOLVED CAR	PLACE CAR STRUCK TRACTOR-TRAILER	NUMBER OF INCIDENTS
Running stop sign or failing to yield	Rear	2
	Side	9
Crossing centerline or lane line	Rear	2
	Side	13

D   Field Manufacturing Company produces three major waste products during its manufacturing process: waste A, waste B, and waste C. The company has developed a 10-year program for treating and refining these wastes and then reusing them in the manufacturing process. To aid in completing this exercise, a summary of the cost of treating and refining each waste product appears in Table 7-10 (assume that the 1982 fiscal year has just ended).

1   The supervisor of each waste treatment program—Supervisor A, Supervisor B, Supervisor C—wants to request more money. As part of the argument, each supervisor wants to compare the funding for her or his waste treatment program to the funding granted to the other two programs. You are to design the best visual aid(s) for each supervisor to use in arguing that her or his program is underfunded in relation to the others. (You will need at least one visual aid for Supervisor A, one for Supervisor B, and one for Supervisor C.)

2   The Head of the Waste Treatment Department must present a financial status report on the waste treatment programs to the company's president. The report must describe the economic history of the programs, the current economic status of each, and predictions for the future. Design the best visual aid(s) for the presentation.

E   Construct visual aids for the information in Tables 7-11 and 7-12. Define a generalization or claim you want the aid to support and make the aid as *visually* supportive as possible.

F   Read through at least one report you have written and evaluate its use of visual aids. Make sure that you have used a visual aid in each necessary or appropriate spot, that you have chosen the best type of visual aid for your purpose, that your visual aid is as *visual* as possible, that it does not distort your data, and that it is appropriately integrated into the text. Add or revise (or delete) any visual aids as necessary. When you have finished, trade papers with a friend and evaluate each other's work.

TABLE 7-10 COSTS OF REFINING WASTE PRODUCTS A, B, AND C (IN THOUSANDS OF DOLLARS)

WASTE PRODUCT	EXPENDITURES 1979	INCREASE IN EXPENDITURES 1979–80	EXPENDITURES 1980	INCREASE IN EXPENDITURES 1980–81	EXPENDITURES 1981	INCREASE IN EXPENDITURES 1981–82	EXPENDITURES 1982	TOTAL (10-YEAR) COST OF PROJECT
A	153.4	16.6 (10.8%)	170.0	19.0 (11.8%)	189.0	20.5 (10.8%)	209.5	2661.3
B	48.5	0.5 (1.03%)	49.0	2.5 (5.1%)	51.5	3.5 (6.8%)	55.0	1237.0
C	178.0	9.0 (5.05%)	187.0	6.0 (3.2%)	193.0	1.2 (0.62%)	194.2	5169.7
Total (A + B + C)							458.7	9068.0

TABLE 7-11 GROSS SALES FOR YOUGAN DAIRY PRODUCTS, 1967–71

YEAR	FRESH MILK PRODUCTS	DRY MILK PRODUCTS	DIVERSIFIED PRODUCTS	DOLLARS (THOUSANDS)
1971	56.7%	15.5%	27.8%	2074
1970	54.0%	17.0%	29.0%	1750
1969	67.0%	20.0%	13.0%	1229
1968	72.0%	23.0%	5.0%	986
1967	74.0%	24.0%	2.0%	932

TABLE 7-12 RELATIVE STEEL PRODUCTION IN THE COMMON MARKET COUNTRIES

	1956	1962	1963	1964	1965	1966
Germany	26.564	32.563	31.597	37.339	36.821	35.316
France	13.442	17.234	17.554	19.781	19.599	19.591
Italy	6.076	9.757	10.157	9.793	12.681	13.639
Belgium	6.367	7.351	7.525	8.725	9.162	8.916
Luxembourg	—	3.456	4.010	4.032	4.559	4.390
Holland	1.000	2.087	2.354	2.659	3.145	3.309

7.6
FORMATTING CONVENTIONS THAT MAKE READING EASIER

There are many features of technical writing that make it look different from most writing we see in newspapers, books, and personal letters. Look, for instance, at Figure 7-29, the beginning of a typical engineering report. (The full report appears in Appendix C.) You will notice that it has some very interesting formatting features:

1   Single-spacing

2   Short paragraphs

3   Lists

4   Headings (underlined titles)

5   Numbers to mark the various paragraphs

6   Liberal use of white space

All of these features occur frequently in scientific and technical writing because they are functional: single-spacing saves space, and the others make a text easier to read, especially for busy and inattentive readers. Headings clearly announce the contents of a section so that busy readers can skip that section if they don't need its details. Short paragraphs and white space make a report easy on the eye, even though it may be single-spaced. The numbering, indentation, and lists provide clues to the organization of the report:

FIGURE 7-29 Formatted version of the Discussion of a technical report.

1. FAILURE DUE TO TEMPERATURE AND MOISTURE AND PROPOSED PREVENTION MEASURES

One of the main problems facing LCDs is their strong sensitivity to temperature and moisture. These factors can change the behavior of LCDs rather drastically and result in gradual degradation of the device. Two basic categories can be recognized in this case: damages due to combination of high temperature and high humidity and damages due to high temperature only.

1.1 Damages Due to a Combination of Temperature and Moisture

The most damaging environment that an LCD is likely to experience is produced by the combination of high temperature and high humidity. Depending on how the display is constructed, three failure modes can occur under this condition: failure of the polarizers, failure of the adhesive on the polarizer and reflector, and, finally, failure of the liquid crystal mixture.

1.1.1. Failure of the Polarizers

The most common failure occurs with the polarizers. Regular polarizing material will begin to degrade after a short period of time (4 days) when subjected to a temperature of 50°C and 95% relative humidity. Once moisture penetrates the plastic backing of the polarizing material, it affects the polarizing properties. The digit area of the display will start to lose contrast, gradually turning from black to brown until it finally disappears completely.

1.1.1.1. Preventing Failure of the Polarizers

To provide protection against the effects of high temperature and humidity, the use of a polarizing material known as "K-sheet," introduced by Polaroid Corporation, is recommended. This material has polarization properties which are relatively unaffected by moisture.

1.1.2. Failure of the Adhesive

Moisture can also attack the adhesive on the polarizer and reflector and cause it to begin peeling away from the glass. If the alignment is lost, dark patches will appear in the viewing area like those as already described.

1.1.2.1. Preventing Failure of the Adhesive

To prevent this kind of failure, the use of a truly hermetic seal is recommended to protect the adhesive.

1.1.3. Failure of the Liquid Crystal Mixture

If any of the components in the liquid crystal mixture (such as Schiff-base compounds) are attacked by moisture, the liquid crystal molecules will start to break down, thereby lowering the clearing point of the mixture and the effective upper operating temperature of the display. Due to the ionic by-products of the chemical reactions, the electrical currents of the display will also show a very significant increase.

FIGURE 7-29 *(continued)*

1.1.3.1. Preventing Failure of the Liquid Crystal Mixture
The procedure described in part 1.1.2.1 should be adopted.

1.2. Failure Due to High Temperature Only
Failure can occur when an LCD is exposed to high temperatures only. The twisted nematic display requires the glass plates to be specially treated to impress a uniform alignment on the liquid crystal molecules next to the glass surface. The quality of this alignment can be degraded by high temperature because of decomposition of the interface layer.

1.2.1. Preventing Failure Due to High Temperature
The procedure used in part 1.1.2.1. should be adopted.

2. FAILURE DUE TO LOSS OF CONTACT
Problems can arise due to loss of contact between the connector and the indium oxide leads on the LCD. The user will often blame the LCD for an electrical failure such as an open segment when in fact it is the connector which is responsible.

2.1. Preventing Failure Due to Loss of Contact
This problem was solved by the introduction of the DIL displays with substrate chips by Liquid Crystal Displays in 1975. They offer a display that can be soldered directly in a circuit board to reduce the possibility of any disconnections.

they help the reader see what sections provide proof for the generalizations, and they allow a reader to skip freely from section to section without reading everything.

To get a good idea of how helpful these simple formatting considerations can be, look at the unformatted version of the Discussion section of the report, presented in Figure 7-30. Do you agree that it is much more difficult to read? Do you agree that formatting makes the version in Figure 7-29 and Appendix C more functional, that is, easier to read and understand?

One thing that new technical writers have to learn is that the formatting conventions just discussed are widespread as well as functional. Short paragraphs, for instance, are common in much good technical writing. This may seem strange to those who were taught that a "good" paragraph should be at least 100 to 200 words long to be fully developed. Notice, however, that short paragraphs are not necessarily undeveloped paragraphs. You can spread out a 200-word proof over several short indented paragraphs and create a fully developed section, as illustrated in Figure 7-29. Such a section is much easier to read and probably more meaningful than the long-paragraph version presented in Figure 7-30.

FIGURE 7-30 Unformatted version of Figure 7-29.

One of the main problems facing LCDs is their strong sensitivity to temperature and moisture. These factors can change the behavior of LCDs rather drastically and result in gradual degradation of the device. Two basic categories can be recognized in this case: damages due to combination of high temperature and high humidity and damages due to high temperature only.

The most damaging environment that an LCD is likely to experience is produced by the combination of high temperature and high humidity. Depending on how the display is constructed, three failure modes can occur under this condition: failure of the polarizers, failure of the adhesive on the polarizer and reflector, and, finally, failure of the liquid crystal mixture. The most common failure occurs with the polarizers. Regular polarizing material will begin to degrade after a short period of time (4 days) when subjected to a temperature of 50°C and 95% relative humidity. Once moisture penetrates the plastic backing of the polarizing material, if affects the polarizing properties. The digit area of the display will start to lose contrast, gradually turning from black to brown until it finally disappears completely. To provide protection against the effects of high temperature and humidity, the use of a polarizing material known as "K-sheet," introduced by Polaroid corporation, is recommended. This material has polarization properties which are relatively unaffected by moisture. Moisture can also attack the adhesive on the polarizer and reflector and cause it to begin peeling away from the glass. If the alignment is lost, dark patches will appear in the viewing area like those as already described. To prevent this kind of failure, the use of a truly hermetic seal is recommended to protect the adhesive. If any of the components in the liquid crystal mixture (such as Schiff-base compounds) are attacked by moisture, the liquid crystal molecules will start to break down, thereby lowering the clearing point of the mixture and the effective upper operating temperature of the display. Due to the ionic by-products of the chemical reactions, the electrical currents of the display will also show a very significant increase. Liquid crystal mixture failure can be prevented in the same way as adhesive failure. Failure can occur when an LCD is exposed to high temperatures only. The twisted nematic display requires the glass plates to be specially treated to impress a uniform alignment on the liquid crystal molecules next to the glass surface. The quality of this alignment can be degraded by high temperature because of decomposition of the interface layer. Failure due to high temperature can be prevented in the same way as adhesive failure.

Problems can arise due to loss of contact between the connector and the indium oxide leads on the LCD. The user will often blame the LCD for an electrical failure such as an open segment, when in fact it is the connector which is responsible. This problem was solved by the introduction of the DIL displays with substrate chips by Liquid Crystal Displays in 1975. They offer a display that can be soldered directly in a circuit board to reduce the possibility of any disconnections.

7.7
FORMATTING CONVENTIONS THAT MAKE WRITING CLEARER

In addition to making reading easier, the two visual elements discussed in Section 7.6, white space and formatting, can also make your writing clearer. For instance, consider the memo in Figure 7-31 and try to write down the list of items the reader is asked to provide. Then compare your list to that in the version of the same memo presented in Figure 7-32. They might not match.

The paragraph structure of the first version allows many readers to miss one or more of the items. However, in the second version, the formatted list and the extra white space separating the items in the list clarify the number of items and force the reader to see them individually. They also force the writer to decide more carefully how the information is to be presented. For instance, does the writer want all of the information specified in paragraph 2 of Figure 7-31 presented together, in a "clump," or does the writer really want the information segmented into categories 1, 2, and 3, as suggested in

FIGURE 7-31   Sample memo—unformatted version.

SPELLBUCH UNIVERSITY
COLLEGE OF ENGINEERING

October 29, 1979

MEMORANDUM

To: Department and Program Heads
            College of Engineering

From: College Curriculum Committee

Subject: Developing Student's Problem-Solving
            Ability: Request for Information

The Curriculum Committee is quite concerned about the emphasis given to developing the student's capabilities in problem solving and synthesis. Virtually all the descriptions of new and revised courses submitted to the Committee for approval identify the technical material to be disseminated but do not identify techniques to be used in problem solving in particular. The Curriculum Committee feels that probably the amount of open-ended problem solving in the College should be increased, but the Committee has limited data on the nature and extent of problem solving now incorporated in College courses.

To more accurately assess the current situation, the Committee is requesting you to provide by November 29, 1979, a list of undergraduate courses (i.e., through 400 level) in your program which incorporate a significant amount of design and open-ended problem solving or which discuss problem-solving techniques for either open-ended or closed problems. An example of such a course is attached. "Significant

FIGURE 7-31 *(continued)*

amount" should be interpreted as at least one or two weeks devoted to the subject including design, analysis, synthesis, and evaluation stages rather than simple assignment of an occasional short problem. Please make a rough estimate of the fraction of effort in these courses which is devoted to open-ended problem solving and design. In addition, please identify which of the listed courses are required in your program and estimate how many are taken by the "average" student.

Please let us have the names of faculty teaching the courses and of faculty who are particularly interested in this topic. The definition of a course as problem solving is obviously not clear-cut, but please make a reasonable interpretation of this classification.

The Committee would also appreciate your giving thought, and possibly some answers, to the following (open-ended!) questions:

What resources could the Curriculum Committee provide to help promote problem-solving courses? Possible examples include organizing seminars, compiling a bibliography of literature related to problem-solving technique and pedagogy, and enlisting the help of the Center for Research on Learning and Teaching. Would involvement be greater if help and advice in developing these courses were available? Are there special computer needs for these courses?

We would appreciate it if we could have your response to the above questions by November 29, 1979. Thank you for your consideration and help on this questionnaire.

HSF:mb

Attachment

Figure 7-32? If the writer wants segmented information, he or she may need to do a lot of analysis and reclassification of the data received from readers of Figure 7-31; presumably this extra work would be avoided by sending out the version shown in Figure 7-32.

Further, a quick or careless reader is also more likely to notice that information is requested in Figure 7-32 because the second paragraph states this in a very obvious way.

In many situations, formatting information into obvious lists or breaking it up into clearly labeled sections helps a reader. When formatting, white space, and labels are used, the reader and writer are more likely to understand each other and to respond appropriately.

FIGURE 7-32 Sample memo illustrating the use of formatting, white space, and headings.

SPELLBUCH UNIVERSITY
COLLEGE OF ENGINEERING

October 29, 1979

MEMORANDUM

To: Department and Program Heads
             College of Engineering

From: College Curriculum Committee

             Developing Student's Problem-Solving
             Ability: Request for Information

The Curriculum Committee is quite concerned about the emphasis given to developing the student's capabilities in problem solving and synthesis. Virtually all the descriptions of new and revised courses submitted to the Committee for approval identify the technical material to be disseminated but do not identify techniques to be used in problem solving in particular. The Curriculum Committee feels that probably the amount of open-ended problem solving in the College should be increased, but the Committee has limited data on the nature and extent of problem solving now incorporated in College courses.

To more accurately assess the current situation, the Committee is requesting you to provide by November 29, 1979, the following information:

1. a list of undergraduate courses in your programs which incorporate a significant amount of design and open-ended problem solving or which discuss problem-solving techniques for either open-ended or closed problems (we realize that the definition of a course as having a "significant amount of problem solving" is not clear-cut, but we have appended an example of such a course and would request that "significant amount" be interpreted as at least one or two weeks devoted to problem solving and its techniques (including focus on design, analysis, synthesis, and evaluation) rather than as the simple assignment of an occasional short problem),

2. a rough estimate of the fraction of effort devoted to open-ended problem solving and design in these courses,

3. the identification of those courses listed in #1 above which are required in your program,

4. an estimate of how many of the courses listed in #1 are taken by the "average" student,

5. the names of faculty teaching these courses who could serve as resource persons within the College, and

FIGURE 7-32 *(continued)*

6. the names of faculty particularly interested in problem solving and its techniques

The Committee would also appreciate your giving thought, and possibly some answers, to the following (open-ended!) questions:

1. What resources could the College Curriculum Committee provide to help promote problem-solving courses? (Possible examples include organizing seminars, compiling a bibliography of literature related to problem-solving technique and pedagogy, and enlisting the help of the Center for Research on Learning and Teaching.)

2. How could the College or the Departments involve more faculty in expanding problem solving in our courses? Would involvement be greater if help, release time, and/or advice in developing these courses were available?

3. What, if any, are the special computer needs for problem-solving courses in your department?

We would appreciate it if we could have your response to the above questions by November 29. Thank you for your consideration and help on this questionnaire.

HSF:mb

Attachment

REFERENCES

1   William Schneider and William D. Green, Jr., "Saving Skylab," *Technology Review* **76**:3 (January 1974), p.44.

2   Wann Langston, Jr., "Pterosaurs," *Scientific American* **244**:2 (February 1981), p. 126.

3   *Technology Review* **81**:8 (August-September 1979), p. 79.

4   Adapted from Daniel J. Minahan and James O'Day, *HSRI Research Review* **8**:3 (The University of Michigan Highway Safety Research Institute, November-December 1977), pp. 4–5, 8.

ADDITIONAL READING

Rudolf Arnheim, *Art and Visual Perception: A Psychology of the Creative Eye,* University of California Press, Berkeley, 1974.

Rudolf Arnheim, *Visual Thinking,* University of California Press, Berkeley, 1969.

Gregg Berryman, *Notes on Graphic Design and Visual Communication,* William Kaufmann, Los Altos, Calif., 1979.

R. N. Haber, "How We Remember What We See," *Scientific American* **222**:5 (May 1970), pp. 104–112.

Michael Macdonald-Ross, "How Numbers Are Shown," *AV Communication Review,* **25** (Winter):4 (1977).

Michael Macdonald-Ross, "Graphics in Text," *Review of Research in Education,* vol. 5, L. Shulman (ed.), F. E. Peacock, Itasca, Ill., 1978.

A. J. MacGregor, *Graphics Simplified: How To Plan and Prepare Effective Charts, Graphs, Illustrations, and Other Visual Aids,* University of Toronto Press, Toronto, 1979.

Cheryl Olkes, "Typography/Graphics," *Document Design: A Review of the Relevant Research,* American Institutes for Research, Washington, D.C., 1980, pp. 103–110, 163–166.

Mary Eleanor Spear, *Practical Charting Techniques,* McGraw-Hill, New York, 1969.

Patricia Wright, "Presenting Technical Information: A Survey of Research Findings," *Instructional Science* **6**:93–134 (1977).

Part V

Major Genres of
Technical Reporting

8

THE ORAL PRESENTATION

In many fields of science and technology, the ability to communicate technical information orally is just as important as the ability to write well. Indeed, those in management positions often find themselves making oral presentations more often than written ones; they find oral presentations faster, easier, and more suited to immediate feedback and clarification. A number of studies suggest that oral communication skills are of crucial importance, in fact, in determining who gets promoted to upper management levels.[1–5]

Managers are not alone, however, in their need to have good oral communication skills: engineers, scientists, and other technical professionals also need these skills. This can be seen, for example, in the results of a recent survey reported in *Engineering Education*.[6] In this survey, 367 engineers in chemical, civil, electrical, geological, mechanical, metallurgical, and mining engineering were asked to rate 30 specific communication tasks for job-related importance. Some of these engineers were working as managers, but most were not. Of the 24 tasks that were rated important by the respondents, 14 were writing tasks and 10 were oral tasks. Of these, five involved the giving of a formal oral presentation (project proposal presentations; oral presentations using graphs, charts, and/or other aids; project progress report presentations; project feasibility study presentations; and formal speeches to technically sophisticated audiences); the remainder were of a more informal nature, involving one-to-one talks or small-group discussions.

This chapter is devoted mainly to giving you advice on how to make an effective formal oral presentation. Preparation is the key: if you plan your presentation properly and practice it beforehand, you'll find that the actual delivering of an oral presentation will be relatively easy; if you don't prepare, it could be a real struggle. So, by all means, give yourself plenty of time to prepare an oral presentation. In particular, follow the procedure we've laid out for you step by step, taking care not to omit any of the steps. That way, you'll not only give an effective presentation but may even *enjoy* doing it!

There are times, however, when you may not be given sufficient time to prepare—when your boss, for example, says that an important group of visitors is arriving in half an hour and you are to give them a briefing. What do you do then? At the end of the chapter, we'll offer some suggestions.

8.1

GIVING A FORMAL ORAL PRESENTATION

Preparation

The basic principle to keep in mind in preparing any kind of oral presentation is that all listeners have a limited attention span and cannot be expected to follow everything you say. Their attention will probably wander from time to time, even if your presentation is only 10 minutes long. So, if you want to make sure that your listeners will come away from your talk with your main points clear in their minds, you must organize your presentation in such a way that these main points stand out. Here is how to do it:

1   *Analyze your audience, and limit your topic accordingly.* What do your listeners already know about the topic? What do they need or want to know about it? How much new information about it can they absorb? If you tell them what they already know, they'll be bored, but if you go to the opposite extreme and give them too much information too fast, they'll be overwhelmed and simply "short circuit."

2   *Determine your primary purpose.* Is there some main point you want to get across? Is there something you want your listeners to believe, or be able to do? Whatever it is, you should have it clear in your mind so that you can build your presentation around it.

3   *Select effective supporting information.* What kind of information will best support your main point? What kind of information will appeal to your audience? They will probably be able to remember no more than three or four main supporting points and no more than two or three supporting details for each of these. So choose wisely!

4   *Choose an appropriate pattern of organization.* Often, your supporting information can be presented according to a single dominant pattern of organization. For example, if you're trying to tell your listeners how to do something, you may want to organize your information into a list of instructions, chronologically ordered. On the other hand, if you're describing an experiment, you'll probably want to use the standard descriptive pattern: Introduction, Materials, Procedure, Results, Discussion. There are many possible patterns of development to choose from, some of which are described in Chapters 2 and 5. Whichever one you choose should be appropriate to the subject matter, to your primary purpose, and to your audience—and it should be followed consistently.

5   *Prepare an outline.* Keep it brief: main points, main supporting points only. Arrange these points according to the pattern of development you chose in Step 4. Do not write out a complete text unless the technical content or circumstances warrant it.

6   *Select appropriate visual aids.* Visual aids, when properly designed, have tremendous power as attention-getters. If you want to emphasize a point, by all means try to do it visually as well as orally. As a general practice, in fact, it's a good idea to emphasize all major points and major supporting points with visual aids. Properly sequenced, the aids will then also serve as cue cards for you. This does not mean simply putting the outline of your speech on flip charts. Words and phrases alone do not really have visual impact, at least not in the way that graphs, sketches, three-dimensional objects, and other pictorial displays do. Visual aids should be truly *visual!* The more visual they are, the more effective they will be in capturing your audience's attention. (Review Chapter 7 on the principles of designing effective visual aids.)

      If you have any reason to believe, however, that your audience may have trouble understanding your pronunciation—for example, if you are a nonnative speaker of English or if you are addressing a group of nonnative speakers—then it would be a good idea to write out your main points and some keywords for your audience.

7   *Prepare a suitable introduction.* It's essential that your listeners have enough background information to understand and appreciate your presentation. Are you addressing some problem? Make sure you define it so that your listeners know exactly what it is and can appreciate your proposed solution. Are you taking sides over an issue and arguing for your point of view? If so, make sure your listeners know exactly what the issue is. In short, if you have any doubts about the audience's background knowledge, be sure to provide a basic orientation and to define important terms.

8   *Prepare a closing summary.* Listeners are typically very attentive at the beginning of a presentation, less attentive as it wears on, and then suddenly more attentive again as it comes to an end. In other words, they perk up at the end, hoping to catch a final summarizing comment or recommendation. This is a well-proven phenomenon, common to all of us: just think about the times you've listened to someone else's presentation and have come away remembering his or her final words best of all.

      In preparing your oral presentation, therefore, you should plan to take advantage of this fact of human psychology—you should prepare a good, solid closing summary. Take this opportunity to repeat and thus reemphasize your most important "bottom-line" conclusions and recommendations, along with the major reasons for them. Make these closing comments crisply and emphatically.

EXAMPLE   You are a systems analyst employed by the Coronado Sugar Company (CSC). For a number of years the company has been using outside help in recording the amount of sugar beets received at its four processing plants. Since this service is quite costly, your supervisor has asked you to investigate the feasibility of using the company's own computer for this purpose. You have done so and are now ready to prepare for a brief oral presentation of your findings. Accordingly, you go through the following eight steps:

1   *Audience analysis.* You expect the audience to include, in addition to your supervisor, the managers of the four processing plants and the company comptroller. All of these people are familiar with the company's system of recording individual deliveries on weight tickets; they are also familiar with the daily and weekly summaries that are produced from these weight tickets. They do not, however, know how these summaries are put together (that's what the outside firm has been doing for $15,000 a year), and they do not know much about computers. Your supervisor and the company comptroller will be concerned mainly with how much money can be saved by adopting a new system. The four plant managers, on the other hand, will want to know mainly whether or not a new system will disrupt plant operations in any way.

2   *Purpose.* Your investigation has revealed that there are three in-house methods that would be substantially more cost-effective than using an outside firm and that any of these methods could be implemented without disrupting company operations. One of these methods would actually save the company more than $100,000 over 8 years, and you would like to see it adopted. However, if you push this one method to the exclusion of the others, your presentation may sound too much like a sales pitch. So it might be a good idea to broaden your scope a bit and promote the idea that any of the three in-house methods would be better than the present system. This will give your listeners a range of options from which to choose. Surely at least one of the options should appeal to them.

    Your primary purpose, then, will be simply to convince your listeners that the present system of compiling beet receiving data should be replaced by an in-house method. Once you've done that, it should be relatively easy to sell your own particular preference from among the three competing options.

3   *Supporting information.* Your supervisor (the decision maker in this case) and the comptroller will be most impressed, of course, with the cost figures, which favor the three in-house options over the present system; be sure to show both the short- and long-term cost advantages. The four plant managers will want to know how the three alternative methods work; in particular, they will want to be reassured that an in-house system will not disrupt their plant operations. All of your listeners

will be pleased to hear that at least one of these in-house methods (the one you prefer) will provide a fringe benefit in the form of making extra equipment available for other uses during the off-season.

4   *Pattern of organization.* Your audience's main concern will clearly be how the four methods compare in terms of cost benefits and impact on plant operations. So the best way to present your supporting information would be according to a comparison-and-contrast pattern of organization.

5   *Outline*   I   Problem statement
                            Need for compiling summaries
                            Present method

                    II   Alternative methods
                            Method 1
                                Costs
                                Advantages and disadvantages

                            Method 2
                                Costs
                                Advantages and disadvantages

                            Method 3
                                Costs
                                Advantages and disadvantages

                    III   Summary
                            All three alternative methods are more cost-effective
                                than the present system.
                            Method 2 is the most cost-effective in the long run.

6   *Visual aids.* Your audience will want to see a projected cost breakdown for each of the three alternative methods, as well as a summary graph comparing all four (Figure 8-1). The summary graph is particularly important, as it will allow you not only to emphasize your main point (that any of the three alternative methods would be cheaper than the present system) but also to show how Method 2 is ultimately the most cost-effective.

7   *Introduction.* You'll want to engage your audience's interest from the outset by explaining the full nature of the problem: CSC needs to compile regular summaries of beet receiving data for various purposes, but the present system of doing so is far too expensive. Don't get bogged down in too much background information: your listeners already know a lot about the present system, and much of what they don't know (for example, how the outside firm puts together the data) they don't really need to know. Also, this is supposed to be a *brief*

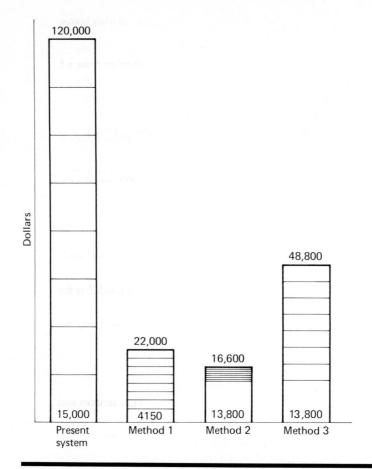

FIGURE 8-1   Comparison of processing methods: total cost for
8 years.

presentation, so don't waste your audience's time telling long-winded
jokes or anecdotes. Instead, get right to the point. After describing the
problem, tell your listeners that you've investigated various possible
solutions and have found three that deserve their consideration.
    Your introduction might go something like this:

As you all know, every fall we collect an enormous number of weight
tickets from our four processing plants. In fact, we've been averaging
about 100,000 of them over the past five years. From all of these tickets
we have to compile regular summaries—daily summaries and weekly
summaries of beets received for each plant, and then final summaries of
delivered amounts for each grower. The daily summaries are particularly
important because we need these to maintain smooth plant operations,

keep to our shipping schedule, and allow for any sudden breakdowns. This means, of course, that the data have to be processed fast and accurately, and in the past we've always felt we needed the MFC people to do it. After all, they're the pros. And indeed, they've done a fine job for us. The problem is that their service is very expensive: $15,000, to be exact, for just the two months we use them.

So last month Glenn asked me to look into the possibility of using our own company computer to do the job. I've now completed my investigation and am happy to report that there are indeed three different ways we could use our own computer to get the job done without spending nearly as much money or sacrificing any efficiency. One of these methods, in fact, could save us as much as $100,000 over the next eight years, without any adverse impact on plant operations. There are even some fringe benefits to be gained. . .

8   *Summary.* You'll want to reiterate the fact that all three in-house methods are more cost-effective than the present system, and you'll want to use your summary chart (Figure 8-1) to emphasize this point. More specifically, you'll want to repeat your claim that Method 2 is the most cost-effective of all, noting that not only would it save CSC the most money in the long run but it would also allow the company to invest in equipment which could be used for other valuable purposes in the off-season. Be sure to make it clear to your listeners that these are not only your conclusions but also your strong recommendations:

. . .and so I recommend strongly that we stop using MFC's services and create instead our own in-house program, using our own equipment and our own people. It would take us a little time to get it up and running, but we have the time and will start saving money the minute it's going. It will also increase our computing capability in all sorts of other ways. If you have any questions, I'd be glad to try to answer them for you.

Practice

Nothing is more helpful to the ultimate success of an oral presentation than practice. Not even the best of speakers can give a totally effective presentation without practicing it first. Practice allows you to spot the flaws in a presentation and eradicate them. It enables you to work on making smooth transitions from section to section, instead of awkward stops and starts. And practice gives you an idea of how long your presentation will take; if it's too long, you still have time to make changes so that you can ultimately deliver it at a tempo that's comfortable for you. All of these benefits promote greater self-confidence, which in turn leads to a more convincing, emphatic, effective style of delivery.

The best way to practice a talk is by rounding up a few friends and trying it out on them. Ask them to hear you all the way through, taking as many notes as possible but not raising any questions until you've finished. Then ask them for a complete "postmortem"; take note of spots where they had trouble following you, and immediately try out some other approach to see if it makes things clearer. Note also what the strong parts of your talk are: maybe you can use the techniques used in the strong parts in other places.

If you'd prefer not to force your friends to listen to you stumble through a practice session, a good alternative would be to use a videotape recorder and then critique your own performance. This is particularly effective in allowing you to spot nervous mannerisms that you might not be conscious of while actually performing. An audiotape recorder, while not as useful as a videotape recorder, can also be used, especially for the purpose of listening to yourself read from a manuscript (see point 6 below).

Here are some specific things to work on while practicing your oral presentation:

1 *Devise ways of repeating your important points without being too repetitive.* Since your important points should all contribute to a single cumulative effect, it's a good idea to reiterate these points occasionally as you go along—especially in summary form at the end of your talk. However, since exact repetition of wording can become annoyingly monotonous the third or fourth time around, try to vary your wording.

2 *Create smooth transitions between sections.* Take note of places where the flow of your presentation seems to break down, and see if you can't insert a phrase or two to act as a bridge. (If you can't, there may be a fundamental flaw in the overall structure of your presentation; in that case, try to reorganize it.)

3 *Familiarize yourself with the equipment you'll be using.* It's embarrassing—and annoying to the audience—to waste precious time by fumbling around with slide projectors, television monitors, microphones, and other equipment when you're supposed to be giving your presentation. It's even worse if your presentation depends crucially on some piece of equipment and you can't get it to operate at all. Therefore, *if you plan on using any equipment, check it out ahead of time and become familiar with it!* Learn how to use it so that it will help your presentation, not hinder it, and be prepared with some backup system just in case.

4 *Prepare yourself for questions.* Listeners may raise questions at any point in your presentation, and it is vitally important that you answer them satisfactorily. If you don't, your most precious asset as a speaker— your credibility—may be endangered. So be sure you know your topic *well.* To test your knowledge, have some friends listen to you while you practice and have them deliberately throw tough questions at you;

if they succeed in stumping you, go do some research. It's not necessary that you have an answer for *every* conceivable question the audience might raise; there are questions for which *no one* has an answer (and you should certainly never try to fake an answer!). But if you have researched a topic well and have state-of-the-art knowledge about it, you'll be able to answer most questions and will be able to say confidently about the others, "We don't have an answer to that question yet."

5   *Develop your own speaking style.* Practice telling stories or jokes to a few friends at a time in an informal setting. Stand tall and face them as in an oral presentation; engage your listeners' interest. Use natural, animated gestures, and vary your intonation and rate of speech. In short, *be expressive! Let your enthusiasm show!* At the same time, take note of any distracting habits you might have: leaning against something, biting your nails, playing with a pencil, and so on. If you are aware of such habits, you can often take measures to keep them under control while you're "on stage."

6   *If you are going to be reading from a manuscript, work on giving it a lively intonation.* There is a strong tendency when reading aloud to adopt a monotonous style of delivery that is very boring for an audience to listen to. So practice varying your intonation as you read aloud. Raise your voice pitch for the more important words and allow it to drop for the less important ones. Pause occasionally at appropriate places. Put special emphasis on contrasting words, as in the following example:

> System 2986 is an experimental system with two processors which can work *independently* or work *together as coprocessors. One* processor is a commercially available single-board computer (Intel's ISBC 86/12A), which is also referred to as System *86. The other* processor is a custombuilt microprogrammable special-purpose processor which is built around AMD 2900 series components and referred to as System *29.* The two processors can talk to each other over the Multibus using the Multibus protocols.

Delivery

As the time draws near for actually delivering your oral presentation, you will experience what all speakers experience: nervousness! If you have adequately prepared and practiced your presentation, of course, you will probably be less nervous than if you have not. No matter how well you prepare yourself, however, you will still be at least a little bit on edge.

The key to an effective delivery is to convert your nervousness into the kind of energy that injects liveliness, enthusiasm, and animation into your speech. If you try to either suppress your nervousness or let it dominate you, the result will be either a listless, boring, "laid back" style or a jittery, key-jingling, pacing-back-and-forth one—neither of which will appeal to your audience.

How can you control your nervousness? Basically, by not being self-conscious. The best way to do this is by establishing a "conversation" with your listeners. If you concentrate on your listeners and on what you're trying to tell them—just as you do in small-group conversation—you'll forget about yourself and lose your self-consciousness. And, in the process, you'll lose your nervousness.

Here are some specific suggestions:

1   *Establish some personal contact with at least some of your listeners before you begin your talk.* Mingle with them. See what they're interested in, what kind of mood they're in. That way, you'll have some friendly faces to look at during your talk.

2   *As you give your presentation, concentrate your full attention on what you want to say to your listeners.* Stick to your outline, make sure you cover all your main points and main supporting points. Plan ahead; know where you're going, what you'll say next. Remember, your listeners are interested more in your topic than in you. If you provide a steady stream of well-organized information, they'll pay attention to you and give you support. If you don't, they'll lose interest.

EXERCISE 8-1

A   Convert one of your written reports into a 5- or 10-minute oral presentation suitable for an audience of nonspecialists.

B   Convert the same report into a 5- or 10-minute oral presentation suitable for an audience of specialists.

C   Take a technical discussion section (300 to 500 words) from one of your reports and prepare it for oral reading to a large audience.

8.2

GIVING AN INFORMAL ORAL PRESENTATION

Occasionally you may be asked to give an oral presentation on short notice and may not have time to practice or prepare any visual aids. For example, as a member of a project team you may be asked at any time to give a spur-of-the-moment progress report. Or perhaps an orientation tour is taking place and you are asked to describe your laboratory. Perhaps your group leader has been asked to present a feasibility study but has been taken ill, and you are a last-minute replacement for her.

In such cases, the best thing to do is to concentrate on the first five steps in preparing any oral presentation, formal or informal: (1) audience analysis, (2) determination of primary purpose, (3) selection of supporting information, (4) selection of a pattern of development, and (5) outlining. Each of these steps is vitally important and should be performed as carefully and thoroughly as time permits. But *keep it simple!* Don't let yourself get bogged down in complicating details. Time does not allow it, and it will only result in disorganization if not outright confusion. Instead, keep your mind on your main points and how you can best and most clearly support them.

EXERCISE 8-2 Take no more than 30 minutes to prepare a 10-minute informal oral presentation on a familiar topic. Then give it. (If you're a student, you might arrange to give the presentation to one of your teachers during office hours. If you're working for a company, try your supervisor or some colleagues or perhaps a technical writer or editor.)

REFERENCES

1   William H. Middendorf, "Academic Programs and Industrial Needs," *Engineering Education,* May 1980, pp. 835–837.

2   Richard M. Davis, "Technical Writing: Who Needs It?" *Engineering Education,* November 1977, pp. 209–211.

3   Ed Gilbert, speech to the Conference on Teaching Technical and Professional Communication, co-chaired by Dwight W. Stevenson and J. C. Mathes, Department of Humanities, College of Engineering, The University of Michigan, 1975.

4   Nicholas D. Sylvester, "Engineering Education Must Improve the Communication Skills of Its Graduates," *Engineering Education,* April 1980, p. 739.

5   Peter M. Schiff, "Speech: Another Facet of Technical Communication," *Engineering Education,* November 1980, p. 181.

6   William R. Kimel and Melford E. Monsees, "Engineering Graduates: How Good Are They?" *Engineering Education,* November 1979, pp. 210–212.

ADDITIONAL READING

Aristotle, *The Rhetoric of Aristotle,* translated by Lane Cooper, Prentice-Hall, Englewood Cliffs, N.J., 1932, 1960. Another good edition of this work is *The "Art" of Rhetoric,* translated by J. H. Freese, Harvard University Press, Cambridge, Mass., 1926, 1967.

Kevin R. Daley, "Presenting Your Viewpoint," *Chemical Engineering*, November 17, 1980, p. 287.

William E. McCarron, "Oral Briefing versus Technical Report: Two Approaches to Communication Problems," *Courses, Components, and Exercises in Technical Communication*, D. W. Stevenson, (ed.), National Council of Teachers of English, Urbana, Ill., 1981, pp. 144–156.

Thomas M. Sawyer, "Preparing and Delivering an Oral Presentation," *Technical Communication*, First Quarter 1979, pp. 4–7.

Peter M. Schiff, "Speech: Another Facet of Technical Communication," *Engineering Education*, November 1980, pp. 180–181.

9

TECHNICAL REPORTS

As noted in Chapter 1, experienced scientists and engineers rate many writing and speaking skills as being very important or even critically important to job performance. Written communications cited for their importance and frequency include project proposals, progress reports, technical descriptions, memos and short reports, feasibility studies, and, depending upon the organization, business letters, long formal reports, and technical articles. Important oral communications cited include one-on-one talks with nontechnical personnel, project proposal presentations, progress reports, feasibility study presentations, other oral presentations using visual aids, and committee work. Since these communication forms are so important, these next four chapters will describe some organizing and formatting conventions for important written communications. Chapter 8 focuses on important oral ones.

Technical reports, proposals, and business letters are read by a variety of audiences: by technical experts in the field as well as by managers and other nonexperts—the busy, inattentive, and perhaps uninterested reader described in Chapter 4. Thus, the report, letter, or proposal should be accessible to all these readers and responsive to their needs and reading habits. In part, such responsiveness appears in two important structural features: (1) the Foreword and Summary (or their analogues) in the technical report and Abstract and Statement of Problem in the proposal and (2) the placement of generalizations or claims before their support in the part of the report, proposal, or letter written for expert readers.

The technical article, which is read almost exclusively by expert readers, uses the second feature in addition to others described in Chapter 12.

9.1
FOREWORD AND SUMMARY: ORGANIZING MAIN POINTS FOR NONSPECIALIST READERS

The Foreword and Summary, which are discussed in Chapter 6, are placed at the beginning of a technical report to be accessible to the busy managerial or nonexpert reader. The Foreword orients the reader and shows that the report addresses a problem or issue important to the reader. The Summary provides a compact statement of results, conclusions, and recommendations to help managerial readers make decisions: it tells what the writer discovered, the implications of the discovery, and the recommendations for action based on the writer's special knowledge—all in terms that the nonspecialist manager will find useful. By placing the Foreword and Summary first in a technical report, the writer allows the managerial reader to stop after the first page or so and still have all of the necessary information.

9.2

CLAIMS BEFORE PROOF: ORGANIZING DETAILS
FOR SPECIALIST READERS

The Abstract

The Abstract is a very important document for the specialist reader. In combination with the title, the Abstract is the vehicle through which readers

1   Identify and locate a report

2   Decide if they need to read it or not

In the second case, the Abstract is a sorting device; it should sort out all readers who don't need to read a report from those who do. To fulfill the identifying and locating function, the Abstract must contain the bibliographic information needed by libraries and information retrieval systems for cataloguing. To fulfill the sorting function, the Abstract must in addition contain a condensation of the report.

The bibliographic information needed in an Abstract varies depending on the report's purpose, audience, and expected circulation. For instance, a report which will circulate only inside the organization which produced it (an in-house document) might need only

1   The name of the author

2   The title of the project

3   The date it was written

4   The project under which the document was produced

5   Perhaps a special organizational number

On the other hand, a report available to people outside the organization which produced it might require all of that information plus

6   The name of the organization which produced it

7   The name of the organization which commissioned it (if any)

8   The contract number under which the document was produced (if any)

9   A security classification

10   Key words and special numbers for cataloguing in an information retrieval system

Items 7 through 10, for instance, are necessary in reports on government-sponsored research.

As a report writer, you will have to find out what types of information are required for your report. You can do this by looking at similar reports which have already been published to see what kinds of information they include, but the best methods are (1) to consult the "information or instructions to authors" published by the organization for which you are writing or (2) to consult with your librarian.

The condensation of the report is probably what we consider to be the heart of the Abstract. It reviews the investigation, main conclusions, and recommendations in the report. All of this needs to be done in 50 to 250 words, depending upon the writing situation and the word limit set on the Abstract. The Abstract needs to be totally self-contained because it will often be listed in abstracting or retrieval sources totally separate from the report.

Since so much needs to be done in so few words, there is a great tendency to write very general condensations, such as the following:

G. E. Lambert and D. F. Bryan. 1978.
The Influence of Fleet Variability on Crack Growth Tracking
Procedures for Transport/Bomber Aircraft
Air Force Flight Dynamics Lab. Rep. AFFDL-TR-78-158
198 pp., illus., Boeing Witchita Co., Witchita, Kan. Unclass.

The purpose of this program is to provide generalized crack growth tracking procedures for transport/bomber aircraft. A study and tests were conducted on crack growth rate characteristics, analysis schemes were developed, and tracking procedures were evaluated.

KEY WORDS: Crack Growth Tracking, Individual Aircraft Tracking, Force Management, Damage Tolerance, Durability, Spectrum Development, Crack Growth Analysis, Crack Growth Retardation, Retardation Models, Data Acquisition Systems.

This type of Abstract is called a descriptive abstract, because it merely describes what the report is about. It does not give any specific information about the problem, method, results, or conclusions. Such an Abstract is not very useful to many readers since it lacks the specific information which would tell them (1) what they need to know about the report and its results or (2) if they really need to get the report to more closely inspect its contents. A more useful type of Abstract is the informative abstract, which gives as much of the important particular information as can be crammed into the word limit. Consider, for instance, an informative version of the descriptive abstract presented above:

The purpose of this program is to provide generalized crack growth tracking procedures for transport/bomber aircraft. The study was composed of three major tasks: (1) an evaluation of the effects of usage parameters on crack

TABLE 9-1 THE RELATIONSHIP BETWEEN SUMMARY AND ABSTRACT*

	SUMMARY	ABSTRACT
Bibliographic data		XX
Overview of problem		XX
Review of method	x	XX
Review of results	XX	XX
Review of conclusions	XX	XX
Review of recommendations	XX	x

*x indicates optional material; XX indicates necessary material.

growth, (2) the development of generalized tracking procedures and (3) an evaluation of the techniques for implementing the individual aircraft tracking program. The KC-135A tanker was selected as the baseline aircraft for this study. Approximately thirty tests were conducted to experimentally verify predictions made from analyses and to determine the crack growth rate characteristics of the specimen material. The results of the parametric and variability studies were used to develop analysis schemes for predicting the effects of usage variations on crack growth. Four tracking procedures were evaluated: (1) pilot logs with the use of parametric crack growth rate data, (2) pilot logs with the use of parametric stress exceedance data, (3) the Mechanical Strain Recorder (MSR) and (4) the crack growth gauge. The implementation of each tracking procedure was evaluated by developing a cost model to study relative life cycle costs.

Notice that this informative version of the Abstract tells the reader much more about the report than did the descriptive version. It tells the reader specifically what the major tasks were in the investigation, what particular aircraft was used in the study, how many tests were conducted and why they were conducted, what uses were made of the results of the studies, which tracking procedures were evaluated, and how the implementation of the tracking procedures were evaluated. It did all of this in 172 words, but more important, it did this by giving some of the principal details of the report.

One last point that should be considered is the relationship between the Summary and the Abstract. In a technical report, the Summary is written for the nonspecialist managerial reader, whereas the Abstract is written for the specialist technical reader. In addition, the Abstract covers more ground than does the Summary, since the Abstract briefly presents bibliographic data and an overview of the problem, which are missing in the Summary. In comparison to the Summary, the Abstract also stresses the methodology and slights the recommendations. The relationship between the two is outlined in Table 9-1.

EXERCISE 9-1
A   Three Abstracts follow, two of them different versions for the same report. See if you can decide which ones are informative and which are descriptive.

B   Rate the Abstracts from best to worst from a user's point of view. Be
     prepared to justify your choices.

C   Write an informative Abstract for a report you are writing or one which
     you have read.

ABSTRACTS 1 and 2

Rinalducci, Edward J., and Arthur N. Beare, "Visibility Losses Caused by Transient Adaptation at Low Luminance Levels," *Driver Visual Needs in Night Driving,* Transportation Research Board Special Report 156, 1975, pp. 11–22.

Transient adaptation refers to the rapid fluctuations in the sensitivity of the eye that result from sudden changes in luminance level. The research reported here examines the effects of transient adaptation and resultant losses in visibility by using luminance levels comparable to nighttime highway lighting conditions. Research has been initiated on the problem of nonuniformities in roadway luminances in the motorist's visual environment. Experiments to examine multiple nonuniformities and the effect of nonuniformities at various distances from the line of sight on transient adaptation are planned.

Transient adaptation refers to the rapid fluctuations in the sensitivity of the eye that result from sudden changes in luminance level. The research reported here examines the effects of transient adaptation and resultant losses in visibility by using luminance levels comparable to nighttime highway lighting conditions. At low luminance levels, sudden increases produce losses in visibility equivalent to those previously found at higher levels. However, at low luminance levels, decreases produce smaller losses than those observed at higher luminance levels. The results also suggest that there is a preadapting level or range of levels below which there is little or no difference between visibility losses for 10- and 100-fold decreases and above which there is a difference. The transition appears to be a gradual one and is complete at about 8 ft-L. The findings of these investigations suggest that visibility loss depends more on the ratio of steady state thresholds, particularly at low luminance levels, than on the ratio of luminance change as previously supposed. Research has been initiated on the problem of nonuniformities in roadway luminances in the motorist's visual environment. Results indicate that the size of a nonuniformity may have little effect on transient adaptation. However, experiments to examine multiple nonuniformities and the effect of nonuniformities at various distances from the line of sight on transient adaptation are planned.

ABSTRACT 3

McKenzie, Robert Lawrence, *"Vibration-Translation Energy Transfer in Vibrationally Excited Diatomic Molecules,"* NASA/Technical Report NASA TR- R-466, October 1976, unclass.

A semiclassical collision model is applied to the study of energy transfer rates between a vibrationally excited diatomic molecule and a structureless atom. The molecule is modeled as an anharmonic oscillator with a multitude of dynamically coupled vibrational states. Three main aspects in the prediction of vibrational energy transfer rates are considered. The applicability of the semi-classical model to an anharmonic oscillator is first evaluated for collinear en-counters. Second, the collinear semiclassical model is applied to obtain numerical predictions of the vibrational energy transfer rate dependence on the initial vibrational state quantum number. Thermally averaged vibration-translation rate coefficients are predicted and compared with CO-He experimental values for both ground and excited initial states. The numerical model is also used as a basis for evaluating several less complete but analytic models. Third, the role of rotational motion in the dynamics of vibrational energy transfer is examined. A three-dimensional semiclassical collision model is constructed with coupled rotational motion included. Energy transfer within the molecule is shown to be dominated by vibration-rotation transitions with small changes in angular mo-mentum. The rates of vibrational energy transfer in molecules with rotational frequencies that are very small in comparison to their vibrational frequency are shown to be adequately treated by the preceding collinear models.

Structuring Proof and Technical Discussions

As just discussed, a report has a short Foreword and Summary which set up a problem and give a proposed solution and cost in general terms for a manager or nonexpert reader. These are followed by details to support the proposed solution, details which the manager could read or skip depending upon a need to know. Thus, the overall report structure moves from gen-eralizations or claims (the Foreword and Summary) to support (the details):

Generalizations (claims)

↓

Details (support)

The section of the report giving details is written for experts, readers with the technical background and interest to follow the details and fine points of the technical argument. This section is usually called the *Discussion* in a long report and the *Details* or *Discussion* in a short report. Such a section is illustrated in the *Details* section of Figure 6-5, which appears on page 105.

As does the structure of the overall report, the structure of the Discus-sion and its individual paragraphs moves from generalizations and claims to support and details. For example, in the *Details* section of Figure 6-5, the first paragraph poses a generalization or claim and then its support. The first sentence of the *Details* section claims that "*the excessive time required to obtain a suitable vacuum. . . is primarily due to contamination of the ion pump.*" The rest of the paragraph supports this claim; it tells *why* the time required is due to contamination. Given this structure, a busy reader could skim the *Details* section for the kind of information needed without having to read *every* word.

Another example of this movement from generalizations or claims to

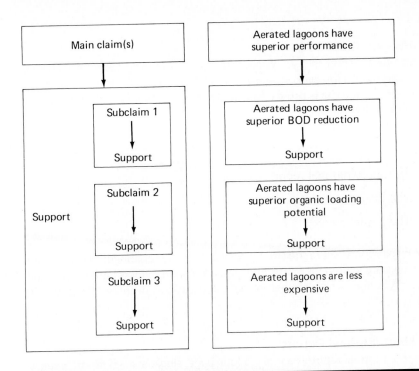

FIGURE 9-1 Organization of information for expert readers.

data and support appears in Figure 6-6, pages 108–110. Note especially the first main section after the Summary, *Performance Superiority of Aerated Lagoons*. This section begins with a heading and then a first paragraph that gives the claim being made: Aerated lagoons have superior performance to activated sludge processes or anaerobic lagoons because aerated lagoons show better BOD reduction and higher organic loading potential. Notice that the main claim (*aerated lagoons have superior performance*) is visible from the heading alone.

The main claim is then supported by the subsections which follow, with each subsection proceeding from claims (or subclaims) to support. In each case, the subclaim appears in the subsection heading—*Superior BOD Reduction by Aerated Lagoon, Superior Organic Loading Potential of Aerated Lagoons, and Cost Superiority of Aerated Lagoons*—and again in the first sentence of the section. This structure is summarized in Figure 9-1.

EXERCISE 9-2 Rewrite the reports in Exercise 6-6, page 107, to strengthen their movement in the Details section from generalizations or claims to data and support. Don't forget to consider the headings.

9.3

TECHNICAL REPORTS

The technical report is one of the most frequently used types of technical communication. However, it is also one of the most difficult for students to master since they have usually had much less experience with it than with any of the other types.

There are three subtypes of technical reports which will be discussed in this chapter:

1   The short informal report

2   The long informal report

3   The formal report

All of these reports share the structural features just mentioned: the Foreword and Summary written for managerial and nonexpert readers and the ordering of generalizations or claims before data and support in the sections written for experts.

The Short Informal Report

The short informal report is a one- to four-page, single-spaced report, usually written to someone within the writer's own company or organization. It may note the existence of a problem, propose some course of action, describe a procedure, or report the results of a test or investigation.

The term *informal* here does not mean sloppy or casual or carelessly done; an informal report should be as carefully prepared as possible— thoughtfully written, neatly typed, and thoroughly proofread for errors. The term *informal* refers to the format in which the report is presented.

The Heading of the Short Informal Report   The short informal technical report begins with a *heading section*. Typically, the heading gives the following information, where *Distribution, Enclosures,* and *References* are optional:

TO:	Name, job title
	Department
	Name of organization
	Address of organization (sometimes included)
FROM:	Name, job title
	Department
	Name of organization
SUBJECT:	Title of report (*Subject* may be replaced by *Re* or *RE,* for *Regarding*)
DATE:	Date

DIST: Distribution list of other people receiving the report; omitted if there is no distribution list

ENCL: Enclosures; other documents which are included with the report; omitted if there are no enclosures

REF: References; list of particularly important background documents; omitted if there are no such documents

Note that the heading is easy to read because of its format: liberal use of white space and aligned columns. It is also functional because the first information it gives is the information readers need first.

It is obvious why writers need to put the subject and date on the report and why they need to give the name of the reader and writer. It may be less clear why they need to give the job title, department, and organization of the reader and writer. These are provided because, at one level, documents are written not to people but to and from the occupants of particular jobs. Thus, it needs to be clear what job or position a person holds. Note that none of the information is presented more than once at the beginning of the report. If the name of the organization and/or the department are included in the letterhead, then these do not reappear under the *To* and *From* headings. If they do not appear in the letterhead, then they must appear under the headings.

Figures 9-2 and 9-3 present two different heading situations with ap-

FIGURE 9-2 Heading segment for short informal report on organizational letterhead (no department or division specified).

American
Electronics
Incorporated

To: Mr. John Nicol
           Chief Optoelectronics Engineer
           Optoelectronics Department

From: Khalil Najafi
           Assistant Optoelectronics Engineer
           Optoelectronics Department

Date: October 30, 1979

Subject: Liquid Crystal Displays: Analysis of Failure and Recommendations for Solution

Dist: Mr. Edward Jones, President
           American Electronics Incorporated

FIGURE 9-3 Heading segment for short informal report on departmental letterhead when reader and writer are in the same department.

American
Electronics
Incorporated

Optoelectronics Department

To: Mr. John Nicol
            Chief Optoelectronics Engineer

From: Khalil Najafi
            Assistant Optoelectronics Engineer

Date: October 30, 1979

Re: Liquid Crystal Displays: Analysis of Failure and Proposed Solutions

Dist: Mr. Edward Jones, President
            American Electronics Incorporated

propriate organization of information. Note that Figure 9-2 is written on organizational letterhead—no department or division is specified in the letterhead—and so *Optoelectronics Department* must appear under the names of the reader and writer. In contrast, Figure 9-3 is written on departmental letterhead, and so *Optoelectronics Department* does not need to be repeated. If the reader had been in a different department—say, in the Finance Department—then the appropriate heading section would be that of Figure 9-4. Notice that the writer's department appears in the letterhead but not under his or her name, whereas the reader's department appears under his or her name and not in the letterhead. The rest of the information remains the same.

Formatting the Short Informal Report Each main section of the report—the Foreword, the Summary, and the Discussion (or Details)—follows the previous section with only a double or triple space between sections. A clearly visible heading indicates the beginning of each section and the contents of the section. The report is organized so that a busy managerial reader (a reader who wants to read no more words than are absolutely necessary) can stop after the Foreword (the statement of the problem) and the Summary (the main results and other important information a manager needs to know). The Discussion or Details section gives the extra information needed by technically involved readers: support for the claims in the Summary or extra

FIGURE 9-4 Heading segment for short informal report on departmental letterhead when reader and writer are in different departments.

American
Electronics
Incorporated

Optoelectronics Department

To: Ms. Karen Smithfield
               Chief Financial Analyst
               Finance Department

From: Khalil Najafi
               Assistant Optoelectronics Engineer

Date: October 30, 1979

RE: Liquid Crystal Displays: Analysis of Failure and Proposed Solution

Dist: Mr. Edward Jones, President
               American Electronics Incorporated

FIGURE 9-5 Structure of the short informal report.

| Foreword | For all readers, especially managerial and nonexpert readers |
| Summary | |

| Details or Discussion | Additional material for technical or expert readers |

details needed to implement or fully understand the solution proposed. This structure is summarized in Figure 9-5. Sample short informal reports appear in Figures 6-5 and 6-6.

The Long Informal Report
The long informal report is quite similar to the short informal report. Both are written for readers within the writer's organization; both use the *To, From,*

FIGURE 9-6   Structure of the long informal report.

Foreword

For all readers, especially managerial
and nonexpert readers

Summary

Discussion

Introduction to the Discussion

Proof and development of claims

For expert or technical readers

Conclusions and Recommendations

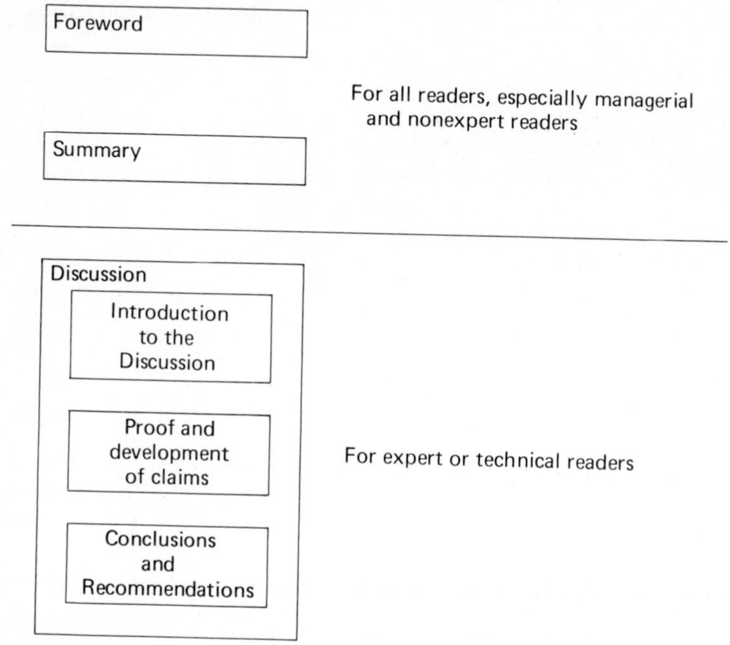

Subject, Date heading; and both are formatted so that each section follows the previous section with only a heading and a double or triple space between sections. However, the long informal report is different from the short report in two important ways: the long report is longer, obviously, and because of its length, its Discussion section is differently organized to make accessible the much greater amount of technical material it contains.

If you examine the long informal report in the Appendix, the similarities and differences between it and the short report are obvious. Let us focus on the differences. As illustrated in Figure 9-6, the Discussion section has three main subsections: the Introduction to the Discussion, the main Proof or Development section, and the Conclusions and Recommendations section. As a whole, the Discussion restates the material in the Foreword and Summary, but for a technically competent and interested reader. The Discussion gives the technical details, technical reasoning, and data that the technical reader needs—in contrast to the general orientation to the subject that the managerial reader needs. How does the Discussion do this?

The Introduction to the Discussion   The Introduction to the Discussion is the first part of the Discussion; it restates the Foreword in terms meaningful to the technical reader. Remember that the Foreword focuses on the implications of the problem for the managerial reader; it presents the problem as one of safety, cost, efficiency, quality control, labor relations, or compliance with legal regulations. In contrast, the Introduction to the Discussion focuses on the implications of the problem for the technical reader; it defines the technical nature of the problem or outlines technical assumptions, values, subproblems, background, or resources relevant to the problem.

For example, consider the following two examples, the first written by an electrical engineer at a company making heaters, fans, and portable lights and the second written by a harried inventory controller. In the first, the Introduction focuses on the technical nature and details of the problem; in the second, it focuses on the technical background.

EXAMPLE 1
FOREWORD
Because of increasing competition in the home appliance market, Electronic Design Division is evaluating the operating convenience and marketability of its less successful products. As part of this effort, I was asked to evaluate a proposed control circuit for our combined heater, light, and fan; to document any problems with this circuit; and if necessary to design a new control circuit. This report presents my evaluation and new design.

INTRODUCTION TO THE DISCUSSION
Ms. Durkis of Marketing has requested a simpler control switch for our combined heater, light, and fan. I have evaluated a preliminary design for a circuit which could turn on all three units by flipping a single switch and then remove power from one or more of the units by depressing another switch. The design for the control circuit was built around three relays and claimed to operate by picking these three relays in different combinations to produce the desired switching capabilities. The following sections detail the preliminary design of the circuit and its problems, a revised design, and a cost analysis of the revised design.

EXAMPLE 2
FOREWORD
During Photon's recent companywide expansion program, the inadequacy of our inventory numbering system became apparent. Thus, I was asked to design a more efficient and systematic numbering system based on our departmental needs and expected growth. This report provides an overview of the system and its response to our needs.

TABLE 9-2 DISTRIBUTION OF CONTENTS IN FOREWORD AND IN INTRODUCTION TO THE DISCUSSION

	FOREWORD	INTRODUCTION TO THE DISCUSSION
Managerial problem	35–70%	0–20%
Technical problem or assignment	5–35%	10–80%
Forecast of contents	10–30%	0–25%

INTRODUCTION TO THE DISCUSSION

The present inventory (parts) numbering system has been in use for almost five years. At the time of its conception, Photon did not have the computer capabilities it has now, nor did it envision the growth it has recently experienced. Because of these factors, the current parts numbering system is quite basic and limited. As the quantity of parts increased over the years, the numbering system could not efficiently keep track of them nor could it assign new part numbers on a systematic basis. (In fact, new part numbers were assigned on a purely random basis.)

Last month, Photon underwent one of its most comprehensive expansion programs to date and suffered critically from the present numbering system. Often customers had to wait for us to reorder unexpectedly depleted supplies, and our inventory department logged hundreds of hours of overtime. Thus, management decided to do away with the present system and develop a new one. This report documents the various departmental needs identified, the computer-assisted inventory control system I have designed, and features of the system directly responding to the needs.

The different emphases of the Foreword and the Introduction to the Discussion are reflected by the relative percentages of each section devoted to the managerial problem, the technical problem or assignment, and the forecast of contents. A rough comparative distribution for the two sections appears in Table 9-2. Notice that the Foreword devotes proportionally more space to the managerial problem, whereas the Introduction devotes more space to the technical problem.

EXERCISE 9-3 A student complained that she was having trouble with the following Foreword, that it was more like an Introduction to the Discussion than a Foreword (and she was right). Turn it into a Foreword by eliminating details unnecessary for a managerial reader. Feel free to rewrite sentences or to move information around.

FOREWORD 1

Flutter is a phenomenon in which structural instabilities arise from aerodynamic loads. In aircraft, these instabilities are often characterized by divergent oscillations of the wings, resulting in structural failure and possible loss of the aircraft. Flight flutter testing involves the tracking of damping estimates of the excited vibration modes at different flight conditions.

The violent nature of flutter makes safety an important concern during flight flutter testing. As with any flight test program, cost is also a major concern. Our objective at Dryden is to provide safe testing at the lowest cost. Costs can be lowered by decreasing flight time. This requires utilizing the fastest analysis technique that will meet the accuracy requirements that safety demands.

Presently, we have available two software packages for data analysis during flight flutter testing. They are

1)   the Power Spectral Density (PSD) package, and

2)   the Random Decrement (RDEC) package

The main objective of each package is to determine damping estimates of the structural modes.

I was assigned to do a comparison study of the two packages looking at two questions:

1)   How accurate are the damping estimates?

2)   How much time is required for the analysis?

These questions are concerned with safety and cost, respectively.

The purpose of this report is to present the findings of my study and to make some recommendations concerning future testing.

Conclusions and Recommendations   The Conclusions and Recommendations section bears the same relationship to the Summary as the Introduction to the Discussion bears to the Foreword. The Summary is written for the managerial reader; it focuses on the conclusions and recommendations relevant to that reader and gives the information a manager must know to act. The Conclusions and Recommendations section, which appears at the end of the Discussion, gives information a technical reader might want in addition to that needed by the manager. It might include the most important technical results which led to the recommendations and statements technically too complex or difficult for the managerial reader. For instance, consider the *Summary* and the *Conclusions and Recommendations* presented below from a report written by a geologist to the Milton Township Council. (The *Foreword* is given first to orient you.) Unless you are a geologist, you probably won't understand all the information presented in the *Conclusions and Recommendations,* though you should easily find your needed information in the *Summary.* Remember that where possible, the *Conclusions and Recommendations* should move from generalizations or claims to data and support.

FOREWORD

Within Milton Township, the greatest industrial and residential development has occurred in areas with limited groundwater resources. This development has already created water shortages and a need to deepen existing wells; more seriously, it has greatly handicapped or prevented the development of new industry. Thus, the township council authorized Moody and Associates to locate, if possible, several high-yield water wells to provide a sufficient municipal water supply. The purpose of this report is to describe the successful location of sand and gravel materials with excellent groundwater yielding capability.

SUMMARY

We conducted studies and six test drillings to locate an area in which a reliable, high-quality, municipal water well field could be developed. We have located such a site.

Results: Four of the six wells drilled were productive and penetrated sand and gravel suitable for development of municipal wells. Each of the four productive wells showed very good water-producing capability, yielding 300 to 500 gallons per minute. Further, the wells produced water of generally excellent quality, though with a potentially high but treatable concentration of manganese. One test well showed manganese concentration now high enough to require treatment, and there is concern that manganese levels in all the wells might increase to such a level. However, this would not prevent development of the well field, and additional study will be required to determine if manganese removal is necessary at all.

Conclusions: Production wells completed at the test well sites should more than supply the township's current and future needs. Analysis of test data indicates that such production wells could produce over 2 million gallons per day, twice the township's projected requirements. Further, if the water-bearing sand and gravel are as extensive as we project them to be, yields up to 5 million gallons per day could conceivably be developed.

Recommendation: We recommend that the property tested be acquired for development as a municipal well field. The property is unique in its supply capability within Milton Township, and a well field on this property should meet any present and foreseeable water requirements within the township.

CONCLUSIONS AND RECOMMENDATIONS

Conclusions: Six test wells were drilled, four of which penetrated sand and gravel suitable for developing adequate municipal wells.

1. The relationship of geologic units encountered in Test Wells 1 through 6 is interpreted and presented in Plate 1. Two aquifers were identified; an extremely thick aquifer was encountered in Test Wells 2, 5, and 6 ranging in thickness from 78 to 125 feet, and a deep aquifer was en-

countered in Test Well 1 extending from 150 to 185 feet. There is theorized to be limited hydraulic connection through fine-grained units between the two aquifers. Some degree of connection between the deep aquifer and upper aquifer was established by the rapid rise in water levels in Test Well 1 observed during a heavy rain in September 1979.

2. A conservative estimate of the yield from the upper aquifer is 2.0 mgd. Aquifer data indicate this could be produced by pumping any combination of two wells among Test Wells 2, 5, and 6 (assuming manganese in Test Well 2 proves acceptable). Ultimate production from a French Creek well field could prove to be in the 3.0 to 5.0 mgd range.

3. It is difficult to estimate the maximum production available from the lower aquifer encountered in Test Well 1 because this unit was not identified in the other five test wells. Based on the data from the pump test of Test Well 1, the yield of the lower aquifer is estimated to be 0.5 to 1.0 mgd.

4. The water quality from all wells is good, with the exception of the anomalously elevated manganese found in Test Well 2. Iron and manganese were present in all other wells but in acceptable concentrations. If there is a source of manganese (none has been identified to date) in the immediate vicinity of Test Well 2, there is concern that due to the extremely high permeability of the aquifers, the high manganese groundwater could migrate toward the other wells with long-term pumping.

Typically, the water is moderately hard to hard, with hardness ranging from approximately 7 grains in Test Well 1 to 10 grains in Test Well 5 to 14 grains in Test Well 2.

Recommendations: Based on results from drilling and sample analysis, the following tests and land purchases are recommended.

1. A short pump test (8 hours) should be conducted on Test Well 2 to recheck the iron and manganese concentrations in this well.

2. In order to stress the aquifer as heavily as possible and simulate municipal well field operation, a pump test should be conducted on Test Well 6 with an attempted pumping rate of 700 to 1,000 gpm for 48 hours.

3. In order to secure the existing test wells, provide a buffer around these wells, and permit some aquifer test drilling, it is recommended the Township purchase property roughly rectangular in form extending 800 feet north and 800 feet south of Test Well 1 and west to the abandoned railroad tracks. This would be approximately 40 acres. If obtaining this acreage is not feasible, then it is recommended that at a minimum the Township obtain property extending 800 feet north and 800 feet south of Test Well 1 and 600 feet to the west of Test Well 1. This would be approximately 22 acres. It is also recommended that a long-term option be obtained on the Barco property south of Test Well 2 and southwest of Test Well 3 for potential long-range future well field development.

EXERCISE 9-4   The following Foreword and Summary were written by an engineer at Jur-Kay Carriers, a moving company specializing in long-distance hauling. Unfortunately, the engineer provided far too much detail for the busy managerial readers of this report. Your job is to write a new Foreword and Summary which give a proper (and short) overview of the problem and proposed solution. Try to make each section no longer than 100 to 150 words. Then write an Introduction to the Discussion and a Conclusions and Recommendations section. A Table of Contents for the report follows to outline the report's structure.

THREADED FASTENERS: FAILURE ANALYSIS AND PROPOSED IMPROVEMENT

TABLE OF CONTENTS
1   Foreword
2   Summary
3   Discussion
   3.1   Background to the Investigation
   3.2   Possibilities of Fault
   3.3   Rejections of Possibilities
   3.4   Location of Fault
   3.5   Correction of the Fault: An Instruction Class
      3.5.1   Bolt Strength and Preload Specifications
      3.5.2   Torque Requirements
      3.5.3   Proper Selection of the Nut
      3.5.4   Why Fasteners Fail
      3.5.5   What Different Failures Indicate about the Inadequacies of a
            Specific Assembly
4   Conclusions and Recommendations

FOREWORD
On April 3, 1974, a Jur-Kay Carriers semitractor-trailer overturned on I-96 approximately 30 miles east of Grand Rapids, Michigan, and the driver was seriously injured. There were no other vehicles involved in the accident. The exact cause of the accident is unknown. However, it is believed that one of three bolts in the steering mechanism failed, causing the driver to lose control. The truck, trailer, and contents of the trailer were a total loss.

On July 27, 1974, a Jur-Kay Carriers semitractor-trailer lost control on U.S. 101, 52 miles south of Grand Rapids, involving three other vehicles in a crash resulting in two deaths. The driver of our vehicle incurred minor injuries. The accident investigation revealed that a bolt in the power steering linkage of our tractor failed, thus causing the driver to lose control. Maintenance Record #1825, dated July 27, 1974, revealed that five hours before the accident a fleet mechanic replaced the original SAE-grade-8 linkage bolt with an ungraded bolt. It can be assumed that if the bolt had been replaced with an SAE-grade-8 bolt, the accident would not have occurred.

Jur-Kay Carriers is now in litigation with a suit brought against them by the owners of the three vehicles involved in the crash.

As a direct result of these two accidents, and because for the last six months our maintenance records have shown an extremely large number of threaded fastener replacements, Mr. Paul Milani, Senior Engineer, Jur-Kay Carriers, suspected that a problem existed with threaded fastener usage in our maintenance department. On July 29, 1974, Mr. Milani requested that I evaluate this usage of threaded fasteners. The investigation has been completed.

The purpose of this report is to show that our mechanics have a dangerous lack of knowledge about fasteners.

The purpose of this report is also to present the needed material for the recommended instruction class, emphasizing:

1. Bolt strength and preload specifications

2. Torque requirements

3. Proper selection of a nut

4. Why fasteners fail

5. What different failures indicate about the inadequacies of the specific nut-and-bolt assembly in question

SUMMARY

A preliminary examination of maintenance records dated since January 1, 1974, was conducted by Mr. Milani. The examination has shown that a large number of threaded fastener replacements and repairs are being made in the maintenance department of Jur-Kay Carriers. In addition to this, there have been two serious vehicle accidents in the last four months caused by Jur-Kay Carriers. They were both the result of threaded fastener failures. Because of these facts, it is believed that a definite problem exists and an investigation into threaded fastener usage in our maintenance department has been made.

The problem had to lie in one of four areas (or a combination thereof): either (1) our fasteners were of inferior quality, (2) there was such a lack of standardization in the fasteners we used that many types of gradings were present, resulting in utter confusion on the part of the mechanics whose job it was to select the replacement part, (3) there was simply a high degree of carelessness in the mechanics' performance of their jobs, or (4) the mechanics were simply not knowledgeable enough to make the proper grade choice when replacing a bolt.

Research revealed that the problem lay in none of the first three areas:

1. Our fasteners are not of inferior quality.

2. There is no lack of standardization in the grading of threaded fasteners we use.

3. There is no degree of carelessness in our mechanics' performance of their jobs. In fact, the opposite was found to be the case.

The problem was found in the last area:

4.   Our mechanics simply do not know the field of fasteners. Indeed, subsequent observations of activities in the garage and conversations with the mechanics verified this conclusion.

Our mechanics lack necessary knowledge in the threaded fastener field. Thus, they are unable to make the correct choices in replacing a nut-and-bolt assembly that has failed. Low-tensile-strength bolts are being substituted for high-tensile-strength bolts, and bolts are not being used for their designed purposes. This has resulted in an unnecessarily large number of failures and replacements, not to mention the two latest vehicle accidents.

I recommend that an instruction class be initiated to inform and educate our mechanics in basic areas of threaded fastener technology. I estimate that five sessions, each an hour in duration, is sufficient to present the necessary material. Such a class would eliminate the problems involving threaded fasteners present in our maintenance department. That is, the serious problem of substituting low-tensile-strength bolts would be avoided. Our mechanics would begin using bolts for their designed purposes. We would also solve the recurring bracketing and "bolt loosening" problems caused by reusing bolts that have stretched way beyond their optimum strength.

I volunteer my services as instructor if the need is to be met. Such a class will emphasize

1.   Bolt strength and preload specifications

2.   Torque requirements

3.   Proper selection of a nut

4.   Why fasteners fail

5.   What different failures indicate about the inadequacies of the specific nut-and-bolt assembly in question

I also recommend that after our mechanics are taught the basic threaded fastener technology, a number of bolt-grade-marking charts be posted throughout the garage.

The Formal Report

Formal reports are usually relatively long reports written to readers outside the writer's organization. Since such reports represent the organization and spread its reputation, they are sent out with a "dressier" look—often a fancy cover and binding with the organization's seal prominently featured. Instead of the *To, From, Subject, Date* heading of the informal report, a formal report begins with a Title Page, an Abstract, a Table of Contents, and a List of Illustrations. Formal reports also begin the Foreword, Summary, and Discussion on new pages. In all of these senses, the formal report closely resembles a book.

The Title Page of the formal report includes all of the information you have already seen in the heading of the informal report: information about *To, From, Subject,* and *Date.* The major difference is in the arrangement of this information on the page. The Title Page spreads the information over the page in a visually pleasing arrangement with lots of white space between items. Sample Title Pages appear in Figures 9-7 and 9-8, which contain the typical heading information:

1 Name of receiver and/or receiving organization

2 Name of writer and writer's organization

3 Subject of report

4 Date of report

Note that these correspond to the *To, From, Subject,* and *Date* of the informal report. In addition, they also include the following funding information:

5 Funding agency (often this is the same as the receiver or receiving organization)

6 Funding contract number(s)

7 Beginning and ending dates of the project

8 File number of the report or document authorizing the project

The specific information on a given Title Page and its format may vary with the requirements of different organizations. You must find out what the specific requirements are for your situation and provide the required information in the required format.

In other ways, the formal report duplicates the long informal report. It is long and has a Discussion section which includes an Introduction to the Discussion and a Conclusions and Recommendations section. These sections restate the Foreword and the Summary, as described above, and the main part of the Discussion provides support for the generalizations in the Summary and the Conclusions and Recommendations. A sample formal report appears in the Appendix.

9.4
AN EXAMPLE OF ADAPTING FORM TO PURPOSE: THE FEASIBILITY REPORT

In most of the reports discussed in this chapter, someone has noticed a problem or undesirable situation and then proposed or asked someone else to propose a solution. Sometimes, however, it may not be clear that there

FIGURE 9-7 Title Page from a formal report.

A DESCRIPTOR VARIABLE
APPROACH TO MODELING
AND OPTIMIZATION OF
LARGE-SCALE SYSTEMS

} TITLE

FINAL REPORT
MARCH 1975 - FEBRUARY 1979

{ TYPE OF REPORT
PROJECT DATES

DONALD N. STENGEL
DAVID G. LUENBERGER
ROBERT E. LARSON
TERRY S. CLINE

{ AUTHORS

SYSTEMS CONTROL, INC.
1801 PAGE MILL ROAD
PALO ALTO, CA. 94004

{ AUTHORS'
ORGANIZATION

DATE PREPARED - FEBRUARY 1979

{ REPORT DATE

PREPARED FOR THE
DEPARTMENT OF ENERGY
OFFICE OF THE ASSISTANT ADMINISTRATOR FOR ENERGY
TECHNOLOGY
DIVISION OF ELECTRIC ENERGY SYSTEMS

{ FUNDING AGENCY

WORK PERFORMED UNDER CONTRACT NCS EX-75-0-0-2090
ET-73-0-0-2353

{ FUNDING CONTRACT
NUMBERS

FIGURE 9-8 Sample formal report Title Page specifying funding agency.

NATIONAL COOPERATIVE HIGHWAY RESEARCH
PROGRAM REPORT **207**

UPGRADING OF LOW-QUALITY AGGREGATES FOR PCC AND BITUMINOUS PAVEMENTS

P.D. CADY, P.R. BLANKENHORN, D.E. KLINE,
AND D.A. ANDERSON
THE PENNSYLVANIA STATE UNIVERSITY
UNIVERSITY PARK, PENNSYLVANIA

RESEARCH SPONSORED BY THE
AMERICAN ASSOCIATION OF STATE
HIGHWAY AND TRANSPORTATION
OFFICIALS IN COOPERATION WITH
THE FEDERAL HIGHWAY
ADMINISTRATION

AREAS OF INTEREST:
PAVEMENT DESIGN AND PERFORMANCE
BITUMINOUS MATERIALS AND MIXES
GENERAL MATERIALS
MINERAL AGGREGATES
SOIL EXPLORATION AND CLASSIFICATION
(HIGHWAY TRANSPORTATION)

TRANSPORTATION RESEARCH BOARD
NATIONAL RESEARCH COUNCIL
WASHINGTON, D.C.   JULY 1979

is a solution or that any given solution would be worth what it would cost in time, money, and effort. At such times, it is useful to conduct a feasibility study, that is, to analyze the problem and its implications, to evaluate alternative solutions, and finally to decide on one alternative or course of action. The results of such a feasibility study are written up in a feasibility report.

The term *feasibility* has two somewhat different meanings. In its more limited sense it means technically capable of being done, executed, or effected. We shall call this *technical feasibility*. If we were considering building a tunnel through a mountain, this more limited sense of feasibility would refer to such questions as the following:

1   Can you build the tunnel? Do you have the technology to keep the soil from caving in?

2   Do you have enough people available for the job?

3   Do they have the required expertise?

4   Can you get the required building supplies and building permits?

In addition to this sense of technically capable of being done, *feasibility* also has the wider, more inclusive sense of "suitable" or "reasonable." To return to the tunnel example, this wider sense of feasibility encompasses such additional questions as the following:

1   Assuming that we can build this tunnel (that it is technically feasible), should we? Will it produce enough benefits to be worth the cost?

2   Is the cost reasonable?

3   Can we afford it, even if it is reasonable?

4   Will building or having the tunnel produce any undesirable side effects?

5   Will it produce any desirable side effects, such as increased employment in the area?

In this larger sense, the notion of feasibility covers the five general criteria discussed in Chapter 5 that are used to evaluate any proposed solution to a problem:

1   *Effectiveness:* Is the solution effective? Will it solve the problem posed? Why? How do you know?

2   *Technical Feasibility:* Can the solution be implemented? Does it require technology or resources that are unavailable? How do you know?

3 *Desirability:* Would one want to implement the proposed solution? Does it have any undesirable effects? Does it have desirable effects? Why? What are they?

4 *Affordability:* What will the solution cost to implement? To maintain? Is this cost reasonable? Is it affordable? Will it reduce costs in the future? Why?

5 *Preferability:* Is the solution better than or preferred over any other possible solution? Why?

This larger notion of feasibility also covers any special criteria necessary or appropriate for the given situation, for instance, important issues of safety, environmental pollution, conformity to particular laws or traditions, precedent setting, or justice. As a technical expert, you must be able to deal with all of the relevant issues and criteria in a feasibility study.

When all of these criteria enter into the notion of feasibility, they provide the basis for a full-fledged argument of fact or policy. For instance, when you judge a tunnel as being effective, technically feasibile, desirable, affordable, preferable, and safe, and then argue on the basis of these judgments that the tunnel is feasible, you have constructed an argument of fact supported by several subarguments of fact. If you go one step further and argue that the tunnel should be built, you have constructed an argument of policy supported by several subarguments of fact, one of which is the argument of feasibility.

Let us examine the arguments being developed in a sample feasibility report to further illustrate this point. The report concerns flight tests on airplanes—in particular, flight flutter tests. A flight flutter test studies the vibratory motion and resulting instability of aircraft wings during flight. When a plane is in flight, air passing over and around the wings creates some wing motion. If this motion is too great, the wings will vibrate catastrophically, or "flutter." If this flutter becomes too severe, the wings can actually break off. The Foreword to the report sets up the problem being addressed in a feasibility study.

FOREWORD

Flight flutter tests are inherently dangerous. To maintain a margin of safety, we must have fast and accurate in-flight analysis of the vibratory motion in the wings created by the test flight conditions. However, the computer program currently used for in-flight analysis is outdated, and its inadequacies present safety problems. Thus, a new analysis technique must be implemented to improve safety and to maintain our reputation in this area of flight testing.

$10,000 was allotted for this purpose, and a survey of the field was made to determine possible choices for the new analysis technique. I was assigned to review the choices for feasibility, select the best one and, if it seemed feasible, identify initial development and implementation concerns. This report presents my solution to our problem and provides initial information on the development and implementation of that solution.

While doing the analysis and evaluation necessary for this report, the writer realized that any proposed solution to the problem—that is, any computer program which would produce an adequate in-flight analysis—would have to meet the following criteria:

1   *Effectiveness:* The solution must be capable of providing accurate damping estimates of up to four superimposed vibration modes from the accelerometer data available.

2   *Technical feasibility:* The solution must be capable of working with the present computer (HP-5451C Fourier Analyzer), with minimal modifications. The solution should be adaptable by our staff, given their expertise and available time.

3   *Desirability:* The solution should have some flexibility in its software structure to allow modifications to the existing package should a particular test call for them.

4   *Affordability:* The solution purchase, development, and implementation costs should be less than the allotted $10,000.

5   *Preferability:* A solution that provides conservative damping estimates and requires the minimum analysis time is preferred.

The writer then evaluated three possible alternatives—the Random Decrement (Randomdec) Technique, the Time Domain Package, and the Flight Test Data Reduction System—against these criteria, finding that two of the alternatives were ultimately infeasible:

1   The Time Domain Package was rejected for failing to meet the desirability criterion. The package is proprietary, and the producer refuses to disclose the details of the software structure. Therefore, the flexibility in modifying the package is absent.

2   The Flight Test Data Reduction System was rejected for failing to meet the effectiveness criterion under certain conditions. Accurate damping estimates were difficult to obtain when two or more modes fell within a frequency range of $\Delta 5$ Hz. Since this condition is quite possible during testing, this package was rejected.

One remaining alternative, however, was feasible; it met all of the criteria and was ultimately recommended as the solution to the problem of inadequate in-flight testing:

I recommend the Randomdec Technique as the solution to our problem because it meets or exceeds all of the given criteria.

1   *Effectiveness:* This technique provides accurate damping estimates of up to four superimposed modes.

2   *Technical feasibility:* This technique can utilize a software structure which is well suited for use on the available computer. We have the personnel and expertise in-house to develop and implement the necessary software.

3   *Desirability:* This technique is well documented and available for use by the general public. A proprietary software package need not be purchased. The necessary software can be developed in-house and tailored to our specific needs with the flexibility desired.

4   *Affordability:* This technique is estimated to cost under the allotted figure. Our survey and in-house discussions gave an initial estimate for development and implementation costs of $8,000 (400 hours of labor @ $20/hour).

5   *Preferability:* This technique is mathematically derived to provide conservative damping estimates, thus allowing some safety margin (Figure 1). Initial estimates of analysis time were also found to be acceptable.

After conducting the analysis and evaluation, the writer had to present the results in a feasibility report. The feasibility report follows the structure already defined in this chapter for technical reports. It has a Foreword and Summary for managerial and nonspecialist audiences and then a more detailed section outlining the argument for specialists. The argument in its final form is presented in the Summary:

SUMMARY

In an attempt to find a replacement for our inadequate flutter analysis package, a survey of the field turned up three alternatives: the Random Decrement (Randomdec) Technique, the Time Domain Package, and the Flight Test Data Reduction System. Each alternative was studied for effectiveness, technical feasibility, desirability, affordability, and preferability. As a result of this study, I recommend the Randomdec Technique as our solution: it meets all of the criteria whereas the others fail to do so. An initial cost estimate of developing and implementing this technique is $8,000 ($2,000 under budget). Implementing this technique will increase the safety of our flight flutter tests and provide a desired margin of safety; it will also allow us to maintain a prominent position in this area of flight testing.

A Details section follows the Summary and provides backup data and references to support the argument stated in the Summary. The Details section first outlines the problem, then presents the argument for the Randomdec Technique, then addresses the issue of development and implementation (a part of the writer's assignment), and finally presents the conclusions. The outline for this structure is given below and the full report appears as Figure 9-9.

Foreword
Summary
  I The Problem and Its Background
  II Argument for the Randomdec Technique
    A Criteria for Judgment
    B Selection of the Randomdec as the Recommended Solution
    C Rejection of Alternatives
  III Development and Implementation of the Randomdec Technique
    A Brief Explanation of the Technique
    B Cost and Time Required
  IV Conclusions
  V References

From this perspective, the feasibility report is much like the reports discussed earlier in this chapter. It is based on a problem; it presents an analysis of a possible solution, of some possible solutions, or of all possible solutions to a given problem; and, if possible, it recommends the best solution given the circumstances involved. Such a recommendation may constitute an argument of policy (you should adopt this solution), but at the least it makes an argument of fact (this is or seems to be feasible). If a feasibility report does not build all the way to recommendations and an argument of policy, it should at least make, as clearly as possible, the various subarguments of fact for the people who will ultimately have to decide whether to fund or approve the project being studied. If these decision makers are better qualified than the report writer to make recommendations, then leaving the recommending to them may be appropriate. However, if they are not better qualified, then the report writer should attempt to make the recommendations which must ultimately be made. Writers of feasibility reports are hired for their interpretative ability as well as for their fact-gathering ability, and in many instances the writer of the feasibility report is the one person (or one of the few persons) best qualified to make the needed recommendations.

9.5

A CONCLUDING NOTE

This chapter has presented some of the conventional ways of organizing reports. It has stressed the persuasive nature of reports—that they are basically arguments of fact or policy which the writer is presenting to the reader. A feasibility report, for instance, makes a claim—that something is or is not feasible, some course of action should or should not be pursued—and then presents the analysis and data to support that claim.

FIGURE 9-9 A sample feasibility report.

DATE: December 5, 1981

TO: Regina J. Petty
                 Branch Chief
                 Dynamics Branch

FROM: Louis J. Bogart
                 Aerospace Engineer
                 Dynamics Branch

SUBJECT: DEVELOPMENT AND IMPLEMENTATION OF A FLUTTER ANALYSIS
                 PACKAGE: A FEASIBILITY STUDY

ATTACHMENT: Functional diagram of the Random Decrement Analysis Procedure

cc: R. J. Jones, Chief, Safety Office
                 T. R. Wilson, Chief, Flight Operations

Foreword

Flight flutter tests are inherently dangerous. To maintain even a margin of safety, we must have fast and accurate in-flight analysis of the wings' vibratory motion created by the flight test conditions. However, the computer program currently used for in-flight analysis is outdated, and its inadequacies present safety problems. Thus, a new analysis technique must be implemented to improve safety and to maintain our reputation in this area of flight testing.

$10,000 was allotted for this purpose, and a survey of the field was made to determine possible choices for the new analysis technique. I was assigned to review the choices, select the best one, submit it to a feasibility test, and, if it seemed feasible, identify initial development and implementation concerns of the selected technique. This report presents my solution to our problem and provides initial information on development and implementation of that solution.

Summary

In an attempt to find a replacement for our inadequate flutter analysis package, a survey of the field turned up three alternatives: the Random Decrement (Randomdec) Technique, the Time Domain Package, and the Flight Test Data Reduction System. Each alternative was studied for effectiveness, technical feasibility, desirability, affordability, and preferability. As a result of this study, I recommend the Randomdec Technique as our solution: it meets all of the criteria whereas the others fail to do so. An initial cost estimate of developing and implementing this technique is $8,000 ($2,000 under budget). Implementing this technique will increase the safety of our flight flutter tests and provide a desired margin of safety; it will also allow us to maintain a prominent position in this area of flight testing.

I. The Problem and Its Background

The problem arises out of the fact that our software package for flutter analysis is outdated. The Power Spectral Density (PSD) package presently in use was developed over ten years ago, and the level of accuracy it provides is no longer

FIGURE 9-9   *(continued)*

acceptable. The accuracy of the damping estimates obtained by the PSD method is erratic, and several superior analysis techniques have been developed in recent years (reference 1).

The importance of having damping estimates that are as accurate as possible can readily be seen from the potential danger involved in flight flutter testing. An incorrect damping estimate obtained during in-flight analysis could result in the loss of an aircraft due to a wing fluttering off.

An updating of our software is needed to reduce the risk involved and to maintain a prominent position in this area of flight testing. Funds ($10,000) have been allotted to improve our capabilities in this area, and a survey of the analysis techniques presently being used in the field has been made. The survey resulted in three analysis techniques being chosen as the most promising alternatives to our present situation. They are

1. The Randomdec Technique - broadly used by a number of aircraft manufacturers.
2. The Time Domain Package - specific package developed by the Aerotest Corporation and used by the Air Force at Edwards Test Center.
3. The Flight Test Data Reduction System - specific package developed and used by the Northrop Corporation.

Since our present PSD package has already been modified to its limits with unsatisfactory results (reference 1), the solution to our problem lies in choosing one of the above alternatives and implementing it into our flight testing system.

II. Argument for the Randomdec Technique

    A. Criteria for Judgment

        1. Effectiveness: The solution must be capable of providing accurate damping estimates of up to four superimposed vibration modes from the accelerometer data available.

        2. Feasibility: The solution must be capable of working with the present computer (HP-5451C Fourier Analyzer), with minimal modifications.

        3. Desirability: The solution should have some flexibility in its software structure to allow modifications to the existing package should a particular test call for them.

        4. Affordability: The solution purchase, development, and implementation costs should be less than the allotted $10,000.

FIGURE 9-9   *(continued)*

    5. Preferability: A solution that provides conservative damping estimates (Figure 1) and requires the minimum analysis time is preferred.

B.   Selection of the Randomdec Technique as the Recommended Solution

I recommend the Randomdec Technique as the solution to our problem because it meets or exceeds all of the given criteria.

    1. Effectiveness: This technique provides accurate damping estimates of up to four superimposed modes.

    2. Feasibility: This technique can utilize a software structure which is well suited for use on the available computer. We have the personnel and expertise in-house to develop and implement the necessary software.

    3. Desirability: This technique is well documented and available for use by the general public. A proprietary software package need not be purchased. The necessary software can be developed in-house and tailored to our specific needs with the flexibility desired.

    4. Affordability: This technique is estimated to cost under the allotted figure. Our survey and in-house discussions gave an initial estimate for development and implementation costs of $8,000 (400 hours of labor @ $20/hour).

    5. Preferability: This technique is mathematically derived to provide conservative damping estimates, thus allowing some safety margin (Figure 1). Initial estimates of analysis time were also found to be acceptable.

C.   Rejection of Alternatives

    1. The Time Domain Package was rejected for failing to meet the desirability criterion. The package is proprietary, and the producer refuses to disclose the details of the software structure. Therefore the flexibility in modifying the package is absent.

    2. The Flight Test Data Reduction System was rejected for failing to meet the effectiveness criterion under certain conditions. Accurate damping estimates were difficult to obtain when two or more modes fell within a frequency range of $\Delta 5$ Hz. Since this condition is quite possible during testing, this package was rejected.

FIGURE 9-9 *(continued)*

III. Development and Implementation of the Randomdec Technique

    A.   Brief Explanation of the Technique

           The Randomdec Technique was developed by Cole (reference 2) and has been used successfully by a number of aircraft companies (references 3, 4). The proposed functional diagram of the analysis procedure can be seen in Attachment 1. To start the procedure, the Randomdec signature is extracted from the initial vibration data (time history) using the methods detailed in reference 2. The signature is then truncated since only the beginning is needed for further analysis. A curve fitting of the signature is then performed to obtain the frequency and damping characteristics of the individual modes. However, first a power spectral density of the signature is taken to supply the curve fitting program with initial frequency estimates of the modes. After the curve fitting, the individual modes are separated and output with their respective frequency and damping estimates.

    B.   Cost and Time Required

           The familiarity of the engineers in the Dynamics Branch with both the computational procedures involved in the Randomdec Technique and the capabilities of our computer system should facilitate a quick and efficient development and implementation of the new package. Having worked extensively in programming the computer in the past, I would estimate that a maximum of 400 hours would be necessary to complete the development, implementation and check-out phases. This represents $8,000 (@ $20/hour), which is below our allowable budget of $10,000.

IV. Conclusions

        I recommend the development and implementation of the Random-dec Technique as the solution to our problem of an inaccurate and out-dated flutter analysis package. Use of the Randomdec Technique will reduce the risk involved in the potentially dangerous area of flight flutter testing and help Dryden to maintain a prominent position in this flight test field. The Randomdec Technique is being recommended because it meets the following necessary criteria:

        1. Effectiveness
        2. Feasibility
        3. Desirability
        4. Affordability
        5. Preferability

The other alternatives were rejected for failing to meet all of the above criteria.

FIGURE 9-9 *(continued)*

V. <u>References</u>

1. DFRC Memorandum 81-0611-2: Inadequacies of the Power Spectral Density Flutter Analysis Package. June 11, 1981.
2. Henry A. Cole Jr.: On-Line Failure Detection and Damping Measurement of Aerospace Structures by Random Decrement Signatures. NASA CR-2205, 1973.
3. W.J. Brignac, H.B. Ness, and L.M. Smith: The Random Decrement Technique Applied to the YF-16 Flight Flutter Tests. AIAA Paper No. 75-776, May 1975.
4. R.V. Doggett Jr. and C.E. Hammond: Determination of Subcritical Damping By Moving-Block/Randomdec Applications. In "Flutter Testing Techniques," NASA SP-415, 1975, pp. 622-46.

Similarly, other types of reports do the same thing. A progress report claims that the writer (or the writer's group) has made certain progress and then supports that claim with evidence and documentation of the progress. A final report presents the results of an investigation (the writer's claim about some state of affairs) along with the data to prove that the investigation was done carefully and appropriately, that is, with the proof that the writer's vision of "reality" is true or defensible. Thus, in all of these types of reports, a writer is making arguments of fact or policy and providing support.

The writer is doing more than providing support for current readers, however. He or she is also providing crucial documentation for future readers, for those who may need to continue, extend, or test the writer's work. This function is crucial in all reports but is especially pressing in the progress report. Progress reports are written before the entire project is completed. This documentation of what has been done is important since usually the only other sources of information about current work on the project are cryptic notes. If something happens to the writer—an accident, untimely death, job change—then those who must continue the project need to have the earlier work available in a useable form. Since most people keep notes and workbooks in a form that only they can understand, the progress report becomes a convenient vehicle for spelling out conclusions and for assembling data at regular intervals. If the writer has finished some library research and discovered something important for the project, she or he should spell out the discovery and provide a relevant bibliography. If the writer has designed a part or developed a process flow chart or derived an important mathematical relationship, that work should be documented, by including the relevant material either in the body of the report or in an appendix. That work shouldn't have to be done again if something should happen to the writer.

10

THE PROPOSAL

A proposal requests support—usually money—for work that a proposer wants to do. It is written to a funding agency, which may be inside or outside the proposer's organization. This funding agency has its own interests and goals, which may or may not coincide with those of the proposer. The proposal must convince the funding agency that the proposed activity will be a good "investment," that is, that the activity is worthy of support and will advance the agency's goals, produce quality results, and do all this better than other activities competing for the same funds. To make things even more difficult, the proposal must make these arguments to busy readers trying to divide too few resources among too many applicants.

To be successful in this environment, a proposal must quickly and clearly answer the questions a potential funder will bring to any request for support:

What does the proposer want to do?

How much will it cost?

Is the problem important and relevant to the funding agency's interests?

Will the proposed activity solve or reduce the problem?

Can the proposed activity be done? Will it duplicate other work?

Is the method or approach appropriate, clearly defined, and well thought out?

Can the results be adequately evaluated?

Is the proposer qualified to do the research? Better qualified than other proposers?

Will the results of the activity be available to others?

Are the proposed schedule of activity and budget reasonable?

The way these questions are answered will vary as particular projects are adapted to the guidelines of particular funding agencies. However, all of these questions need to be addressed at some level, usually in a format similar to the one outlined below. To be sure that a given proposal meets the expectations of a given funding agency, the proposer should carefully consult any proposal writing guidelines put out by that agency and then interview people familiar with the agency's funding interests. A proposer might even

FIGURE 10-1   Parts of a long proposal.

Title page

Abstract

Table of contents

Introduction

>   Problem addressed
>   Purpose or objectives of research
>   Significance of proposed research

Description of proposed research

>   Plan for accomplishing objectives
>   Plan for evaluating results
>   Schedule for project completion

Institutional resources and commitments

List of references (for six or more references)

Personnel

>   Explanation of proposed staffing
>   Relevant experience of major personnel

Proposed budget

>   Budget in tabular form
>   Justification of budget items

Appendices

>   Letters of endorsement
>   Promises of participation, subcontractor's proposals
>   Biographical data sheets (vita sheets)
>   Reprints of relevant articles, reports, background documents

call or visit the funding agency to get the most recent information on its interests and requirements.

10.1

THE ORGANIZATION OF A FORMAL PROPOSAL

Most proposals follow the general outline presented in Figure 10-1, an outline similar to that given in Chapter 5 for the argument of fact and policy. Indeed, you could argue that a proposal is an argument of policy (the funding agency should fund this proposal) based on several embedded arguments of fact (the problem is important, the activity proposed will alleviate the problem, the proposers are qualified to do the work proposed, etc.).

The general outline will be adapted and modified according to the needs of the readers and the demands of the topic proposed. For instance, long or complicated proposals might well contain all of the sections shown in Figure 10-1 in fully developed form with their own headings corresponding to the sections and subsections of the outline. In contrast, shorter or simpler proposals might contain only the sections noted in Figure 10-2. Short proposals may treat a given subsection very briefly or combine several subsections. A longer proposal to a funding agency outside the writer's organization appears in the Appendix. A shorter proposal, but still a formal one, written to someone in a different unit in the writer's organization is shown in Figure 10-6, and a very short proposal written to someone in the writer's own unit is shown in Figure 10-4.

FIGURE 10-2   Parts of a short proposal.

Title page (ideally, with abstract)

Introduction

    Problem addressed
    Purpose or objectives of proposed work
    Significance of proposed work

Plan for accomplishing objectives

Plan for evaluating results

Schedule for project completion

Institutional resources and commitments

Personnel

    Explanation of proposed staffing
    Relevant experience of major personnel

Budget

    Budget in tabular form
    Justification of budget items, where necessary

Title Page

The Title Page provides the basic *To, From, Subject, Date* information found in headings and Title Pages for other types of technical communications; it also includes financial information relevant to proposals alone. Specific formats for Title Pages vary from one funding agency to another, but most require the following:

1   The title of the proposal (as short and informative as possible)

2   A reference number for the proposal

3   The name of the potential funding agency

4   The name and address of the proposer(s): principal investigator and any co-principal investigators or project director

5   The proposed starting date and duration of the project

6   The total funds requested

7   The proposal's date of submission

8   The signatures of the project director and responsible administrator(s) in the proposer's institution

A sample Title Page is shown in Figure 10-3. It lacks items 7 and 8 because these were included on the institution's cover form which accompanied the proposal.

Abstract

Like other Abstracts, the Abstract of a proposal is short, often 200 words or less. In a short proposal addressed to someone within the writer's institution, the Abstract may be located on the Title Page; in a long proposal or one addressed to a funding agency outside the writer's institution, the Abstract will usually occupy a page by itself following the Title Page.

The Abstract is a critical part of the proposal because it provides a short overview and summary of the entire proposal; it is the only text in the proposal seen by some readers. The Abstract should briefly define the problem and its importance, the objectives of the project, the method of evaluation, and the potential impact of the project. It normally does not define the cost. A sample Abstract appears in Figure 10-6; a longer treatment of Abstracts appears in Chapter 9.

Introduction

The Introduction of the proposal, like the Foreword of the technical report, orients a nonspecialist to the subject and purpose of the document. It should define the problem being addressed (perhaps defining what it isn't as well

FIGURE 10-3   Sample Title Page for a formal proposal.

IN REPLY REFER TO:
DRDA 81-2096-P1

PROPOSAL FOR EXTENSION OF

NASA GRANT NSG 1306

MODELS AND TECHNIQUES FOR EVALUATING THE

EFFECTIVENESS OF AIRCRAFT COMPUTING SYSTEMS

Submitted to the

NATIONAL AERONAUTICS AND SPACE ADMINISTRATION
LANGLEY RESEARCH CENTER
HAMPTON, VIRGINIA 23365

Submitted by the

SYSTEMS ENGINEERING LABORATORY
DEPARTMENT OF ELECTRICAL AND COMPUTER ENGINEERING
THE UNIVERSITY OF MICHIGAN
ANN ARBOR, MICHIGAN 48109

PRINCIPAL INVESTIGATOR: John F. Meyer

PROPOSED STARTING DATE: 1 July 1981

PROPOSED DURATION: 1 year

AMOUNT REQUESTED: $39,932

as what it is) and explain the purpose and significance of the proposed project in terms appropriate for a managerial audience. If the project is simple, the Introduction may also include the few relevant details which would belong in the Background section of a more complicated proposal.

Background

As a separate section, the Background allows you to fill in important technical details inappropriate for the nonspecialist readers of the Abstract and Introduction. The Background provides a place to discuss the history of the problem, to survey previous work on your topic (a survey leading up, of course, to some problem or gap in the previous work), and to place this project in a particular context of previous work you may have done on the problem. If this proposal extends earlier work you have done, be sure to show why your previous work needs to be continued and how the proposed work differs from it. Do not spend a great deal of time justifying your earlier efforts and budgets; concentrate on the new work proposed.

If there has been previous work done on your project, you need to demonstrate that you are aware of it and understand its importance and limitations. (Sponsors don't want to pay you to reinvent the wheel.) You usually do this in a section called "Literature Survey" or "Previous Work," and you will demonstrate your competence by carefully selecting and evaluating the works you cite. If you can't select the most crucial items for your project and briefly show why they're crucial, you're probably not expert enough to know what you need to do and how best to do it. Also, you won't be able to show how your work fits into the larger scheme of things, how it builds on previous work and goes beyond it, how it is original and contributes to knowledge in the field.

Description of Proposed Research

The Description of Proposed Research is the most important section of the whole proposal. It describes what you want to do and how you intend to do it:

The objectives to be achieved

The plan for reaching those objectives

The plan for evaluating the results

The schedule for completing the work

This section will be evaluated carefully by the proposal's reviewers, who will be knowledgeable in the field. Their job is to eliminate all proposals whose objectives or plans are inappropriate or unclear or not well thought out. Thus, your job as a writer is to convince them that you are doing what needs to be done and are doing it in the most careful and thorough way.

When preparing the Description section you should assume that you are writing to a critical, hostile audience, for you are. You should provide all of the details a knowledgeable critic would need to assess your argument—the assumptions on which your research is based, the hypothesis you are following, the specific question(s) you are trying to answer, the particular research and evaluation methods you are using, the appropriateness of your methods for the problem proposed. In addition to providing these items, you may need to justify them, especially if there might be any question or controversy about them. You will probably want to overtly demonstrate the appropriateness of your method for solving the problem posed. If this isn't clear, probably nothing else will matter.

You also need to convince the critical reader that your proposed schedule is appropriate and realistic. You don't want to propose to do too much, given your time and resources, or it will seem that you have a poor assessment of the project and don't really know what you are doing. One way to demonstrate that your schedule is realistic is to spell it out very specifically so that the critical reader can easily see its merits and the care and thoroughness you put into determining it. If you demonstrate that you've really thought of everything, carefully, you're halfway to success.

Description of Relevant Institutional Resources

If you are proposing a project that requires special equipment, one important factor in your ability to do the proposed work is having access to that equipment. Having this equipment already available at your institution is a big plus for your proposal since a funding agency could pay you much less to do the work than it would have to pay someone who had to buy the equipment. Thus, it is to your advantage to list relevant institutional resources. Further, funding agencies often feel that proposers work harder (and institutions monitor them more carefully) if the proposer's institution has resources invested in the project.

List of References

If you have enough references to interrupt the text if you insert them as you go along, you may want to set them up in a separate section. You may also want to do this if the previous work is especially important and you want your reviewers to see that you have cited all the "right" items. As a rule of thumb, if you have more than six references, you might consider a List of References, placed before the sections on Personnel and Budget. The references are listed consecutively as they appear in the text and are listed with the author's name in normal order (first name or initials first).

Personnel

The purpose of this section is to explain who will be doing what and to demonstrate that the people listed for a proposed activity are competent to do it. This is normally accomplished in two subsections, one outlining the responsibilities of the individual participants and the structure to coordinate them and one providing short biographical sheets for the main participants. Both sections should focus on only the *relevant* qualifications of the participants.

Budget

Like the Personnel section, the Budget section has two purposes: to explain what things will cost and to justify and explain individual expenditures, especially where these are not obvious. The budget is usually summarized in a table, such as Table 10-1. (A simple proposal may have a much simpler budget.) The typical headings in a budget are Personnel, Equipment, Supplies, Travel, Computer Time (if relevant), and Indirect Costs. Items typically included under these headings are listed in Table 10-2. Note that not all proposals will need to include all items, but where appropriate the items listed in Table 10-2 should be treated in a proposal's budget.

Appendices

Appendices are reserved for necessary supporting documents which, because of their length or type, would disrupt the "flow" of the proposal. The most common Appendix items are biography sheets for more than six to eight people, letters of endorsement for the proposal, and promises of participation from important participants. Other materials may be pertinent to a given proposal, but the proposal writer should consider carefully any item included in the Appendix and eliminate anything not really needed to support the importance of the topic, the credentials of the proposers, or the ability of the proposers to carry out their work.

10.2

THE ORGANIZATION OF A SHORT INFORMAL PROPOSAL

The outlines for proposals presented in Figures 10-1 and 10-2 are relatively full outlines; they allow writers to argue convincingly to someone who doesn't know them that their problem is important, that their proposed activity will alleviate the problem, and that they are qualified to do the proposed work. Sometimes, however, proposers need to write very short, almost routine proposals meant for people in their own unit. Such proposals might concern some small thing to be done, some small problem to be noticed, some small piece of equipment to be purchased. In such a situation, a proposal using a compressed form of the outline shown in Figure 10-2 might be very appropriate.

Such a proposal appears in Figure 10-4. This proposal omits the sections on Personnel and Budget. These are, however, implied in Sections 5 and 6: one person, the writer, will work for 4 weeks, and perhaps someone

TABLE 10-1 SAMPLE 12-MONTH BUDGET

	CONTRIBUTED BY SPONSOR	CONTRIBUTED BY OUR COMPANY	TOTAL
PERSONNEL			
Project director, half-time	$ 8,000	$ –0–	$ 8,000
Project associate, ¼-time	–0–	3,000	3,000
Research assistant, full-time	6,000	–0–	6,000
Clerk-typist, full-time	5,000	–0–	5,000
Subtotal	$19,000	$3,000	$22,000
Staff benefits (15% of salaries and wages)	2,850	450	3,300
Subtotal	$21,850	$3,450	$25,300
CONSULTANTS			
Warren Duval, $100/day, 2 days	$ 200	$ –0–	$ 200
EQUIPMENT			
Methometer	$ 2,000	$ –0–	$ 2,000
MATERIALS AND SUPPLIES			
Miscellaneous office supplies	100	–0–	100
Glassware	100	–0–	100
Chemicals	100	–0–	100
Subtotal	$ 300	$ –0–	$ 300
TRAVEL			
Project director consultation with sponsor, Chicago to Washington, D.C., and return; 1 person, 2 days			
Air fare	$ 94	$ –0–	$ 94
Per diem © $33/day	66	–0–	66
Local transportation	20	–0–	20
Subtotal	$ 180	$ –0–	$ 180
TOTAL DIRECT COSTS	$24,530	$3,450	$27,980
INDIRECT COSTS (69% of salaries and wages, including staff benefits)	$15,077	$2,381	$17,457
GRAND TOTAL	$39,607	$5,831	$45,437

SOURCE: *Proposal Writer's Guide,* Division of Research Development and Administration, University of Michigan, Ann Arbor, September 1975, p. 9.

from the Hardware Department will work for a short time to provide the necessary hardware. However, the proposal does have the other sections in relatively obvious terms. It has an Introduction which states

1   The problem (we need a catchy demonstration program)

2   The objectives of the proposed work (to provide a demonstration program by simulating the control panel of a nuclear reactor)

TABLE 10-2 CHECKLIST FOR PROPOSAL BUDGET ITEMS

SALARIES AND WAGES
1   Academic personnel
2   Research assistants
3   Stipends (training grants only)
4   Consultants
5   Interviewers
6   Computer programmers
7   Tabulators
8   Secretaries
9   Clerk-typists
10 Editorial assistants
11 Technicians
12 Subjects
13 Hourly personnel
14 Staff benefits
15 Salary increases in proposals that extend into a new year
16 Vacation accrual and/or use

EQUIPMENT
1   Fixed equipment
2   Movable equipment
3   Office equipment
4   Equipment rental
5   Equipment installation

MATERIALS AND SUPPLIES
1   Office supplies
2   Communications
3   Test materials
4   Questionnaire forms
5   Duplicating materials
6   Animals
7   Animal food
8   Laboratory supplies

9   Glassware
10 Chemicals
11 Electronic supplies
12 Report materials and supplies

TRAVEL
1   Administrative
2   Field work
3   Professional meetings
4   Travel for consultation
5   Consultants' travel
6   Subsistence
7   Automobile rental
8   Aircraft rental
9   Ship rental

SERVICES
1   Computer use
2   Duplication services (reports, etc.)
3   Publication costs
4   Photographic services
5   Service contracts
6   ISR services (surveys, etc.)

OTHER
1   Space rental
2   Alterations and renovations
3   Purchase of periodicals and books
4   Patient reimbursement
5   Tuition and fees (training grants)
6   Hospitalization
7   Page charges
8   Subcontracts

INDIRECT COSTS

SOURCE: *Proposal Writer's Guide,* Division of Research Development and Administration, University of Michigan, Ann Arbor, September 1975, p. 10.

3   The significance of the work (to prove the ability of the computer programming language, Industrial Pascal, to communicate with various input-output devices and to test its capabilities to monitor and control several simultaneous tasks)

The proposal also has a plan for accomplishing the objectives, and this is outlined in Sections 2 to 5 of the proposal. It has an implied plan for evaluating the results; an observer would simply see if the demonstration program simulates the nuclear control panel by producing the actions described in Section 4. Finally, the proposal has a schedule (Section 6), and it lists resources already available in the unit (Section 2) as well as needed resources (Section 5).

In what it includes, then, this proposal provides the context and information necessary for those reviewing it while still being an appropriately short proposal for a small project.

FIGURE 10-4 Sample of a short informal proposal.

Chen Computer Systems—Interoffice Memorandum *

To: John Van Roekel
            Software Manager

From: Larry N. Engelhardt
            Senior Programmer

Date: 9 August 1979

Subject: Proposed Demonstration Project for Industrial Pascal: Specifications
            and Description

1. Introduction

The Programming Division has just finished developing an Industrial Pascal language system for our new 3935 RacPac minicomputer system. We now need a computer program that will allow our sales people to demonstrate the capabilities of the RacPac system running Industrial Pascal in a "real" environment.

I propose to develop a demonstration program that will control a simple but realistic representation of a control panel of a nuclear reactor. This demonstration program will be written entirely in Pascal and will run on a RacPac system in an Industrial Pascal environment.

The RacPac will be hooked into a 3910A-3920A simulator, and together the RacPac and simulator will represent the control panel of a nuclear reactor. The program will read and write digital and analog inputs and outputs to and from the controls and indicators on the simulator front panel and respond in an interesting and amusing manner. This demonstration will both prove the ability of Industrial Pascal to communicate with various input/output devices and test its capabilities to monitor and control several simultaneous tasks.

2. Hardware Capability

The hardware configuration on which the demonstration program will run consists of a simulator built by Bill Lepior and a 3935-A RacPac system. The simulator is a cabinet which features two potentiometers and a meter connected to a 3910A analog I/O controller and eight lamps and eight switches connected to a 3920A digital I/O controller. Both controllers are plugged into the chassis of the RacPac. Furthermore, the RacPac has an integral plastic keyboard and 5-inch CRT screen controlled by an 1812 interface for operator I/O. This mix of digital and analog I/O devices offers a good sample of control devices.

3. The Demonstration

One assembler language program has already been written to demonstrate the simulator. Of course, a Pascal program could be written to duplicate the functions of the assembler program exactly. However, to be interesting for the customers, the Pascal demo program could be made more realistic. In addition, a dynamic and robust demonstration program is needed to make use of all the I/O devices of the simulator and also to exercise the multitasking features in IP in a real-time control situation. This document proposes that the IP demo program be designed to be a nuclear reactor control panel.

FIGURE 10-4   *(continued)*

4. Operator Controls

The IP demo program would turn the simulator, keyboard, and screen into a simplistic representation of a control panel of a nuclear reactor. For example, the eight panel switches could move damper rods into and out of the make-believe reactor. The corresponding panel lights would go on when a rod is removed. As damping rods are removed, the core temperature would rise over time. The core temperature can be displayed by the large analog meter. As more rods are withdrawn the temperature would rise at a faster rate. The "operator" can reduce the rate of temperature rise by increasing the cooling water feed rate with one potentiometer. The second potentiometer could set an alarm point on a temperature which will cause the reactor to "scram" automatically. The goal of the "operator" is to keep the reactor running in a range between letting the core cool off and scramming, or going critical.

The screen of the RacPac can display values for all of the activity of the simulation. The screen can also display helpful but ominous warning messages such as "Warning: Evacuate the state of Pennsylvania!" While the screen duplicates the analog and digital inputs and outputs, the keyboard will accept some simple commands to control execution of the program.

5. Hardware Required

The development of the demonstration program will require a 3800B for editing and compiling. Intermediate debugging could be done by plugging the simulator into a 3800B if it is equipped with an 1850 controller board in addition to its standard 1824. Final checkout will require either down-line loading of the program from a 3800B to the simulator's RacPac or burning of the program into PROMs for installation into the RacPac. The Hardware Department has both a 3800B and an 1850 controller board, and it could burn the program into PROMs and install the PROMs.

Description	Weeks Required	Total Time
Research drivers	1	1
Design program	1	2
Code	1	3
Debug	1	4

---

*Note to readers: Pascal is a computer language, and Industrial Pascal (IP) is a version of Pascal used on small computers in industrial settings to control machinery or to monitor various processes. Software is a computer program or programs. I/O is input/output, that is, the text or information put into a computer (input) or that given out by it (output). A CRT screen is a cathode ray tube screen, a screen similar to a television, which prints out the letters and numbers the operator types and which prints anything the operator commands the computer to calculate and print. Hardware is equipment, computers or the "hard" parts of computers, and a PROM (A programmable, read-only memory) is a small part in a computer.

10.3

EDITING THE PROPOSAL

Once you have written a proposal, it is wise to analyze it for weak spots, areas needing more proof or detail. As a guide, you might refer to the analysis below of the problems detected in 605 proposals rejected by the National Institutes of Health.[1] More than one item may have been cited in rejecting a proposal.

A	PROBLEM (58 percent)	
1	The problem is not of sufficient importance or is unlikely to produce any new or useful information.	33.1
2	The proposed research is based on a hypothesis that rests on insufficient evidence, is doubtful, or is unsound.	8.9
3	The problem is more complex than the investigator appears to realize.	8.1
4	The problem has only local significance, or is one of production or control, or otherwise fails to fall sufficiently clearly within the general field of health-related research.	4.8
5	The problem is scientifically premature and warrants, at most, only a pilot study.	3.1
6	The research as proposed is overly involved, with too many elements under simultaneous investigation.	3.0
7	The description of the nature of the research and of its significance leaves the proposal nebulous and diffuse and without a clear research aim.	2.6
B	APPROACH (73 percent)	
1	The proposed tests, methods, or scientific procedures are unsuited to the stated objective.	34.7
2	The description of the approach is too nebulous, diffuse, and lacking in clarity to permit adequate evaluation.	28.8
3	The overall design of the study has not been carefully thought out.	14.7
4	The statistical aspects of the approach have not been given sufficient consideration.	8.1
5	The approach lacks scientific imagination.	7.4
6	Controls are either inadequately conceived or inadequately described.	6.8
7	The material the investigator proposes to use is unsuited to the objective of the study or is difficult to obtain.	3.8
8	The number of observations is unsuitable.	2.5
9	The equipment contemplated is outmoded or otherwise unsuitable.	1.0

C   INVESTIGATOR (55 percent)

1   The investigator does not have adequate experience or   32.6
      training for this research.

2   The investigator appears to be unfamiliar with recent pertinent   13.7
      literature or methods.

3   The investigator's previously published work in this field does   12.6
      not inspire confidence.

4   The investigator proposes to rely too heavily on insufficiently   5.0
      experienced associates.

5   The investigator is spreading himself too thin; he will be more   3.8
      productive if he concentrates on fewer projects.

6   The investigator needs more liaison with colleagues in this   1.7
      field or in collateral fields.

D   OTHER (16 percent)

1   The requirements for equipment or personnel are unrealistic.   10.1

2   It appears that other responsibilities would prevent devotion of   3.0
      sufficient time and attention to this research.

3   The institutional setting is unfavorable.   2.3

4   Research grants to the investigator, now in force, are adequate   1.5
      in scope and amount to cover the proposed research.

10.4

GETTING THE PROPOSAL APPROVED FOR SUBMISSION

Different organizations have different procedures for handling proposals. Often proposals to someone within the writer's organization can be submitted without much red tape. You may need to get your supervisor's approval before sending out a proposal asking that you work for another department or unit, but a verbal OK may be all the approval you need. (Note, however, that you should at least inform your supervisor if you propose to work on a project outside your unit.)

In contrast to the simple procedures for approving internal proposals, the procedures for approving proposals to funding agencies outside the proposer's organization may be quite involved. Organizations often insist on formally reviewing and approving any proposal to external funders. This allows the organization to monitor commitments made to outside organizations; it also allows the organization to eliminate undesirable competition between members of its own staff who may be applying for funding in competition with each other. The problem with this organizational review and approval—from the proposer's point of view—is that it takes extra time, and the proposer must plan this time into the writing process. In addition, the proposer may need to get special forms filled out and signed by administrative officers in the proposer's organization.

For instance, the sample approval form in Figure 10-5 for proposals

FIGURE 10-5 Sample approval form for a proposal seeking
external funds.

Proposal no. DRDA-81-2096-P1
Date 5/21/81

The University of Michigan

APPROVAL OF APPLICATION FOR GRANT OR CONTRACT
(See back of form for instructions.)

<table>
<tr><td>__X__ Research</td><td>_____ New</td></tr>
<tr><td>_____ Instruction</td><td>__X__ Continuation of acct. no.: 014524</td></tr>
<tr><td>_____ Facilities</td><td>_____ Competing renewal</td></tr>
<tr><td>_____ Fellowship/traineeship</td><td>_____ Non-competing renewal</td></tr>
<tr><td>_____ Equipment/materials</td><td>__X__ Supplement</td></tr>
<tr><td>_____ Service</td><td></td></tr>
</table>

To be submitted to: National Aeronautics and Space Administration
                                        (proposed sponsor)

Project Director   Prof. John F. Meyer
Department or unit   Elec. & Comp. Eng.
                                (Limit to One)

Organization code   2160

1. Major fields of study to which project is related:
   Code nos.   11.80.50
                              (Please Limit to 3 Codes)

2. PROJECT TITLE (60 spaces max.)   Models and Techniques for Evaluating
   the Effectiveness of Aircraft Computing Systems

3. Proposed period (Must agree with period for which budget in Item 9 is
   given):
   From 7/1/81 through 6/30/82   Total months: 12   Total cost: $39,932

4. Does the proposed activity involve:
   a. Use of human sub- Yes _____ No __X__ If yes, date of committee approval ____
      jects
   b. Use of vertebrate   Yes _____ No __X__
      animals
   c. Lease or purchase Yes _____ No __X__
      of computer equip-
      ment
   d. Recombinant DNA Yes _____ No __X__ If yes, date of committee approval ____
   e. Classified research Yes _____ No __X__ If yes, date of committee approval ____

5. Participating faculty and students:
   a. The following faculty members other than the project director will partici-
      pate.

FIGURE 10-5 (*continued*)

Name	Rank and Unit	Nature of Participation
_____	_____	_____
_____	_____	_____
_____	_____	_____
_____	_____	_____

  b. Number of students to participate: Postdoctoral: _____ Graduate: __2___
      Undergraduate: _____

6. University facilities required:
    a. Will adequate space be available for the period proposed?
        Yes __X___ No _____
        If yes, Bldg.: E. Eng. Room: 2523, 2506, 1   Approved: _____
        If no, space required: _____ sq. ft. Source: _____ Approved: _____
    b. Space renovation $_____ UM account no.: _____ Approved: _____
    c. Will acquisition of major equipment items require installation and building
        modification at cost to the University?
        Yes _____ No __X___.

7. Is additional office equipment required for this project? Yes _____ No __X___
    Is it provided for in the proposal? Yes _____ No _____
    If not, attach list and indicate University unit to provide: _____

    _____
        (approved)

8. Is work to be performed off University property? Yes _____ No __X___
    If yes, indicate location and duration: _____

9. Summary of proposed budget:

	UM sources:	Sponsor:	Total:	Acct. no. for UM sources:	Approved:
a. Salaries and Wages (S&W)	_____	21,405	21,405	_____	_____
b. Staff Benefits (SB)	_____	3,853	3,853	_____	_____
c. Fellowships/Stipends	_____	_____	_____	_____	_____
d. Tuition	_____	_____	_____	_____	_____
e. Consultants	_____	_____	_____	_____	_____
f. Travel	_____	3,100	3,100	_____	_____
g. Supplies/Materials	_____	_____	_____	_____	_____
h. Equipment*	_____	_____	_____	_____	_____
i. Computer costs	_____	1,000	1,000	_____	_____
j. Subcontracts	_____	_____	_____	_____	_____
k. Other (Itemize under notes)***	_____	600	600	_____	_____
l. Total Direct Costs (TDC)	_____	29,958	29,958	_____	_____
m. Indirect Costs**	_____	16,974	16,974	_____	_____

FIGURE 10-5 *(continued)*

<u>17.8%</u> Computer = $178
<u>58%</u> MTDC = $16,796
Less carryover _____ (−7,000) (−7,000)
n. TOTAL $39,932 $39,932

10. NOTES:
   *Itemize items of equipment costing more than $10,000.
   **Explanation if other than negotiated rates.
   ***Publication costs/supplies.

(Requested by: Project Director)

J. F. Meyer

(Fiscal/Legal review)

(Approved by: Dept. or Unit Head)

G. I. Haddad

(Approved for DRDA by: Project Representative)

Mary L. Egger

(Approved for School/College by: Dean)

M. J. Sinnott

(Approved for the University by: Vice-President)

C. G. Overberger

---

seeking external funds had to be filled out and signed by four administrative officers in addition to the proposer. Just finding these people can take some time, and theoretically each person should have some time to review the contents of the proposal he or she is approving.

If the proposer's organization requires such approval, the proposer should be aware of this fact long before the proposal's deadline and should plan to allow sufficient time for each reviewer.

EXERCISE 10-1

A   Compare the short formal proposal shown as Figure 10-6 to the proposal outlines given in Figures 10-1 and 10-2. See if you can identify the various sections outlined in Figure 10-2. Compare the treatment of these sections with the comparable treatment of sections in the short informal proposal in Figure 10-4 and then to that in the formal proposal in the Appendix.

B   Write a short formal or informal proposal for some project you want to do. Be sure to include all relevant information, and be prepared to justify the information you have included or excluded.

FIGURE 10-6 A short formal proposal.

THE DEVELOPMENT OF INSTRUCTIONAL MODULES IN LECTURE
COMPREHENSION FOR FOREIGN STUDENTS

A proposal submitted to the Center for Research on Learning
and Teaching in competition for a Faculty Development Fund
Teaching Grant

Project Director: Leslie A. Olsen, Associate Professor, Department of Humanities, College of Engineering

Co-Principal Investigator: Thomas N. Huckin, Assistant Professor, Department of Humanities, College of Engineering

Date of Submission: February 15, 1979

Proposed Starting Date: May 1, 1979

Proposed Duration: 1 year

Amount Requested: $5000

ABSTRACT

Many foreign students attending the University of Michigan have serious problems understanding course lectures. The objective of this proposal is to improve the University English instruction for such foreign students by developing expertise and instructional materials in lecture comprehension. Specifically, we propose (1) to compile experimental evidence from several disciplines to provide a solid explanatory base for teaching lecture comprehension and (2) to develop a set of graded video tapes and handout materials on lecture comprehension aimed at science and engineering students.

The results of this project, which will be class tested and reviewed by outside reviewers, could affect potentially all of the foreign science and engineering students at the University who need instruction in lecture comprehension. It is estimated that one third of the approximately 1630 foreign technical students here now need such instruction and that about 300 of the students entering each year would profit from this instruction when they first arrive. The instructional materials will be made available for both class and individual use.

For years the English Language Institute (ELI), the Department of Linguistics, and the Department of Humanities in the Engineering College have conducted a program to improve the English of foreign students at the University. The ELI program has been oriented toward basic, sentence-level skills and spoken "survival" English, while the programs in the Linguistics and Humanities Departments have been oriented toward improving writing and conversational ability.

FIGURE 10-6 *(continued)*

Recently we in the Humanities Department have become aware of a strong need for another component in our programs for science and engineering students, a component on lecture comprehension. This need has been expressed by the Oster-Brown report on the state of English-as-a-second-language (ESL) instruction at the University, by specific requests from Dean Sussman at Rackham (stimulated by complaints from various professors), by complaints from professors in our college to us, and by comments and requests from our students, even those students who did well on the various English tests.

Science and engineering professors often complain that foreign students cannot follow lectures, especially the interpretive material added by the professor in the lecture. On the other hand, science and engineering students complain that they can't remember all of the material (they often try to write down and memorize a lecture word for word rather than to look for generalizations). Further, the students often complain that they can't tell what is important and what is unimportant material (they can't tell when an aside or a digression occurs in contrast to an important gloss or a shift in the line of argument) and they can't follow a lecturer who is disorganized, who mutters, or who digresses. In addition, we and other researchers have noted that students often do not know how to read visual aids presented to summarize or to illustrate a point, and they often don't see the redundancy we put in our speech to signal our various shifts in purpose, mood, or importance of subject. In short, there is a strong consensus that something needs to be done and done quickly.

We are thus requesting $5000 to develop some materials in lecture comprehension appropriate for the foreign science and engineering students at the University. (These students account for approximately 70% of the 2100 foreign students at the University of Michigan.) In the pages which follow, we would like to argue (1) that there is a need for new materials to be developed because there are no adequate materials available and (2) that we are qualified to develop such materials.

NEED FOR NEW MATERIALS

An important recent study from the University of Lancaster (England), one of the few studies of "real" lecture situations and the only serious in-depth study of the problems foreign students have with science and engineering lectures, has identified all of the problems we have just described. This study has also stressed the critical importance of using lecture comprehension materials (1) which incorporate a real lecture situation and (2) which really represent the total communication situation with all of its visuals, movements, digressions, etc. We agree completely with this conclusion.

Unfortunately, the materials currently available to teach listening comprehension are inadequate from both points of view. First, most materials are inadequate because they are too basic; they do not represent real lecture situations even at the end of an instructional sequence. (People seem to have assumed that a student living in this country would readily assimilate listening skills and thus would

FIGURE 10-6 *(continued)*

not need more advanced, real lecture materials. We have already tried to demonstrate that this assumption is false.) Second, the available materials are inadequate because they simply do not simulate the total communication situation in a classroom with its asides, glosses, hesitations, digressions, visual aids, blackboard notes, etc.; the available materials are usually audiotapes of texts carefully read from written or highly polished material.

Thus, if we are to develop adequate instructional modules in lecture comprehension, we need to prepare a series of videotapes of lectures of varying difficulty and to develop class materials to accompany these tapes. We also plan to survey the literature in cognitive psychology, psycholinguistics, and speech for further insights and experimental support for our current hypotheses on lecture comprehension for foreign students.

OBJECTIVE

The general objective of this proposal is to improve the Humanities Department's (and the University's) offerings in the area of foreign student instruction by developing expertise and instructional materials in the now neglected area of lecture comprehension. We have already done some basic reading devoted to lecture comprehension in scientific and engineering disciplines and have some ideas about those features of lectures in science and engineering which must receive special emphasis—such as kinesics (movement and how it is tied to the lecture's structure), visual aids, and ways of signaling transitions within a unit and transitions to peripheral or explanatory material. Thus, we propose for this project to

1) compile a body of experimental evidence from cognitive psychology, psycholinguistics, and speech and rhetoric to test and support or refine our analysis of those features of lectures needing special attention for foreign students. This will develop our expertise as well as provide a solid explanatory base for our teaching materials;

2) develop a set of graded videotapes and handout materials for a class unit on lecture comprehension.

ACCOMPLISHING THE OBJECTIVES

To accomplish the objectives listed above, we plan to

1) survey the literature in speech, psychology, and psycholinguistics for evidence describing the relative importance of various features in establishing coherence in lectures or in oral discourse;

2) define additional areas or features not covered in the literature above which we feel might be important in establishing (or breaking) coherence for foreign students;

FIGURE 10-6   *(continued)*

3) survey tapes of lectures already available to find at least some segments of lectures we could use to describe and teach the important
features identified in steps 1 and 2 above;

4) secure permission to use appropriate sequences from prerecorded
tapes for instruction in lecture comprehension;

5) produce tapes as necessary to provide a modularized series of 5 or 6
tapes (3 tapes 5-10 minutes long and 2 or 3 tapes 25-30 minutes
long);

6) develop for each tape written handout materials and questions focusing on the important features of the tape.

Any materials developed would be made available to teachers in any unit of
the University and would be placed in the Sight and Sound Room of the Undergraduate Library for individual use.

EVALUATING THE OUTCOME

In general, we have been successfully utilizing—and plan to continue utilizing—two different forms of evaluation for our project: testing of students and evaluation by peers. Student-testing is provided mainly through class assignments
(written reports and papers) and subsequent classroom critique sessions and editing exercises; we observe the students' performance to see if our instruction has
been learned and put to use. We also use class exercises aimed directly at the
teaching points and solicit student response and suggestions on these class exercises. Peer evaluation will be provided as indicated by the following sources:

1) on a monthly basis by colleagues in the English Language Institute,
the Linguistics Department, the Speech Department, and the communications group of the Humanities Department;

2) at the project's end by researchers at Carnegie-Mellon University
working on discourse analysis from the perspectives of cognitive psychology (John Richard Hayes) and rhetoric (Richard E. Young and
Linda Flower); and

3) by researchers at the University of Lancaster (England), Christopher
Candlin and Dermot Murphy, who have done the most significant research to date on listening comprehension in technical discourse.

SCHEDULE FOR PROJECT COMPLETION

If our complete budget request is allocated, our proposed team could complete the pre-evaluation part of it within 12 weeks. We could begin as early as May
1 and complete this part by the end of July. Evaluation and subsequent refining
would begin in early September and continue through the academic year.

FIGURE 10-6 *(continued)*

PROPOSED BUDGET

Since there are three of us working on this project, including a research associate, we are asking for the maximum amount of support the CRLT can provide, namely, $5000. This total is broken down as follows:

Partial salary support for the project director, Leslie Olsen, during May-July	$1550.00
Partial salary support for the co-principal investigator, Thomas Huckin, during May-July	1550.00
Research Associate II salary for Russell S. Tomlin: 200 hours @ $7.75	1550.00
Supplies, photocopying, typing, 2 videotapes	350.00
	$5000.00

We are especially requesting support to work on the project during the summer because we have heavy teaching loads and thus limited research time during the school year. (Our normal course load is 3 courses per term per instructor, and this winter term Olsen and Huckin are each teaching 4 courses.)

QUALIFICATIONS OF RESEARCHERS

Any teacher trying to develop materials on lecture comprehension for students in science and engineering should be able:

1) to apply the principles of discourse analysis developed in linguistics to analyze texts (in order to be able to adequately describe important features of the material students need to study);

2) to describe specifically the features of the English of science and technology (EST) and especially to describe the special ways in which EST establishes coherence and marks transitions;

3) to describe and analyze features of intonation and phonology and to describe such features to students having trouble with the sounds of words;

4) to explain the relationship between speech, movement, and gestures in a lecture (be able to explain why the speaker moved when and where he/she did);

5) to understand enough about the subjects chosen for the lectures to select appropriate material for the videotapes (this would be done in consultation with specialists in the field).

Our proposed team meets all of these qualifications. To address the first and second criteria, all three of us have done research and teaching in discourse analysis.

FIGURE 10-6 *(continued)*

Olsen and Huckin have been teaching scientific and technical writing for five and two years, respectively; have published in the area; and are finishing up a text on the subject. Further, all three team members are currently teaching a graduate-level course in the area. To address the third and fourth criteria, one of our team, Huckin, wrote his dissertation on phonology and intonation and all three members of our team have started studying the relationship between speech, movement, and gestures in videotaped lectures on technical subjects. Finally, to address the fifth criterion, Olsen and Huckin have had up to 12 years of experience with technical students and subjects and Olsen has worked as a technical writer and editor and has a degree in chemistry with significant work in math, physics, and biochemistry.

POTENTIAL IMPACT OF PROJECT

The expertise and materials developed in this project could potentially affect all of the foreign science and engineering students at the University who need instruction in lecture comprehension. We don't know exactly how many students there would be in this category, since none of the standard English tests for foreign students (including the Michigan Test and the TOEFL) measure lecture comprehension. However, from all of our informal measures, experience, and anecdotal evidence, we estimate that one third, or 550, of the approximately 1650 foreign technical students here at both the undergraduate and graduate levels need such instruction in lecture comprehension now and that about 300 students entering each year would profit from this instruction when they first arrive. The instructional materials developed on lecture comprehension will be made available for both class and individual use.

Further, the results of this project, including the materials and the evaluations of the materials, should increase our ability to secure funding from the National Science Foundation for a large, multidisciplinary research project on the discourse analysis and comprehension of lectures.

REFERENCE

1 *Proposal Writer's Guide,* Division of Research Development and Administration, University of Michigan, Ann Arbor, 1975. From E. M. Allen, *Science,* November 25, 1960, pp. 1532-1534.

11

THE BUSINESS LETTER

There are a number of activities which a business does frequently. These include requesting information or equipment, ordering supplies, praising or thanking someone for a job well done, complaining about a job badly done, and responding to someone else's request, order, praise, or complaint. While some of these activities may be accomplished orally or by filling out a form, many of them require writing a letter. A letter provides a record of the activity for someone's file, it allows the writer to provide more context or explanation than is usually possible on a form, and it helps the audience remember what is to be done. While letters may be written more frequently in some jobs than in others, all technical people should be able to write a good letter when it is needed.

Writing a good letter is an art. It basically requires a writer to produce a one-sided conversation with the reader. Of course, all writing does this to some degree in that the writer anticipates the reader's questions and provides answers to those questions where the reader might ask them. However, letters differ from most other forms of writing: they are often more personal, even emphasizing the reader/writer relationship with the generous use of such pronouns as *I, we,* and *you.*

Despite this frequent personal sense, letters share several of the organizational features already described for the less personal technical report. Like reports, letters need to first orient the reader to the topic at hand, then explain why the writer is writing, and then provide enough information so that the reader can easily understand what he or she is to do.

The orientation of the reader is especially important when the reader does not expect a letter or might not remember the subject the letter refers to. For instance, consider the letters in Figures 11-1 and 11-2. The first letter has no orientation to the topic being discussed. Its writer jumps right into the letter as if she were simply continuing the phone conversation which initiated it. Unfortunately, however, the conversation occurred almost 3 weeks before the letter was written, and the reader may well have a hard time remembering which shroud is being discussed in the first paragraph and what he needed to know about it. In contrast, the version of the letter in Figure 11-2 gives a good orientation to the topic; it reminds the reader of the conversation and the main questions raised. It also provides a rationale for the organization of the letter's details: they are grouped as answers to the three questions.

FIGURE 11-1 Sample business letter without a good orientation to the topic.

ADVANCED TECHNOLOGIES, INC.
40 Technology Park, Milford,
Massachusetts 01757
617-555-4553

October 4, 1981

Mr. Frank E. Lee
Equipment Department
Harrison Radiator, Inc.
1451 Murray Avenue
Pittsburgh, PA 15217

Dear Mr. Lee:

The shroud you called about on September 15 has the same shape, construction, and functions as the shroud previously purchased by Harrison Radiator Division from A.T.I. However, that machine was designed to balance a series of blower assemblies having only one fan. With the present balancer you wish to have the capability to balance a new blower assembly having two fans, one on each end of the motor shaft. This new blower assembly is longer than all previous single-fan blower assemblies. The only dimensional change needed to account for this is to widen the shroud from 10″ to 16-½″. Also, a new clamping fixture is needed to hold the double-fan assembly, necessitating that a 3″-diameter hole be added to allow a clamping cylinder to protrude. These changes are illustrated on the enclosed print.

I have found that the change of the focal point of the fan speed sensors, due to the new blower fan's position within the shroud, in relation to the sensors, is well within the limits of the adjustments designed into the original shroud, as is shown on the print. This should not present a problem.

The estimated weight of the shroud is 6-½ lbs., only ½ lb. heavier than the previous one. This results in a cost difference of an additional $5.75 for the new shroud.

I refer you to the enclosed print; it might answer additional questions.

Sincerely,

Barbara C. Benson

Barbara C. Benson
Engineering Department

BCB/rs

Enclosure

FIGURE 11-2 Sample business letter with a good orientation to the topic.

ADVANCED TECHNOLOGIES, INC.
40 Technology Park, Milford,
Massachusetts 01757
617-555-4553

October 4, 1981

Mr. Frank E. Lee
Equipment Department
Harrison Radiator, Inc.
1451 Murray Avenue
Pittsburgh, PA 15217

Dear Mr. Lee:

In our phone conversation of 28 September 1981, you wanted to know:

--What the differences will be between the original shroud for Harrison Radiator Division's D-25-PS balancing machine and a similar one purchased from A.T.I. on September 15, 1981 (Ref.: Sales Order #N5682).

--How these differences affect the cost of the shroud.

--If the introduction of our double-fan blower assemblies will create any problems with the fan speed sensors.

I have listed this information below and enclosed a blueprint of our new shroud for your reference (#25508).

Basically, the two shrouds have the same shape, construction, and functions. However, there are three major differences between the Harrison and A.T.I. shrouds:

1) The A.T.I. shroud has a new double-fan blower assembly.

2) The A.T.I. shroud is 6-½" wider.

3) The A.T.I. shroud has an additional 3"-diameter hole in the top of the shroud.

These changes will result in an additional cost of $5.75. Also, the new blower assemblies will not present any problems with regards to the fan speed sensors.

The original Harrison shroud was designed to work only with a series of single-fan blower assemblies. The introduction of longer double-fan blower assemblies (fans on both ends of the motor shaft) on the A.T.I. shroud requires two things.

FIGURE 11-2   *(Continued)*

Frank E. Lee
October 4, 1981
Page Two

The shroud must be lengthened from 10″ to 16-½″. Also, a new cradle fixture is required to hold the blower assemblies, resulting in the addition of a 3″-diameter hole in the tip of the shroud, to allow a clamping cylinder to protrude.

Since the two shrouds are the same construction--same material, same method of assembly--the only cost difference will result from the greater weight of the new shroud. Widening the shroud 6-½″ adds ½ lb. to the total weight. This translates to an additional cost of $5.75.

As can be seen on the enclosed print, the focal point of the new blower fan is well within the range of the electro-optical fan sensors, as provided for by the sensor adjustments designed into the original shroud.

I refer you to the enclosed print. It may answer additional questions.

Sincerely,

Barbara C. Benson

Barbara C. Benson
Engineering Department

BCB/rs

Enclosure

11.1

BASIC LETTER FORMATS

Even though letters do not have labels such as *To, From, Subject,* and *Date,* they do have conventional places where readers look for such background information. Figure 11-3 indicates these places, and Figure 11-4 gives an example of a short letter arranged in a conventional format for business letters. Notice that the subject of the letter is mentioned in the first sentence to orient the reader.

The conventional format illustrated in Figure 11-4 is only one of three quite common formats for business letters: the unblocked format, the semi-blocked format, and the blocked format. These formats are presented in skeletal form in Figures 11-5 to 11-7 for letters written on plain paper, that is, not on letterhead stationery. It should be noted, however, that when a writer is representing a company or organization, the writer should use the

FIGURE 11-3 Location of background information in a letter, unblocked format. (On letterhead stationery, the writer's address will already be printed.)

Writer's street address
Writer's city, state, zip code
Date

Reader's name
Job title
Reader's company/organization
Organization's street address
Organization's city, state, zip code

Subject:

Dear name of reader:

Subject of letter and references to past correspondence, if any

Sincerely yours,

Writer's signature

Writer's name typed

WRITER'S INITIALS: typist's initials
(if writer did not type letter)

Enclosure (if appropriate)

cc: Name of recipient of carbon copy of letter
     (if appropriate, or xc: Name of recipient of photocopy of letter)

FIGURE 11-4   Example of a short business letter in the unblocked format.

205 State Street
Livonia, Alabama 35851
June 20, 1982

Mr. R. B. Sparks
Sales Representative
Computer Connection
384 Grand Drive
Harrington, Georgia 30018

Subject: BASIC Software for Apple II System

Dear Mr. Sparks:

The Computing Club of Brewster High School will be sponsoring a series of computing activities during the fall of 1982. Since our current computing equipment is inadequate for our projected needs, we are considering buying an Apple II system to expand our equipment line.

Last week I wrote to you requesting information on the cost of the Apple II Plus system and its peripheral equipment as well as information on the PASCAL software available with the system.

I would now like to request additional information on

1)   the BASIC software currently available with the system, especially for plotting graphs such as that enclosed with this letter; and

2)   any BASIC software expected to be released in the next year for the APPLE system.

Thank you for your attention. We are looking forward to hearing from you in the near future.

Sincerely yours,

Ann Jones

Ann Jones
President, Computing Club

AJ:vg

Enclosure

cc: Mr. Peter Murray
    Ms. Mary Corvina

FIGURE 11-5 Unblocked format for business letter.

(Number of Blank Lines
Between Sections)

205 State Street
Livonia, Alabama 35851
June 20, 1982

(3)

Mr. R. B. Sparks
Sales Representative
Computer Connection
384 Grand Drive
Harrington, Georgia 30018

(1)

Subject: BASIC Software for Apple II Systems

(1)

Dear Mr. Sparks:

(1)

_____

_____
_____
_____

(1)

_____
_____
_____

(1)

_____
_____
_____

(1)

Sincerely yours,

(3)

Ann Jones
President, Computing Club

(1)

AJ:vg

(1)

cc: Mr. Peter Murray
Ms. Mary Corvina

FIGURE 11-6 Semiblocked format for business letter.

*(Number of Blank Lines
Between Sections)*

205 State Street
Livonia, Alabama 35851
June 20, 1982

(3)

Mr. R. B. Sparks
Sales Representative
Computer Connection
384 Grand Drive
Harrington, Georgia 30018

(1)

Subject: BASIC Software for Apple II Systems

(1)

Dear Mr. Sparks:

(1)

(1)

(1)

(1)

Sincerely yours,

(3)

Ann Jones
President, Computing Club

(1)

Enclosure

(1)

cc: Mr. Peter Murray
Ms. Mary Corvina

FIGURE 11-7 Blocked format for business letter.

*(Number of Blank Lines
Between Sections)*

205 State Street
Livonia, Alabama 35851
June 20, 1982

(3)

Mr. R. B. Sparks
Sales Representative
Computer Connection
384 Grand Drive
Harrington, Georgia 30018

(1)

Subject: BASIC Software for Apple II Systems

(1)

Dear Mr. Sparks:

(1)

_____

(1)

_____

(1)

Sincerely yours,

(3)

Ann Jones
President, Computing Club

(1)

AJ:vg

(1)

Enclosure

(1)

cc: Mr. Peter Murray
     Ms. Mary Corvina

FIGURE 11-8 Sample continuation page, blocked format.

*(Number of Blank Lines
Between Sections)*

Mr. R. B. Sparks   -2-   June 20, 1982

(3)

_____
_____
_____
_____

(1)

Sincerely yours,

(3)   *Ann Jones*

Ann Jones
President, Computing Club

(1)

AJ: vg

(1)

Enclosure

(1)

cc: Mr. Peter Murray
    Ms. Mary Corvina

organization's letterhead stationery for any correspondence with people outside the organization. When such letterhead is used, the only change in the formats presented in Figures 11-5 to 11-7 involves the location of the writer's address, city, and state. These are usually given, along with the organization's name and telephone number, in the letterhead printed at the top of the page, as shown in Figure 11-1, for example.

If a letter requires more than one page, the additional pages are called continuation pages. They are typed on plain paper, not letterhead. A sample continuation page appears in Figure 11-8. Notice that it contains enough information at the top of the page to identify it if it gets separated from the first page.

In Figures 11-5 to 11-8 the number of blank lines separating the various sections of the letter are indicated in the left margin. One blank line separates most sections, but three blank lines separate the writer's address at the top of the page from the reader's name and address and three blank lines separate the closing (*Sincerely yours*) from the typed version of the writer's name. The latter spacing should leave plenty of room for the writer's signature.

The letters in Figures 11-1 to 11-8 all have some additional bits of information the function of which may not be as obvious as that of other parts of the letter. This information appears at the left of the page below the typed version of the writer's name. Such lines should be added to a letter whenever they are appropriate; if any one of these lines is omitted, the remaining lines should be moved up, with one blank line separating them from each other and from the typed version of the writer's name. The following information is included in these lines:

Initials line. The first line under the writer's name is the initials line (*AJ:vg*), which indicates that someone other than the writer typed the letter. The capital letters are the initials of the writer, the lower case letters the initials of the typist. If the writer types the letter, the initials line is omitted.

Enclosure line. The second line under the writer's name (*Enclosure*) indicates that something has been enclosed with the letter and alerts the reader to look for this item. The enclosure may be a whole document, a single page, a picture, a drawing, or anything else. If nothing is enclosed, the Enclosure line is omitted.

Copy line. The third line under the writer's name (*cc: Mr. Peter Murray*) indicates that a carbon copy or photocopy of the letter has been sent to the person (or persons) whose name appears after the *cc:*. Sometimes *xc:* is used to indicate that a photocopy has been sent. If no one has been sent a copy of the letter, the *cc:* line is omitted.

11.2

LETTER OF TRANSMITTAL

When a document of more than two or three pages—a report, article, or proposal—is sent to some reader, it is often accompanied by a letter of transmittal. The purpose of this letter is to identify the document and to orient the reader to it. It may help the reader decide whether to read the document, file it, or send it on to someone else. Thus, the letter of transmittal should be short, no more than one page if possible, but it should identify the document being sent, the project about which it was written, and any especially important points being made about the project which the reader does not already know.

The form of the letter of transmittal may vary with the audience. For instance, suppose you have been working with a partner on a proposal for the city of Grand Rapids to develop an industrial waste pretreatment program and a modification to a facilities plan your company submitted to Grand Rapids last spring. You and your partner have written the proposal together, and you have agreed to incorporate some last-minute details into a final draft and have that typed. When the final draft is ready, you might put the following short letter on the report and send it to your partner:

> Ann—
>
> Here is the final draft of our Grand Rapids proposal. How do you like it?
>
> Roger

Obviously, this short letter assumes that the reader knows which proposal you are talking about when you mention "our Grand Rapids proposal" and what points are especially important. Since the reader is a coauthor of the proposal, this is probably a safe assumption.

On the other hand, if you were sending this proposal to your supervisor for review and approval before you send it out to the city of Grand Rapids, your letter of transmittal might look like this:

> Dear Phil:
>
> You assigned Ann and me to prepare a proposal for the development of an industrial waste pretreatment program for Grand Rapids as an addendum to the facilities plan we submitted to Grand Rapids last spring. Here's a final draft of our proposal for both subjects. As you suggested earlier, this proposal is based on a cost-plus-fixed-fee-with-a-maximum contract. It does not include the city's involvement by force account in the work other than identifying equipment to be purchased by the city.
>
> Think it's ready to go?
>
> Roger

This letter is longer and provides more information than did the version to your partner. It assumes somewhat less knowledge on the part of the reader about the particular project and thus points out certain assumptions behind the proposal which would have been obvious to your partner. On the other hand, it assumes that the supervisor has a relatively high level of technical knowledge in waste pretreatment programs and facilities plans. It assumes, for instance, that the supervisor will see the appropriateness of basing a proposal of this sort on a cost-plus-fixed-fee-with-a-maximum contract and of not including the city's involvement in the work other than identifying the equipment it will purchase.

Finally, if you were sending the final proposal to the city of Grand Rapids, you would need to provide much more context and information than provided in the two previous letters. Those letters were written to people who had knowledge about the project, who understood the implications of important technical statements, and who were aware of the experience and credentials of your company in developing industrial waste programs for waste pretreatment and facilities modifications. Much of this knowledge needs to be spelled out in more detail for someone outside the company, someone less familiar with the company's credentials and the details of the development plans. Thus, a letter of transmittal accompanying the proposal might look like that presented in Figure 11-9. This letter, written to the city engineer but potentially read by anyone else to whom the document will be routed, identifies the proposal it accompanies and then spells out some of the information the city engineer and city planners might need to know about the proposers. It emphasizes the proposers' willingness to cooperate with the city and stresses their qualifications in the area of the proposed work. This is probably the most important thing to stress, since any community about to spend a large amount of money wants to believe that the money will be well spent, that the community's problem will be solved for the specified amount. This will most probably occur only if the company funded to do the work is knowledgeable, effective, and experienced.

Whatever the audience, the function of the transmittal letter is to orient the audience to the document. This can be done well only if the writer of the letter is sensitive to the needs, interests, and knowledge of the reader.

11.3

LETTER OF COMPLAINT

Sometimes in our dealings with individuals or businesses we find that we have not received what we were promised or what we expected. We may have received a part different from what we ordered, one that is defective, or one that does not perform as advertised. We may have been badly treated by someone, or we may be exasperated by a long-term dangerous situation. Whenever these sorts of things happen, whenever we feel cheated or victimized, especially in important ways, we want to complain. We want to get things "off our chests" and to correct unpleasant situations.

This desire to complain and correct gives rise to the letter of complaint. The goal of such a letter is usually to improve the situation about which one is complaining. To do this, the writer of a letter of complaint should politely but firmly

1   Identify the nature and seriousness of the problem

2   If possible, request or suggest a solution to the problem

This procedure is illustrated in Figure 11-10, a sample letter of complaint.

FIGURE 11-9 Sample letter of transmittal to an
audience outside the writer's organization.

June 15, 1981

Mr. John L. Hornbach, P.E.
City Engineer
300 Monroe Avenue, N.W.
Grand Rapids, MI 49503

RE: Request for Proposals

Dear Mr. Hornbach:

We are pleased to submit four copies of our proposal for the Development of
an Industrial Waste Pretreatment and Non-Domestic Users Program and the Preparation of a 201 Facilities Plan Addendum.

Our proposal is based on a cost plus fixed fee with a maximum type contract
and is our best estimate at this time to complete the work outlined in your request. We are prepared to make an oral presentation of its contents when requested, and, should we be selected as your engineer, we are prepared to discuss
our proposal in greater detail.

Our proposal does not include the city's involvement by force account in the
work other than identifying equipment that is to be purchased by the city. We believe that there will be considerable manpower saving on the city's part resulting
from our understanding of your needs; our team approach; our knowledge of your
wastewater collection system, your wastewater treatment facilities, and your
NPDES requirements; our ongoing working relationship with the Michigan DNR;
and the fact that we did prepare and had Michigan DNR approval of a Step 1
Grant Amendment for this work.

We are aware of the slow-down in the U.S. EPA grant program and the possibility of some of the work not qualifying for federal and state assistance. We are
prepared to discuss a scale-down of this work if so directed by the city.

We value our long association with the city of Grand Rapids and look forward to its continuance. Please contact us if you have any questions on our proposal.

Sincerely,
McNAMEE, PORTER AND SEELEY

By _Philip C. Youngs_____

Philip C. Youngs

FIGURE 11-10 Sample letter of complaint.

HARRISON
RADIATOR, INC. 1451 Murray Avenue
Pittsburgh, Pennsylvania
412-555-4356   November 12, 1981

Ms. Barbara C. Benson
Engineering Department
Advanced Technologies, Inc.
40 Technology Park
Milford, Massachusetts 01757

Dear Ms. Benson:

Two weeks ago we purchased a shroud #22508 from Advanced Technologies, Inc., for our Harrison Radiator Division's D-25-PS balancing machine (reference sales order #N5682). Your sales representative told us that your shroud would fit our Harrison balancing machine. Unfortunately, when we tried to install the shroud, it did not fit.

I am writing now to request instructions for adapting your shroud to our Harrison balancer. If you have such instructions, we would appreciate receiving them as soon as possible. We need to use our balancer within two weeks, if that is possible, and we cannot operate it until we have a suitable shroud.

If we cannot adapt your shroud to our Harrison balancer, we would like to return the shroud for full credit and find a suitable substitute from another supplier.

Thank you for your prompt attention to this matter.

Sincerely,

Frank Lee

Frank Lee

FL:aa

Notice that this letter of complaint asks the reader to provide time, effort, and perhaps money to address the complaint. Since this is typical of many letters of complaint, you can imagine that such letters should be polite but firm. They probably shouldn't threaten the reader, at least not on the first exchange of letters, since they shouldn't alienate the person who may be able to solve the problem. For the same reason, if the letter of complaint suggests a solution to some problem, the solution should be reasonable. From a purely practical standpoint, reasonable solutions have a better chance of being accepted than unreasonable ones. Further, they show the writer to be fair and helpful even in a difficult situation, and thus they may encourage people to do business with the writer in the future.

11.4
RESPONSE TO A LETTER OF COMPLAINT

Responding to a letter of complaint poses its own set of problems. The writer of such a response needs to look knowledgeable, helpful, concerned, and appropriately apologetic. As with other letters, the writer needs to

1   Identify the purpose of the letter, reminding the reader of the source of the complaint as well as of any suggestions the reader originally made for dealing with the complaint

2   Deal with the complaint, outlining whatever the writer can do to help the reader

3   Assure the reader of the writer's goodwill and attention to the problem

These characteristics are illustrated in Figure 11-11. Notice that the writer of this letter stresses the speed with which she is responding and then carefully provides the help the reader needs: she outlines the characteristics of the two shrouds and the few simple modifications which need to be made to the problematic shroud. Finally, the letter ends on a positive, helpful note. It does not grovel, but it stresses the company's commitment to satisfying its customers.

11.5
LETTER OF REQUEST

Writers sometimes need to request information or equipment. Since such requests usually cost the reader time or resources in responding, the writer should very clearly

1   Orient the reader to the topic of the letter

2   Indicate why the writer should be willing to respond

3   Indicate exactly what the writer is requesting the reader to provide

Orienting the reader to the topic was discussed earlier in this chapter and in Chapter 6 and so will not be treated again here. Indicating why the writer

FIGURE 11-11 Sample response to a letter of complaint.

ADVANCED TECHNOLOGIES, INC.
40 Technology Park, Milford,
Massachusetts 01757
617-555-4553   November 16, 1981

Mr. Frank E. Lee
Equipment Department
Harrison Radiator, Inc.
1451 Murray Avenue
Pittsburgh, PA 15217

Dear Mr. Lee:

On November 12, you wrote to me about a problem you were having fitting our shroud #22508 on your Harrison balancing machine. You asked for instructions, as quickly as possible, which would allow you to adapt our shroud to your balancer.

I have just received your letter and am sending by return mail the instructions you requested. I am also enclosing, for your reference, a blueprint of our shroud #22508.

Basically, our shroud and the original Harrison shroud have the same construction, shape, and functions. However, there are three major differences between our shroud and the Harrison model:

1) Our shroud has a new double-fan blower assembly.
2) Our shroud is 6-$\frac{1}{2}$ inches wider.
3) Our shroud has an additional 3-inch-diameter hole at the top of the shroud.

These differences are probably causing your problem with fit.

Fortunately, it will be very simple to accommodate our shroud to your Harrison balancer. The original Harrison shroud was designed to work only with a series of single-fan blower assemblies. Our shroud introduces longer double-fan blower assemblies (fans on both ends of the motor shaft), and this requires only two small modifications.

1) Our shroud must be lengthened from 10″ to 16-$\frac{1}{2}$″.

2) Our shroud will require a new cradle fixture to hold the blower assembly. You can insert a cradle fixture by drilling an additional 3″-

FIGURE 11-11 *(continued)*

Mr. Frank E. Lee   - 2 -   November 16, 1981

diameter hole in the top of the shroud, inserting the fixture, and allowing the clamping cylinder to protrude.

If you make these simple modifications, I am sure that your new shroud will fit perfectly.

If there is anything else we can do to help you in this matter, please feel free to call us. We firmly believe that customer satisfaction is our most important goal.

Sincerely yours,

Barbara C. Benson

Barbara C. Benson
Engineering Department

BCB/rs

Enclosure

should be willing to respond may be done along with the orientation or in addition to it. An example of a letter which provides a motive for responding along with an orientation appears in Figure 11-12; one which provides a motive for responding in addition to the orientation appears in Figure 11-10, a letter of complaint as well as a letter of request.

11.6

RESPONSE TO A LETTER OF REQUEST

The response to a letter of request, like any of the other letters already described, should (1) orient the reader, (2) identify the purpose of the letter, reminding the reader of his or her request and any conditions upon it, and (3) provide the information requested. If the reader requested a physical object that will be sent in a separate package or at a later date, the writer should include any available information on the package's status and projected date of arrival. If the reader requested information which is now available, the letter writer should try to include all of the relevant information the reader requested without overwhelming the reader with irrelevant or unnecessary detail. A sample response to a letter of request appears in Figure 11-13. Notice that in form this letter resembles a very short technical report, minus the *To, From, Subject, Date* format and headings to indicate the Foreword and Summary.

FIGURE 11-12 Sample letter of request.

ANDERSON COLLEGE

400 Haggerty Road
San Francisco, California 94105
(415) 555-6400

May 28, 1982

Mr. R. B. Sparks
Sales Representative
Computer Connection
384 Grand Drive
Harrington, Georgia 30018

Dear Mr. Sparks:

The Computing Science Club of St. Mary's School will be sponsoring a new course
in the PASCAL programming language for the Fall Term of 1982. Due to the over-
crowded conditions of our present computer system, we are going to purchase several microcomputers rather than add PASCAL software to our present system.

We are considering purchasing ten of your Apple II Plus systems for our PASCAL
course. However, before we can make a final decision, we must review the following:

1. Information on the cost of the Apple II Plus system and its peripheral
   equipment.

2. Information on the PASCAL software available with the system.

3. Specifications for the computer system and its microprocessor.

Could you please send us whatever information you have which addresses these
questions?

Thank you for your attention. I am looking forward to hearing from you in the near
future.

Very truly yours,

C. Beth Norther

C. Beth Norther
Instructor, Data Processing

CBN/vr

FIGURE 11-13 Sample response to a letter of request.

Pine Dunes Country Club
Orchard Dunes, New Jersey 07081

August 16, 1980

Mr. Thomas F. Wilson
Club Manager
Pine Dunes Country Club
Orchard Dunes, New Jersey 07081

Dear Mr. Wilson:

After receiving numerous complaints from some of the club members about the unsafe conditions of the wooden boat docks on the club's west shoreline, you asked me to investigate the condition of the docks. In particular, you asked me to find out what parts were broken or missing from the dock (if any) and what it would cost to make the necessary repairs. I am now writing to report my findings and actions.

I inspected the boat docks on August 12. In general, they seemed to be in good condition, but they did have a few minor trouble spots. Six crossarm brackets were missing bolts, and three stanchion braces had pulled away from the dock and required some minor welding.

After receiving authorization from Virginia Lette, the Beach Director, I bought the required bolts for $4.75, replaced the missing bolts in the crossarm brackets, and welded the three stanchion braces. The docks are now in good working order.

Details on Repairs

On August 13, the docks which were causing the problem were removed from the water. The missing bolts from the crossarm brackets were replaced with new bolts (stock #83443.7AG) that were bought from Commerce Dock Supply for $4.57. They were tightened to the recommended torque of 40 ft-lb. as stated in the installation instructions.

The three stanchion braces that needed welding were brought over to the Maintenance Department, where they were rewelded by Roger Sands. No problems were encountered during the welding, and the stanchion braces seemed to be completely repaired. There was no cost for the welding.

On August 14, the repaired dock sections were reinstalled and the dock system is now in good working order.

Sincerely,

Steven M. Berggruen

Steven M. Berggruen
Assistant Beach Director

EXERCISE 11-1

A   Write a letter of request for information that can be provided by one
     of your classmates, friends, or colleagues. Then ask that classmate,
     friend, or colleague to write a letter responding to your letter of request.

B   Write a letter of request or complaint to someone you do not know.
     Try to pick a situation about which you really need information or about
     which you really want to complain. Send your letter and see if it is
     effective.

C   Write a letter of transmittal for a report you have produced, found in
     the library, or encountered on the job. If you are a student, you may
     choose a report you have produced in one of your technical classes or
     in your technical writing class. Define the audience and situation for
     your letter, and write at least one version that might accompany your
     report to total strangers.

11.7
LETTER OF APPLICATION FOR EMPLOYMENT
OR ADMISSION AND RESUME

If you need to write a letter of application for employment or admission and
a resume, it is probably important that you do them well. If they are impressive enough, you may get an interview with a prospective employer or
admission to a desired graduate school or other training program. If they are
poorly done, you will probably be screened out of the pool of applicants and
into the ranks of those still looking. (Most companies and programs have
more applicants than positions and thus are looking for ways to eliminate
applicants; a bad letter or resume makes elimination easy.)

What Makes a Good Applicant?

If you were hiring an employee, you would probably want someone who
filled a need in the company and who was (not necessarily in this order) well
trained, technically competent, smart, hardworking, reliable, honest, loyal,
well organized, disciplined, tactful, helpful, resourceful, and able to get along
with others, to take initiative, and to communicate well orally and in writing.
Therefore, when you are applying for a position your letter of application
and resume need to say as many of these things as possible about you as
an applicant.

     This can be done by following a few general principles. First, you should
understand that applications for jobs or training programs are usually read
by at least several readers coming from different perspectives and having
different "pet peeves," assumptions, and preferences. For instance, many
readers believe that competent, educated, serious writers should not make
grammatical errors; most readers are offended by form letters; many readers

prefer a conservative look in a letter or applicant, while others prefer a "different" look. Thus, you should design your letter with multiple audiences in mind. Second, you should stress what you can do for the company or program, not what it can do for you. This means showing that you are aware of *its* needs, that you meet these needs, and that you want to be a part of *this* company or program, not just any one. Third, you should address the person having the authority to arrange an interview or to admit you into a program; don't waste your letter on someone who can't help you. Fourth, you should stress your accomplishments and responsibilities, since these will show off many of the desirable qualities listed above. Fifth, you should set up your letter and resume in expected formats, but still highlight your best qualifications.

Applying the first principle is easy: to be successful, you must write a letter and resume which are short (no more than one page each, if possible) and which please as many and offend as few readers as possible. We suggest that you look as competent, careful, and "proper" as possible: have each letter individually typed by a professional secretary with a professional-quality typewriter on good white bond paper. This may seem expensive, but it really isn't. You're preparing a $20,000 to $50,000 letter (your salary for a year or so if you get the job or the training program that prepares you for the job). Applying the next three principles is difficult unless you know what you want to do, what you can offer, what the company or program does, and what it offers. Thus, writing a good letter involves doing some serious thinking and research.

To find out what you want to do, you need to consider your short-term and long-term goals, your previous training and interests, and your financial, geographical, or personal constraints. Do you eventually want to be a technical person or a manager? (Managers are often technical people who, because of the demands of their jobs, have developed managerial skills but grown away from their technical skills.) What type of training and experience do you need to reach your goal? Is there a conflict between your short-term and long-term goals? If so, what is the best way to resolve the conflict? What are the constraints on your plans and goals? Be sure to consider financial and geographical constraints, as well as such personal constraints as limits on your time, attention, concentration, and energy. You may not be able to answer all of these questions without consulting other people, but if you need some outside advice, go get it. You are making important decisions and should do so with the best knowledge you can get.

To assess your previous training and interests, you might refer to the matrix in Table 11-1. Be thorough and immodest at this stage; you are generating information for yourself, not for an offendable outsider. And don't forget your special talents and accomplishments; if you speak a language besides English or play the violin, write that down. Most Americans do not have such skills, and what is unexceptional to you may be very exceptional

TABLE 11-1   A MATRIX TO HELP YOU ASSESS YOURSELF AND YOUR INTERESTS

RESULTS / ACTIVITIES	PRODUCTS OR ACCOMPLISHMENTS	SKILLS	KNOWLEDGE	REWARDS OR RECOGNITION
Working				
Studying				
Playing				
Living				

and interesting to an employer. Note that not everything you write down in the matrix will appear in your letter and resume. You are writing to ensure that you don't leave out something important; you still have to select the information you will finally put in the letter. Be bold and thorough.

Designing a Good Letter of Application

To write a successful application letter, you need to know the conventions for such things. This is easy. The letter is ideally only one page long and has four main sections:

1   The heading, which includes the writer's address, the reader's address, and the date

2   A first paragraph, which introduces you (if possible by citing an impressive recommender) and then establishes the company's need and your ability to fill it (or your reason for writing and what you are requesting of the reader)

3   A second paragraph, or series of paragraphs, which establish your *most relevant* experience and qualifications and which stress your accomplishments, responsibilities, and work quality, not just your activities. This section should include some proof that you did a good job, by citing either a reference ("X will provide a letter of reference") or some objective measure of a job's quality. (If you have designed an oven as a senior engineering project, distinguish it from an oven an English major would design: "My oven maintained a constant temperature within a 1 percent error.") This section should be very selective and focus on only your *most relevant* qualifications.

4   A closing paragraph, which gives any other pertinent data, asks for an interview, provides your telephone number and the hours you may be reached, and thanks the readers for their attention.

Two sample letters are shown in Figures 11-14 and 11-15. The letter in Figure 11-14 was written by an engineering senior and has been nicely

FIGURE 11-14 Letter of application for employment and resume.

6601 Bursley-Lewis
1931 Duffield Street
Ann Arbor, MI 48109

May 16, 1980

Mr. Lewis R. Tassellski
Technical Group Coordinator
Pacific Power & Light Company
1634 S.W. Columbia
Portland, OR 97201

Dear Mr. Tassellski:

At the suggestion of Mr. S. Feldman, Instrumentation Supervisor, Fermi II Project Pioneer Corporation, I am writing to apply for a position in the Instrumentation and Control Group at the Astoria Point Power Project. My work experience in both instrumentation and start-up groups should be of value to the project, as should my engineering experience at the University of Michigan in automatic control and power plant systems. My credentials include:

Employment as an engineering aide with the Detroit Edison Company at the Enrico Fermi II Nuclear Power Project. While there, I independently researched in-service inspection problems of certain ASME Class II welds, reporting to the Production Department Lead Superintendent. I also worked on the initial rough draft of the In-Service Lubrication Manual for the plant. (Ref. R. J. Szcotnicki, Technical Group Supervisor)

Employment as a student engineer with Daniels International Corporation at the Enrico Fermi II Nuclear Power Project. I researched material for input into the Component Control System, checking data for authenticity and inputting the data to update the system. (Ref. T. R. Bietsch, I. & C. Lead Superintendent)

I expect to receive my bachelor of science in mechanical engineering from the University of Michigan in April, 1981, and my future goals include a desire to advance in the area of technical management. I am very interested in discussing my credentials with you at your convenience. If this is possible, I can be reached at (313) 555-1857 during the afternoon and evening.

Thank you for your consideration.

Sincerely,

Steven J. Kaercher

Steven J. Kaercher

PERSONAL DATA RESUME

Steven J. Kaercher
6601 Bursley-Lewis
Ann Arbor, Michigan 48109
(313) 555-1857

Career Plans and Objectives

I am seeking employment in the consumer power industry, especially in the area of instrumentation and control of power systems. I feel that my experience and education will enable me to contribute to such projects now; later I plan to advance into management.

Qualifications and Experience

Summer 1979 — Engineering Aide (student) with Detroit Edison Company at Enrico Fermi II Nuclear Power Station. While working for the production department, I 1) researched potential problems in the In-Service Weld Inspection program and 2) worked on the Plant Lubrication Manual, documenting equipment and specified procedures.

Summer 1978 — Engineering Aide (student) with Daniels International Corporation at Enrico Fermi II Nuclear Power Station. While there, I worked on a Component Control System for the Instrumentation and Control Group. This involved monitoring all sources of information and doing in-plant inspections to keep the system updated.

College Work — Employed by University of Michigan, Ann Arbor, during all terms in school. I worked at the School of Public Health as a coder one year and at the Bursley Hall Cafeteria for the last two years. This employment paid for twenty-five percent of my college costs.

Academic Experience

Education — Bachelor's Degree in Mechanical Engineering, College of Engineering, University of Michigan, expected in April, 1981. Emphasis on instrumentation and control of power system equipment.

Societies — American Society of Mechanical Engineers, student member.

References

References and transcript will be supplied upon request.

FIGURE 11-15 Letter of application for admission and resume.

717 Dartmoor
Ann Arbor, MI 48103
(313) 555-9872

February 15, 1978

Dean Arjay Miller
Graduate School of Business
Canston University
Bolton, Connecticut 06431

Dear Dean Miller:

At the suggestion of Mr. Richard E. Spaid, I am writing to request your support for my application to the Canston Graduate School of Business. My qualifications include a bachelor of science in engineering from the University of Michigan and the following abilities:

Ability to Solve Problems: I have concentrated in mechanics, a strongly problem-oriented field, while maintaining a diverse background in engineering and science. The human aspects of engineering have received the majority of my attention; I have independently researched current advances in the fields of stress-related anxiety, materials for implantation in humans, and nerve conduction velocities in the hand. (Ref. - Prof. D. A. Sonstegard)

Ability to Lead and Motivate Others: I had 24-hour direct counseling responsibility for 20 teenage boys at the National Music Camp at Interlochen (summer, 1976). I used my motivational and leadership skills to keep the boys interested in their musical and physical activities as well as to deal with homesickness, personality conflicts, group problems, and personal problems. Additionally, I supervised many of the waterfront, recreational, and evening activities. (Ref. - L. E. Dittmar)

Ability to Achieve Desired Goals: I have been able to simultaneously attain my goals in swimming and school over the past twelve years through self-motivation, my competitive spirit, and my organizational abilities. I stand in the top fifth of my class scholastically, and I have won three varsity letters as a member of the University of Michigan swim team. (Ref. - A. P. Stager)

Ability to Maintain Diverse Interests: I have kept my interests diversified through activities such as being Rush Chairman of my fraternity and joining Tau Beta Pi, National Engineering Honorary Society. I have been able to share my interests with others as a counselor, scholastic tutor, swim instructor, and music student.

It is essential that the manager master these abilities if he is to contribute to the business world. My perspective of these abilities differs from that of the typical applicant with a B.B.A., and I believe it is a perspective that would be valuable to the M.B.A. program at Canston. I would be happy to provide any further information you might wish. Thank you for your consideration.

Sincerely,

Brian Wylie

Brian D. Wylie

Brian D. Wylie
717 Dartmoor
Ann Arbor, MI 48103
(313) 555-9872

Career Goals

My career goal is to become an effective and conscientious corporate general manager. My strong engineering background, a projected M.B.A., and my diverse activities will have formed a solid foundation for this goal.

Education and Work Experience

Current	Bachelor's degree in engineering-science from the University of Michigan (expected completion: May 1978). Emphasis on biomechanics and industrial human performance. Independent studies on stress-related anxiety, materials used for human implantation, and nerve conduction velocities in the hand.
Summer 1977	Assistant Pool Manager, Travis Pointe Country Club, Saline, Michigan. Responsibilities included maintaining effective relations with the membership, organizing the physical plant, and teaching group and private swim lessons.
Summer 1976	Cabin Counselor, National Music Camp, Interlochen, Michigan. 24-hour counseling responsibility for 20 teenage boys with homesickness, personality conflicts, group problems, and personal problems. This job enabled me to use my motivational and leadership skills to keep the boys interested in their musical and physical activities. I also had responsibility for the waterfront, the recreation program, and special events.

Activities and Honors

Activities	Michigan swim team, 3 varsity letters (1974-78) Michigan varsity club water polo (1974-77) Psi Upsilon fraternity (1976-78), Rush Chairman (1978)
Honors	Tau Beta Pi National Engineering Honorary (1977-78) Dean's List (3 semesters); Regents' scholar (1974) Academic scholarship (1974-75, 1976-78) Athletic scholarship (1975-78)

References

Richard E. Spaid, President, Hillman Manufacturing Company
David M. Sonstegard, Ph.D., Assoc. Prof. of Applied Mech., U.M.
Larry E. Dittmar, All-State Boy's Director, National Music Camp
Augustus P. Stager, U.M. Varsity Men's Swim Coach

presented at many levels. It is addressed to the person in charge of screening applicants, written specifically to the company, and individually typed on a professional-quality typewriter. This gets the letter to the right reader(s) and shows that the writer cares enough about the job to prepare the letter carefully. It first cites a reference known and respected by the readers and then defines the job being sought and the writer's general qualifications for the job. The two indented paragraphs in the middle of the letter give the writer's main qualifications, focusing on his accomplishments and on proof that he did a good job. This section is relatively straightforward since the writer comes from a good school and has some obviously good qualifications for the job. Finally, it provides useful information if the employer wants an interview and then ends politely.

The letter in Figure 11-15 presents a more difficult problem. It is a letter of application to an excellent graduate program, written by a good student but one with apparently marginal qualifications for that program: swimming, work as a camp counselor, and an engineering degree as preparation for one of the most competitive business programs in the country. Since the letter was relatively successful in a difficult situation, it is useful to see why.

First, the student has a good reference in the first sentence. (If no one with an impressive title or reputation has offered to recommend you, then politely ask someone to do so; the worst you can get is "no.") Second, the student has assessed his qualifications and activities and reinterpreted them from the point of view of the reader. (What are the characteristics an admissions officer would look for in applicants to business school? In future managers? What kinds of things have I done to demonstrate these characteristics?) Obviously, a writer is limited by the facts of a given case, but she or he can organize those facts in more or less effective ways. This student has chosen a more effective way and highlighted it. He has focused on his qualifications by using indented paragraphs which draw the reader's eye. Further, he has used underlining and parallel construction (the repetition of *Ability to*) to highlight the qualities he wants the reader to note most carefully: Ability to Solve Problems, Ability to Lead and Motivate Others, Ability to Achieve Desired Goals, and Ability to Maintain Diverse Interests.

When you actually write such a letter, try to avoid having a lot of empty space at the bottom of the page. You don't want to suggest that your life has been a blank. Also, try to produce a balanced and visually attractive format. If you have only a little information, spread it out and use wide margins. Finally, make your letter and resume as concise as possible, a model of efficiency.

Designing a Good Resume

The resume is a summary of *all* your activities and experience. It will repeat some (or all) of the information in the letter of application, and it may be more inclusive than the letter. However, since the letter and resume may get separated, the resume should be as strong and impressive as the letter and arranged to highlight your most important qualifications.

What should the resume look like? First, it ought to be easy to read: not too long, not too much material crammed on a page, easily visible headings, and, if you prefer, short phrases rather than full sentences. Second, it ought to give your vital statistics: your name, address, and telephone number and, if necessary, a permanent address and telephone number if you intend to move soon. (Resumes used to include such information as height, weight, sex, and marital and military status, but this information is often irrelevant and prohibited by law, and so it is not included here. Obviously, you can include it if you particularly want to.) Third, the resume ought to suggest where you are headed professionally (perhaps in a section called "Career Goals") and where you have been (in a section called "Qualifications and Experience," including education and work experience, or two sections entitled "Work Experience" and "Education").

The ordering of information is important in the Qualifications and Experience section. Generally, you want to put your most relevant and impressive qualifications first. If you have a lot of relevant work experience, you should list that before your educational experience. If you have only a little work experience, you will have to emphasize your education and its special features. What makes you different from any other student with your degree? Have you had any research or design courses which simulate a job situation? Do you have a number of honors and extracurricular activities? You might want to highlight them in a separate section entitled "Honors and Activities," since such features show that you are organized enough to handle several activities at one time.

Finally, you need a section entitled "References," which either states that references are available on request or lists your references' names (and addresses) if these are particularly impressive or if you need to use up some extra space. (Before you list someone as a reference, ask that person if she or he is willing to write letters of recommendation for you. It is impolite and potentially disastrous to list someone who is annoyed at being taken for granted or unwilling to write a good letter.)

One note about the placement of information on the page. As with your letter of application, try to avoid a lot of empty space at the bottom of the page. Again, use wide margins if necessary, spread the information out evenly on the page, and give the full names and addresses of your references to use up more space.

Sample resumes appear in Figures 11-14 and 11-15, and other resumes for discussion are presented in the following pages as Figures 11-16 to 11-19 for Exercise 11-2 (page 274).

FIGURE 11-16   Letter of application for admission and resume:
example 1 for analysis.

1320 South University
New Brunswick, New Jersey 07032
January 23, 1981

Prof. Walter H. Arnaz
Assistant Professor of Electrical
and Biological Engineering
University of Illinois, Urbana-Champaign
Urbana, Illinois 60537

Dear Professor Arnaz:

At the suggestion of Professor Lin of Rutgers University, I am writing to express my desire to become a part of the Bio-Engineering Department at the University of Illinois in the capacity of a teaching or research assistant. Because of my financial situation, it will be necessary for me to obtain an assistantship while working for my master's and possibly my doctorate degree. Considering the current research and educational activities at your University, including biological computation and instrumentation, I am sure that my educational background and activities would be of great value to the Bio-Engineering Department. My background includes:

Six months of employment in the Kubik Hydradrive Company of Troy, New Jersey. In the Design Department of this company my duties included the design of control mechanisms for hydraulic machinery. I was also assigned individual design projects in which I worked in the field without direct supervision.

My studies have given me an extensive background in instrumentation, biology, and chemistry. In May, I will receive a combined degree in Electrical Engineering-Instrumentation and Engineering Science-Biology.

Active participation in extracurricular activities, including serving as Vice President, Athletic Director, and Educational Programming Chairman for a University dormitory. I am currently an instructor for the New Jersey Heart Association in Cardio-Pulmonary Resuscitation. I am also a member of the Engineering Honor Society, Tau Beta Pi.

Ranking in the upper quarter of my class at Rutgers University.

I believe my background and abilities will make me an asset to the Bio-Engineering Department. I would appreciate your contacting me as to the projected needs of your department as they apply to my experience and background. I can be contacted at the above address or most evenings at the enclosed number.

Sincerely,

Patricia J. Lingri

Patricia J. Lingri

<u>PERSONAL DATA RESUME</u>

Patricia Jensen Lingri
1320 South University #4
New Brunswick, New Jersey 07032
(201) 555-9064

Career Plans and Objectives:

I am seeking employment as a University Research Assistant or closely related position. My particular area of interest is in the application of the rapidly increasing technology in instrumentation to the area of biological
measurements and diagnostics. My experience in the areas of teaching and
leadership will enable me to make significant contributions to the reputation
and quality of the Bio-Engineering Department.

Academic Summary:

September 1977 to present	Bachelor's degree in Engineering Science-Bio-Engineering (Rutgers University; expected completion: December 1981). In this engineering program I obtained an exten-sive background in most areas of engineering, chemistry, and biology.
September 1977 to present	Bachelor's degree in Electrical Engineering-Instrumenta-tion (Rutgers University; expected completion: May 1982).

Qualifications and Experience:

May 1980 to January 1981	Employed by the Kubik Hydradrive Company of Troy, New Jersey, in the Design Department. My duties included the design of control mechanisms for hydraulic machinery. I was also assigned individual design projects in which I worked in the field without direct supervision.
September 1979 to present	Instructor for the New Jersey Heart Association in Car-dio-Pulmonary Resuscitation. In this capacity, through the New Jersey Heart Association, I teach groups of individ-uals who request instruction in this life-saving technique.

Organizations

While attending Rutgers University	Vanner Hall House Council, Vice President 1980 Vanner Hall Educational Programming Committee, Chair-person 1980 Vanner Hall Athletic Director, 1979-1980 Tau Beta Pi, Rutgers University Chapter

References and transcript will be supplied upon request.

FIGURE 11-17 Letter of application for employment and resume: example 2 for analysis.

921 Second Street
Charleston Heights, South Carolina 29405
September 21, 1981

Mr. Donald Abelson
Senior Engineer-in-Charge: Works Engineering Activity
Fisher Body Central Engineering
General Motors Technical Center
Warren, Michigan 48019

Dear Mr. Abelson,

Clara Kinnie, who is employed in the Mechanical Engineering section of your department, has informed me that you require an engineering student to fill the temporary vacancies which will occur when vacations begin next summer. I will be graduating from the University of South Carolina this April with a bachelor of science degree in engineering, and I have prior training which would assist my performance in this position.

Three years ago, I attended General Motors Institute under the sponsorship of Fisher Body. While there I obtained work experience in many different central engineering activities which included:

Works Engineering Activity--Plant Layout and Mechanical Product Testing
Laboratories--Elastomers
Manufacturing and Development--Resistance Welding
Body Engineering Activity--1980 Detroit Pilot
Production Engineering Activity--Fixture Design
Product Research and Safety--Seats and Headliners
Seat and Wagon Load Floor--B-C-E Body Rear Seats

While assigned to the Works Engineering Activity I worked on the following projects:

Plant Layout--Worked with Hugh Reins at Fisher Body Fleetwood during the rework for the 1980 model change-over. Also interacted with the Architecture group to produce a collection of plant elevation cross sections to be utilized in layout work.

Mechanical--Aided Clara Kinnie in the preparation of a study concerning effluent guidelines on chrome plating operations, requested by the EPA. Drew the new drainage system for the plant in Syracuse, N.Y.

My schedule is very flexible, and I would be glad to come to Warren for an interview, at your convenience. I will be at the number listed in my resume on Monday and Friday afternoons and also most nights if you require any additional information.

Sincerely,

Alexander Herrod

Alexander Herrod

PERSONAL DATA RESUME

Alexander Herrod
921 Second Street
Charleston Heights, South Carolina 29405
(805) 555-5879

Career Plans and Objectives:

I am seeking summer employment in a company that requires an engineer with a diverse background. My experience and training will enable me to make contributions in many different work situations.

Qualifications and Experience:

August 1978 to 1980 — Cooperative student, General Motors Institute, employed by Fisher Body Central Engineering, Warren, Michigan. While there I was rotated for six-week periods throughout the many different departments and acquired experience in Production Engineering, Designing, Layout, Product Testing, and Product Development. I saved the corporation $15,000 by locating a design error in a plant drainage system construction project. I also designed a screwless installation sunshade support which would save an estimated $1 per car on installation costs. This support was under patent investigation at the time I left the company.

Academic Summary:

Education — College of Engineering, University of South Carolina (enrolled September 1980). BS Engineering degree expected April 1982, in interdisciplinary program, pre-biomedical-engineering option. This option is geared toward the obtaining of a master's degree and involves engineering applications in the medical field.

Engineering program, General Motors Institute, electrical engineering option (enrolled August 1978-August 1980). Left program to enter University of South Carolina.

Am applying for admission to Samuels School of Graduate Studies, Yale University, where I plan to obtain a master's degree in biomedical engineering.

Societies Honors — Considered for Management Honor Society (GMI). Withdrew name from consideration because of transfer to USC. Dean's List (GMI).

References:

Will be supplied upon request.

309 Packard Road
Embry, Ohio 44107
September 22, 1981

Dr. Rodger L. Krakau
Biomedical Engineering Research
Johns Hopkins University
Applied Physics Laboratory
8621 Georgia Avenue
Silver Spring, Maryland 20910

Dear Dr. Krakau:

Pursuant to the advice of Dr. Roberta Shofer, Department of Neurology, Einstein Medical College, New York, I am writing to you concerning potential employment. I will be graduating in May, 1982 from the University of Ohio, College of Engineering, with a degree in electrical engineering and with an emphasis in biological sciences and bio-engineering. I believe that I could be a valuable addition to your research staff. My qualifications and background include the following.

> For five summers I have been employed as a refrigeration and air conditioning mechanic and technician. I hold an unlimited Refrigeration Journeyman's License. While not directly related to my field of study and interest, I believe this employment (and the invitations to return each year) are indicative of my integrity and initiative in a commercial situation and my ability to solve technical problems through a logical diagnostic approach. I have letters of recommendation from my two employers, which I would be happy to forward should you so wish.
>
> During the course of my education at the University of Ohio, I have studied in both the College of Engineering (Department of Electrical and Computer Engineering) and the College of Literature, Science, and the Arts (Department of Zoology). I believe that my background in zoology, supplemented with course work in bio-engineering, gives me a solid and adequate basis for work in biomedical engineering.
>
> Outside of the classroom, I have been active on many fronts. I am currently serving as treasurer of the U of O chapter of IEEE. I served on the executive board of the local Democratic Party and did research on issues and policy for an unsuccessful bid for a Congressional seat (Ohio 2nd district) in 1974. I have also performed in Washington, D.C., for the President with an Ohio-sponsored vocal jazz ensemble.

I would like very much to discuss your research projects and my potential employment and am confident that a convenient date and time can be arranged. At present, I am planning to be in the Maryland area for the Thanksgiving and Christmas recesses, or I could arrange to come earlier should it prove necessary. As I am difficult to reach, I will contact your office shortly to arrange a suitable and convenient meeting.

I thank you for your time and consideration, and am looking forward to meeting you.

Sincerely yours,

Lisa C. Caron

Lisa C. Caron

PERSONAL DATA RESUME

Lisa C. Caron
309 Packard Road
Embry, Ohio 44107
(614) 555-8932

Career Plans and Objectives:

I am seeking employment in biomedical engineering research and development. My education and background in electrical engineering, biology, and biophysics will enable me to contribute significantly to the growth of knowledge in this vast and growing field and thereby eventually help to improve the quality and technology of medical care.

Work Experience:

Summer 1977, 1978, 1979, 1980
Bishop Equipment Company, Fairfax, Virginia

Summer 1981
The Alma Corporation, College Park, Maryland

I worked for these five summers, to finance my education, as an air-conditioning mechanic, beginning as an apprentice and working my way up to master mechanic. I worked in all phases of residential and commercial air conditioning.

Education:

University of Ohio
College of Literature, Science, and the Arts
Department of Zoology 1977-1979

University of Ohio
College of Engineering
Department of Electrical and Computer Engineering 1979-1982
B.S.-E.E. May, 1982

Language fluency: French

Societies and Honors:

IEEE 1980-1982; presently serving as treasurer, U of O chapter
Invitation to join Eta Kappa Nu

Professional License:

Unlimited Refrigeration Journeyman's License

References will be supplied upon request.

FIGURE 11-19 Letter of application for employment and resume:
example 4 for analysis.

---

806 Packard
Ann Arbor, Michigan 48104
September 24, 1976

Mr. Edwin L. Russell
Manager, Planning Department
FMC Corporation
Prudential Plaza
Chicago, Illinois 60701

Dear Mr. Russell:

My job placement investigation of careers combining engineering and finance has led me to your Business Planning Department of FMC Corporation. I
believe that I would be of value to your department and firm, as my education,
work experience, and personal objectives closely match your preferred qualifications as outlined in your recent Business Planner job bulletin. My qualifications
include:

Over three years of work experience in the research and development divisions of General Motors Corporation. Functioning as a development engineer, I redesigned the GM rotary engine coolant flow distribution, directed
development tests on rotary engines, and worked with instrumentation and
computers on engines and suspensions.

My overall experience in many engineering departments has made me familiar with the functions of an R&D division within a major corporation.

Controllership of a college student body and a social fraternity, giving me
three years of practical experience in designing budgets, formulating financial policies, and resolving financial problems.

Graduate studies in the area of financial management, including managerial
accounting, management planning, a seminar in corporate financial policy,
and hopefully a seminar in management problems in large corporations.

Undergraduate studies in the area of automotive mechanical engineering.

I will graduate with an MBA from the University of Michigan in the beginning of May, 1977, and be available for employment by the end of the month. I
would like to meet with you and your department on the first Monday or Friday
convenient for you. If you have any questions, please call me any evening at the
number listed in my resume.

Sincerely,

Lincoln H. McGhee

Lincoln H. McGhee

LINCOLN H. McGHEE
806 Packard
Ann Arbor, Michigan 48104
(313) 555-6279

OBJECTIVE

A position in corporate financial management requiring the use of my engineering experience and analytical abilities in the formulation and solution of problems in the areas of financial, product, and corporate planning. Long-term goal is corporate management.

EDUCATION
1976 - 77

Graduate School of Business Administration   The University of Michigan
MBA emphasizing financial management. Primary supportive coursework: managerial accounting, management planning and control systems, seminar in corporate financial policy.

1970 - 75

General Motors Institute (GMI)   Flint, Michigan
B.S. in Mechanical Engineering. Ranked in top five percent of class.

Offices: Comptroller, GMI student body (annual budget $125,000).
Comptroller, Phi Delta Theta Fraternity (annual budget $90,000).
Treasurer, national convention and local chapter of Tau Beta Pi.
President and Founder, Karate Club.

Honors: Sobey Scholar (school's highest honor). GMI Alumni travelship.
Robot Society (leadership). Tau Beta Pi (engineering honor society).
Listed in Who's Who in American Colleges.

EMPLOYMENT
1970 - 75

General Motors Corporation   Warren, Michigan
Co-operative student engineer accumulating 3.5 years' work experience at GM Technical Center, Milford Proving Grounds, and Chevrolet Flint Assembly.
Responsible for:

Redesign of the GM rotary engine coolant flow distribution.
Computer simulation of a high-performance rotary engine.
Test and evaluation of anti-drive on Monte Carlo front suspensions.

Exposed to testing, development, computer simulation and design of automotive engines, transmissions, and suspensions. Worked at European engineering facilities, summer 1974.

1968 - 80
(summers)

Manual laborer for a construction company and assembler for a mobile home corporation.

FIGURE 11-19   *(continued)*

PERSONAL BACKGROUND	Born in Tokyo, Japan; raised in several locations in United States and Europe.
	Graduated top five percent in high school; member National Honor Society.
	Volunteer fire fighter and ambulance assistant for a hometown fire company (1968-70).
	Self-supporting since 1970.
	Interests: cars, golf, karate, skiing, camping, and tennis.
REFERENCES	Will be supplied upon request.

October 1976

EXERCISE 11-2   Evaluate and revise where necessary the letter and re-sume combinations of Figures 11-16 through 11-19. Note the order of categories and information in both the letters and the resumes. Do the most important qualifications come first? Are they appropriately highlighted? In each pair, are the letter and resume of equal strength? (If they get separated, you do not want someone important to see only the weaker unit.) Are they concise and smooth enough?

12

THE JOURNAL ARTICLE

A good journal article is like other good technical communications: it addresses a problem or issue important to its audience and presents its argument clearly and coherently. In this sense, it is not exempt from the principles of communication presented in this text. However, it differs somewhat from the other types of technical communication because it is written for a very different audience. The other types are written for a wide spectrum of readers with varied backgrounds, attention, interest, and time—audiences including nontechnical people, technical nonspecialists, and specialists. Thus, the other types have to provide features appropriate to this diversity of audience. In contrast, the journal article is written for a narrow audience: specialists in a field, who share assumptions, knowledge, and backgrounds and who have the need and interest to read carefully. This means that the journal article can eliminate some of the features provided primarily for nonexperts, in particular, the Foreword and Summary. However, the article should maintain the features providing clarity and ease of reading for experts, in particular, the movement from generalizations and claims to data and support. Unfortunately, this movement is sometimes reduced by writers relying perhaps too strongly on the expert reader's interest and determination to understand.

Articles in different fields and journals vary somewhat in style. If you are writing an article for a particular journal, you should read that journal's instructions on organization and format and then look at sample articles in several issues to become familiar with the journal's style. (Here, style includes such things as format for footnoting, ways of setting up headings and visual aids, length of paragraphs, and other special features.)

Although they may differ in style, most articles share a remarkably uniform purpose and structure. The purpose of an article in any field is to advance an argument of fact or policy: (1) an argument of fact that the results reported are valid, that previously reported results are supported (or not), that a given theory is supported (or not), that other observations are necessary to resolve some debate in the field; or (2) an argument of policy that previous results should be questioned or reinterpreted, that a given theory should be abandoned, recast, or extended. These arguments are made in a structure that is quite consistent over many fields and includes the following sections:

1   *Introduction,* which defines the problem and describes its importance

2   *Materials and Method,* which describes how the research arrived at the results

3   *Results,* which describes what was discovered

4   *Discussion,* which analyzes the importance of the results and their implication(s)

Let us examine these sections in more detail.

12.1

INTRODUCTION

The Introduction is an especially tricky part of the article since it must present a great deal of information and orientation in a short space. Robert Day, an experienced journal editor and author of a helpful and entertaining book on writing articles, has provided four rules for a good Introduction:

> (i) It should present first, with all possible clarity, the nature and scope of the problem investigated. (ii) To orient the reader, the pertinent literature should be reviewed. (iii) The method of the investigation should be stated. If deemed necessary, the reasons for the choice of a particular method should be stated. (iv) The principal results of the investigation should be stated. Do not keep the reader in suspense; let him follow the development of the evidence. An O. Henry surprise ending might make good literature, but it hardly fits the mold that we like to call the scientific method.[1]

Articles aimed at specialists may begin with an Introduction based on either a long-form or a short-form problem statement. (Since specialists share assumptions, methods, and knowledge of their field, a writer can often assume this shared experience and not state it.) In contrast to other types of Introductions, article Introductions aimed at specialists include technical details and a short review of previous work on the topic. These establish a context against which the author contrasts an observation or an inadequacy in theory or method. This is illustrated in the following example.

**THE THERMAL CONDUCTIVITY AND SPECIFIC HEAT
OF EPOXY-RESIN FROM 0.1-8.0 K**

A term	The thermal properties of glassy materials at low temperatures are still not completely understood. The
Amplification of or proof for A term	thermal conductivity has a plateau which is usually in the range 5 to 10 K, and below this temperature it has a temperature dependence which varies approximately as T^2. The specific heat below 4 K is much larger than that which would be expected from the Debye theory,

(Unstated however)	and it often has an additional term which is proportional to T. Some progress has been made
B term	towards understanding the thermal behaviour by assuming that there is a cutoff in the phonon spectrum at high frequencies (Zaitlin and Anderson 1975 a,b) and that there is an additional system of low-lying two-level states (Anderson 1975, Phillips 1972).
Missing information	Nevertheless, more experimental data are required, and in particular it would seem desirable to make experiments on glassy samples whose properties can be varied slightly from one to the other. The present
Purpose	investigation reports attempts to do this by using various samples of the same epoxy-resin which have been subjected to different curing cycles. Measurements
Assignment	of the specific heat (or the diffusity) and the thermal conductivity have been taken in the temperature range 0.1 to 8.0 K for a set of specimens which covered up to nine different curing cycles.[2]

As suggested by this example, article Introductions and especially their problem statements can be quite complex and varied. They have been studied in some detail by John Swales, who has proposed the amplified structure for them presented in Table 12-1; this is based on 16 articles each from physics, biology/medicine, and the social sciences. In Swales's proposed structure, the problem statement extends into or through Move Three, the statement

TABLE 12-1 A POSSIBLE STRUCTURE FOR ARTICLE INTRODUCTIONS

THE FOUR MOVES			NUMBER OF OCCURRENCES IN 48 ARTICLE INTRODUCTIONS
Move One	Establishing the field		43
	A Showing centrality		25
	i By interest	6	
	ii By importance	6	
	iii By topic prominence	7	
	iv By standard procedure	6	
	B Stating current knowledge		11
	C Ascribing key characteristics		7
Move Two	Summarizing previous research		48
Move Three	Preparing for present research		40
	A Indicating a gap	20	
	B Raising questions	14	
	C Extending a finding	6	
Move Four	Introducing present research		46
	A Giving the purpose	23	
	B Describing present research	23	

SOURCE: John Swales, *Aspects of Article Introductions,* Aston ESP Research Report No. 1, University of Aston, Birmingham, England, 1981, p. 22a.

of missing information (the question or task) appears in Moves Three and Four, and—if it appears at all—the purpose is found in Move Four. (Note that a statement of purpose appears in only half of Swales's examples.) The sample article introduction analyzed above is repeated below, this time with Swales's divisions.

THE THERMAL CONDUCTIVITY AND SPECIFIC HEAT
OF EPOXY-RESIN FROM 0.1-8.0 K

1 Establishing the field	The thermal properties of glassy materials at low temperatures are still not completely understood. The thermal conductivity has a plateau which is usually in the range 5 to 10 K, and below this temperature it has a temperature dependence which varies approximately as T^2. The specific heat below 4 K is much larger than that which would be expected from the Debye theory, and it often has an additional term which is proportional to T.
2 Summarizing previous research	Some progress has been made towards understanding the thermal behaviour by assuming that there is a cutoff in the phonon spectrum at high frequencies (Zaitlin and Anderson 1975 a,b) and that there is an additional system of low-lying two-level states (Anderson 1975, Phillips 1972).
3 Preparing the present research	Nevertheless, more experimental data are required, and in particular it would seem desirable to make experiments on glassy samples whose properties can be varied slightly from one to the other.
4 Introducing present research	The present investigation reports attempts to do this by using various samples of the same epoxy-resin which have been subjected to different curing cycles. Measurements of the specific heat (or the diffusity) and the thermal conductivity have been taken in the temperature range 0.1 to 8.0 K for a set of specimens which covered up to nine different curing cycles.[2]

12.2

MATERIALS AND METHOD

The Materials and Method section of any report you write is a critical part of the report's argument since it establishes the validity of your results and allows them to be taken seriously. It demonstrates that you have done everything "the right way": that you have been scrupulously careful and thorough, that you have used an accepted method, that you have made no technical mistakes.

This section also provides the mechanism by which the scientific community can repeat and verify your work. It must contain sufficient detail to allow any relatively experienced researcher in your field to reproduce your results exactly. This means that you must

1   Identify exactly what materials you used to conduct your re-search—reactants, enzymes, catalysts, organisms, experimental subjects (human or animal), etc. You should also identify your materials specifically enough that another researcher could use exactly what you specify to reproduce your results.

2   Identify any special conditions under which you conducted your re-search—special temperatures, irradiation with ultraviolet light, testing with unusually high current or voltage loads.

3   Identify any special criteria you used to select materials, subjects, test apparatus, or test method. (If you chose one gluing material or catalyst over another, why?)

4   Identify the specific method you used to conduct the research. If you followed a standard procedure, you may simply reference it. If you followed an unorthodox or new procedure, you need to describe it fully.

5   Justify, where necessary, any of your choices of criteria, materials, method, or conditions.

12.3

RESULTS

The Results section of an article presents (1) the major generalization(s) you are making about your data and (2), in compact form, the data supporting the generalization(s). The generalizations must be clearly and obviously stated: "This glue has successfully bonded stainless steel artificial joints to human bone" or "UV-irradiated DNA contains 5,6-dihydroxydihydrothymine as well as pyrimidine dimers in ratios dependent upon wavelength."[3] The supporting data must be presented fully enough that the reader can evaluate the strength of your generalizations but succinctly enough that the data do not overwhelm the generalizations.

For instance, consider the relationship between generalizations and data in the following example from an article entitled "The Effect of Graded Doses of Cadmium on Lead, Zinc, and Copper Content of Target and Indicator Organs in Rats." The subheadings in the Results section are "Zn," "Cu," "Pb," and "Correlations." Only the section on zinc (Zn), and one of the tables containing the data are presented here.

Zn

The Zn contents of the target and the indicator organs are presented in Tables I and II. The Tables also state significant differences ($p \leq 0.001$) between the control group and the experimental groups for each tissue. As can be observed several of the soft tissues had increased Zn levels when the Cd dose had been

TABLE I ZN CONTENT IN μG/G WET WEIGHT IN TARGET ORGANS OF RATS HAVING RECEIVED GRADED DOSES OF CdCl$_2$ IN THE DRINKING WATER FROM CONCEPTION UNTIL SACRIFICE AT THE AGE OF TEN WEEKS

CD DOSE IN PPM	KIDNEYS		LIVER		SPLEEN		HEART		MUSCLE		ADRENAL GLANDS		EPIPHYSES		DIAPHYSES	
	$\bar{x}$	SD	$\bar{x}$	SD	$\bar{x}$	SD	$\bar{x}$	SD	$\bar{x}$	SD	$\bar{x}$	SD	$\bar{x}$	SD	$\bar{x}$	SD
0	23.38 ± 1.53		27.71 ± 1.21		17.42 ± 0.98		14.54 ± 1.05		9.86 ± 2.51		13.79 ± 2.61		119.86 ± 10.90		121.73 ± 8.17	
2.5	22.53 ± 5.26		22.98 ± 1.77		17.09 ± 0.54		13.68 ± 0.73		8.35 ± 1.96		11.25 ± 2.18		114.64 ± 11.87		111.45 ± 7.73	
5	24.69 ± 1.30		30.48 ± 3.01		16.81 ± 0.95		14.40 ± 1.25		8.46 ± 1.12		14.61 ± 5.05		123.86 ± 12.89		123.67 ± 6.02	
7.5	28.12 ± 1.14[a]		33.44 ± 2.42[a]		21.71 ± 0.83[a]		18.42 ± 1.21[a]		11.17 ± 1.27		20.31 ± 1.98[a]		125.13 ± 6.63		129.77 ± 6.45	
10	25.04 ± 1.52		31.84 ± 3.33[a]		18.21 ± 1.20		15.31 ± 0.57		7.65 ± 1.04		13.38 ± 0.96		122.39 ± 4.65		123.93 ± 6.04	
12.5	24.35 ± 1.15		32.35 ± 3.04[a]		19.91 ± 1.96[a]		17.90 ± 0.95[a]		10.02 ± 1.79		27.60 ± 13.18		106.66 ± 8.17[a]		112.52 ± 9.59	
15	23.11 ± 1.62		26.43 ± 3.74		18.64 ± 1.24		16.66 ± 0.82[a]		9.79 ± 2.12		17.30 ± 2.38		102.50 ± 8.56[a]		114.80 ± 7.80	
17.5	27.45 ± 0.99[a]		42.67 ± 4.85[a]		21.40 ± 3.47[a]		17.07 ± 0.97[a]		10.96 ± 1.32		17.63 ± 2.14[a]		104.63 ± 10.89[a]		107.85 ± 5.65[a]	
20	27.20 ± 1.53[a]		35.13 ± 2.64[a]		20.30 ± 1.09[a]		16.96 ± 0.74[a]		10.44 ± 1.91		16.44 ± 3.06		106.41 ± 5.17[a]		110.54 ± 7.71	
22.5	25.72 ± 2.07		32.67 ± 1.84[a]		19.05 ± 0.97[a]		15.83 ± 0.89		9.80 ± 1.29		13.43 ± 3.40		92.99 ± 9.11[a]		101.67 ± 6.65[a]	
25	26.77 ± 1.65[a]		31.87 ± 3.37[a]		16.43 ± 1.11		14.63 ± 1.13		16.10 ± 3.81[a]		20.73 ± 3.62[a]		106.98 ± 5.09[a]		99.95 ± 6.01[a]	

[a]The concentration is significantly higher/lower ($p \leq 0.001$) than in the control group of the same tissue.

above 5 ppm, whereas for Cd doses above 12.5 ppm significantly decreased Zn contents in epi's and dia's were registered (see Table I). In addition, Cd supply was associated with a significant decrease in Zn content in incisors and molars of the majority of the experimental groups (see Table II) [not included].

In none of the pooled left molar samples was the Zn content of enamel different from that of dentin. This was valid also for incisor samples through all groups.[4]

Notice that the table contains the condensed results of many experiments, presenting average values for the ten experiments conducted with each dose of cadmium (Cd).

A second example of the relationship between generalizations and supporting data in the Results section comes from an article on tensile and fatigue tests on human cortical bone specimens. The example presents only a part of the Results section, the results of the tensile tests; it does not include the two tables of supporting data referenced in the text.

TENSILE TESTS
The mean values of the physical characteristics of the test specimens are shown in Table 2. The mechanical parameters of ultimate stress, yield stress, yield strain and modulus of elasticity showed no statistically significant correlation with porosity, ash fraction, wet density, or dry density.

Statistically significant ($P < 0.05$) positive linear regressions were found between ultimate stress and elastic modulus and between yield stress and elastic modulus (Table 3). A weak negative linear regression (not statistically significant) was found between yield strain and elastic modulus. The influence of bone elastic modulus of the ultimate stress, yield stress, and yield strain are shown in Fig. 1 [see page 282]. The mean values ($\pm$ s.d.) were ultimate stress 140 ($\pm$ 12) MPa, yield stress 129 ($\pm$ 11) MPa, yield strain 0.0068 ($\pm$ 0.0004), and elastic modulus 17.5 ($\pm$ 1.9) GPa.[5]

12.4

DISCUSSION
The Discussion explains the implications of your results. It fits the results into the context of the field by relating your results to other work, both theoretical and experimental. Along with the Introduction, it explains why your work is important, how it contributes to the advancement of the field. It is critical that this be done carefully and thoroughly, for as Robert Day has noted,

> *Many* papers are rejected by journal editors because of a faulty Discussion, even though the data of the paper might be both valid and interesting. Even more likely, the true meaning of the data may be completely obscured by the interpretation presented in the Discussion, again resulting in rejection.[6]

If you want to show how your work contributes to the advancement of your field, you might consider what circumstances create advancement.

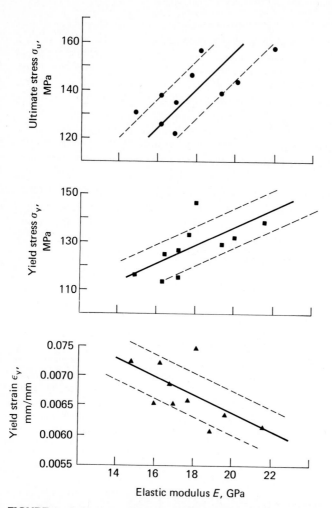

FIGURE 1   Influence of elastic modulus on the ultimate stress, yield stress, and yield strain of the tensile specimens.

One philosopher, John Platt, believes that some fields advance much more rapidly than others because of a rigorous intellectual methodology:

> These rapidly moving fields are fields where a particular method of doing scientific research is systematically used and taught, an accumulative method of inductive inference that is so effective that I think it could be given the name of "strong inference.". . . The steps are familiar to every college student and are practiced, off and on, by every scientist. The difference comes in their systematic application. *Strong inference consists of applying the following steps to every problem in science, formally and explicitly and regularly:*

1 *Devising alternative hypotheses;*

2 *Devising a crucial experiment* (or several of them), *with alternative possible outcomes,* each of which will, as nearly as possible, *exclude one or more of the hypotheses;*

3 *Carrying out the experiment so as to get a clean result; and*

4 *Recycling* the procedure, making subhypotheses or sequential hypotheses to refute the possibilities that remain; and so on.[7] [italics original]

The keys to this system, according to Platt, are logical rigor, the alternative hypotheses, the crucial experiments, and the disproving of hypotheses. These allow scientists in a field to identify those problems crucial to the advancement of the field, then to construct a series of logical options or lines of inquiry for each problem, and then to eliminate quickly unproductive hypotheses and lines of inquiry. This system reduces wasted time because it focuses the theoretical and experimental efforts of a field on the hypotheses and lines of inquiry likely to produce successful results.

Now, given this perspective, what do you say in an adequate Discussion? How do you establish the importance of your work for the advancement of your field? If you can, you argue that your work is a crucial experiment, disproving some hypothesis or supporting another. Notice that this view encourages the reporting and discussion of negative results. "Hypothesis A predicted that I would find B, but I did not. This result questions the validity of hypothesis A or indicates its inadequacy." Such a statement might save researchers and graduate students from wasting years of needless effort. It may also provide the best kind of statement scientists may make about nature:

> As the philosopher Karl Popper says today, there is no such thing as proof in science—because some later alternative explanation may be as good or better—so that science advances only by disproofs. There is no point in making hypotheses that are not falsifiable, because such hypotheses do not say anything; "it must be possible for an empirical scientific system to be refuted by experience."[8]

In arguing that your work is important, ideally because it is a crucial experiment, you typically consider some or all of the following questions.

1 Were your results expected? If not, why not?

2 What generalizations or claims are you making about your results? How do you interpret these generalizations?

3 Do your results contradict or support other experimental results?

4   Do they suggest other observations or experiments which could be done to confirm, refute, or extend your results?

5   Do your results support or contradict existing theory?

6   Do your results suggest that modifications or extensions need to be made to existing theory? What are they?

7   Could your results lead to any practical applications? What are they?

EXERCISE 12-1

A   If you expect to write technical articles, analyze the organization and style of articles in one journal to which you might submit an article. Compare the journal's features with those discussed above and with those presented in Chapter 9 for technical reports.

B   For a topic you are interested in, write an Introduction to an article on a research project you have conducted or would like to conduct.

C   Select a laboratory report from one of your technical courses and write a Materials and Method section for it. Then write a Results section. If you can, write an Introduction and then a Discussion. Try especially hard to produce a Discussion.

REFERENCES

1   Robert A. Day, *How To Write and Publish a Scientific Paper*, ISI Press, Philadelphia, 1979, p. 24.

2   John Swales, *Aspects of Article Introductions*, Aston ESP Research Report No. 1, University of Aston, Birmingham, England, 1981. p. 16.

3   Philip C. Hanawalt, Priscilla K. Cooper, Ann K. Ganesan, and Charles Allen Smith, "DNA Repair in Bacteria and Mammalian Cells," *Annual Review of Biochemistry* **48**:787 (1979).

4   G.B.R. Wesenberg, G. Fosse, and P. Rasmussen, "The Effect of Graded Doses of Cadmium on Lead, Zinc, and Copper Content of Target and Indicator Organs in Rats," *International Journal of Environmental Studies* **17**:193–194.

5   Dennis R. Carter and William E. Caler, "Uniaxial Fatigue of Human Cortical Bone: The Influence of Tissue Physical Characteristics," *Journal of Biomechanics* **14**:463–464 (1981).

6   Day, op. cit., p. 33.

7   John R. Platt, *The Step to Man*, Wiley, New York, 1967, p. 20.

8   Platt, op. cit., p. 27.

ADDITIONAL READING

American National Standards Institute, *American National Standard for the Preparation of Scientific Papers for Written or Oral Presentation,* ANSI Z39.16-1972, New York, 1972.

V. Booth, *Writing a Scientific Paper,* 4th ed., Biochemical Society, London, 1978.

CBE Style Manual Committee, *Council of Biology Editors Style Manual: A Guide for Authors, Editors, and Publishers in the Biological Sciences,* 4th ed., Council of Biology Editors, Washington, D.C., 1978.

Robert A. Day, *How To Write and Publish a Scientific Paper,* ISI Press, Philadelphia, 1979.

L. DeBakey, *Guidelines for Editors, Reviewers, and Authors,* C. V. Mosby, St. Louis, 1976.

J. H. Mitchell, *Writing for Professional and Technical Journals,* Wiley, New York, 1968.

John R. Platt, "Strong Inference," in *The Step to Man,* Wiley, New York, 1967, pp. 16–36.

John Swales, *Aspects of Article Introductions,* Aston ESP Research Report No. 1, University of Aston, Birmingham, England, 1981.

PART VI

Readability

13

MAKING YOUR WRITING READABLE: INFORMATION SELECTION

As emphasized in earlier chapters, most readers of scientific or technical writing do not have as much time for reading as they would like to have and therefore must read selectively. This is especially true of managers, supervisors, executives, senior scientists, and other busy decision makers, who often skim-read for main points and ideas. However, it is also true of professionals who often need to read more closely and slowly, for thorough understanding, and it is true of technicians, workers, and consumers who may need to read and follow operating instructions. These different types of readers are selective in different ways: the skim-reading decision maker may be looking for "bottom-line" cost figures and performance data; the professional may be looking for the main thread of an argument; the technician, worker, or consumer may need to use operating instructions only as a checklist.

For such readers, writing is readable to the extent that it provides the information they need, located where they can quickly find it, in a form in which they can easily use it. This takes considerable effort on the writer's part; as the old adage has it, "It's easy to make things difficult, but difficult to make things easy." Nonetheless, it's *worth* the effort. If you can make your writing readable, you'll greatly increase its chances of being read and used; in short, you'll greatly increase its *effectiveness*. Conversely, if you don't make the effort, your readers may not either.

How can you make your writing more readable? Unfortunately, there is no simple formula to follow. (The so-called readability formulas are not designed to guide the writing process, and so prescriptions based on such formulas, such as "Use short sentences and short words," are not very reliable.) There *are* steps you can take, however, that should be of some help, and these are laid out in the next three chapters. First, in this chapter, we make suggestions for selecting appropriate information and for making this information accessible to the reader. Then, in Chapters 14 and 15, we suggest a number of things you can do to make it easier for the reader to absorb details; the focus will be on sentences, phrases, and individual words.

BASIC PRINCIPLE
Put information that is new to the reader into a framework of information already known to the reader.

Communication involves the transmitting of information to someone who does not already possess that information, in other words, the transmitting of new information. When new information is absent from a message, the message does not communicate anything: it is noninformative. To most of us, the statement "The Pope is a devout Catholic" is noninformative: it tells us nothing we don't already know. In general, new information is a necessary ingredient of meaningful communication.

New information is not the only ingredient in successful communication, however. If statements consisted of nothing but new information, they would be incomprehensible. Try this, for example: "On big-wall Grade VI's, the second usually jumars all off-widths and overhangs." Got that—or did it go sailing right over your head? It's filled with technical rockclimbing jargon, and so unless you're familiar with rockclimbing, you probably found it totally incomprehensible. In general, for new information to be comprehensible to someone, it must be couched in a meaningful framework of information already known to that person, i.e., a framework of given information. Given information is the background knowledge that we call upon in trying to make sense of new information.

Most of what we know about the world, however, is stored in distant recesses of our mind, in what psychologists call long-term memory; it is not consciously on our mind. Thus, before this background knowledge can be put to use in interpreting new information, it must first (to use another psychological term) be activated, that is, brought to conscious awareness. Fortunately, our background knowledge exists not in isolated bits and pieces but rather in clusters and networks of associated concepts. Thus, mention of even a single word may trigger in the reader's mind a whole host of related images, facts, beliefs, etc. Some of these related concepts may be useful in helping the reader interpret a particular piece of new information; others may not be.

While there is almost no limit to the amount of knowledge we can store in long-term memory, there are very severe limits to the amount of knowledge we can maintain in active consciousness. We can only think about one thing at a time, as the saying goes. So, it's important for you as a writer to activate the *right kind* of given information in the reader's mind—the kind of given information that will help most in understanding the new information. It should pertain directly to the topic of discussion so that it fits into the context already established and allows the reader to anticipate what's coming next. The word or words you use to refer to this given information should be familiar, concise, and loaded with imagery so as to maximize the amount of background knowledge you can bring to the reader's conscious awareness. Consider the following negative example and its rewritten version:

NEGATIVE EXAMPLE

Modern cryptanalysis has attracted some of the most capable mathematical minds. In recent years the growing prospect that postal and diplomatic communication will soon be replaced by other forms of communication has furnished increased incentive for mathematicians and engineers to invent an unbreakable cipher. Messages will be able to be typed, delivered, and quickly typed on the other end. But unbreakable codes will be needed to protect senders against snoopers.

No doubt you can understand the basic meaning of this passage. But is it vividly memorable? Do you have a clear impression of what kind of unbreakable cipher or what kinds of snoopers the writer is talking about? If not, you may find this rewritten version more to your liking:

REVISED VERSION

Modern cryptanalysis has attracted some of the most capable mathematical minds. In recent years the growing prospect that postal and diplomatic communication will soon be replaced by electronic communication has furnished increased incentive for mathematicians and engineers to invent an unbreakable cipher. Electronic devices will permit messages to be telephoned, delivered, and quickly typed on the other end. But unbreakable codes will be needed to protect senders against electronic snoopers.[1]

The only difference between the two versions is the use of the word *electronic* in the second one. This one word does not really add any new information to the passage; someone reading the first version, for example, could probably figure out that the phrase *other forms of communication* most likely includes electronic communication. But this word does bring to mind the kind of given information that's needed to fully understand what the writer is talking about. The word *electronic* is familiar and concise. Even more important, it activates in our minds a number of vivid images which we can associate with words like *communication, devices,* and *snoopers,* thus giving much greater meaning to these concepts. Notice, too, how the writer has used the word repeatedly so as to keep it in the reader's active consciousness and thus help maintain continuous focus on the topic of the passage:

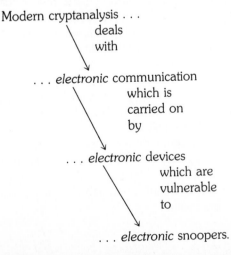

Modern cryptanalysis . . .
      deals
      with

   . . . *electronic* communication
        which is
        carried on
        by

    . . . *electronic* devices
         which are
         vulnerable
         to

     . . . *electronic* snoopers.

As can be seen from this example, the use of key words can greatly enhance the coherence and memorability—and hence the readability—of a passage. Key words are most effective when they (1) trigger vivid imagery in the reader's mind, (2) are related in an obvious way to the topic of the passage, and (3) are related in an obvious way to the reader's purpose in reading the passage. If you select key words that satisfy these criteria, and if you put them in prominent positions in the text, you will be doing much to activate the right kind of given information in the reader's mind.

What are the prominent positions in a text? Basically, these are the places that are visually prominent by virtue of having more white space around them and being more easily located: titles, headings, subheadings, captions, labels, paragraph topic statements, and sentence subjects. Some of these text features have already been discussed (see Chapters 2 and 7 especially); the remainder of this chapter and Chapter 14 will go into further detail.

13.1

ESTABLISH YOUR TOPIC AND PURPOSE

Make it clear what the main topic of the report or the section is. Then state your purpose explicitly, so that your readers can anticipate how you will be dealing with the topic. Readers of scientific and technical writing are typically purpose-directed and pressed for time. So, rather than reading word for word and cover to cover, they often prefer to merely "consult" a document, looking only for the information they need. When you define your topic and state your purpose, you make it easier for the reader to determine right away how to process the document: whether to read it closely, skim-read it, pass it on to someone else, or disregard it. A clear statement of topic and purpose allows the reader to form certain expectations about the rest of the text, specifically, how the topic is likely to be developed. It is a well-known fact that we process information most quickly and efficiently when it accords with our preconceptions; this is why it's important to create the right preconceptions in the reader's mind in the first place.

Scientific and technical writing genres customarily have various features designed to announce the topic and set up initial expectations: titles, abstracts, summaries, overviews, etc. Use these to full advantage by loading them with key words and main ideas instead of vague phrases. If you're writing a report dealing with some problematic issue—as is the case with most reports—be sure to include a well-written problem statement at the beginning. Engineering and other applied sciences are fundamentally problem-oriented, and so, as discussed in Chapter 6, a good problem statement usually has important orientational value.

13.2

USE KEY WORDS PROMINENTLY

Build sections and paragraphs around key words related to the main topic. If possible, make these key words visually prominent by using them in headings, subheadings, topic statements, and sentence subjects. Once you've established a conceptual framework at the beginning of your text, you can turn your attention to filling it in with appropriate details. To make sure that your discussion is a coherent one, you should strive to link these details as directly as possible to the main topic. The best way to do this is to establish a hierarchy of intermediate topics and subtopics for the various units and subunits of your text, with each being directly related to the immediately higher topic or subtopic. (If you've taken the time to outline your text before writing the draft, this should be a fairly straightforward matter.) These intermediate topics and subtopics should consist of appropriate key words, as discussed above.

Figure 13-1 is an excerpt from a scientific article illustrating this technique. Notice how key words (circled) are used to weld the headings, subheadings, and topic sentences together into a hierarchy of topics and subtopics; notice, too, how the writer has given these key words prominence by putting them in the subject position of the sentences.

A well-structured discussion is highly functional in at least two respects. First, it builds on the basic framework established at the beginning of the text, allowing for easier interpretation and promoting greater coherence at the same time. As new information is progressively added to the initial framework, it is interpreted in terms of this framework and integrated into it. As such, this new information is transformed into given information and can then be used to help interpret succeeding pieces of new information. Second, a hierarchically structured text facilitates selective reading. Since the sections and subsections are arranged in a general-to-specific order, the reader can quite easily zero in on desired levels of detail—especially if the respective topics of these sections and subsections are made visually prominent through the use of headings and subheadings. (Paragraph-level topics can be made visually prominent by introducing them in topic statements at the beginnings of paragraphs, as discussed in Chapter 3.)

13.3

EXPLAIN IMPORTANT CONCEPTS WHEN WRITING
FOR NONSPECIALIST READERS

When writing for nonspecialists, be sure to clarify the important technical concepts in your text by using examples, analogies, visual aids, or other forms of verbal or visual illustration. Research by information theorists in the past few decades suggests that communication proceeds best when there is a

FIGURE 13-1 Building sections and paragraphs around key words related to the main topic. The key words are circled. [From Ben Patrusky, "The Social Cell," *Mosaic* **12**(4):9 (1981).]

Reach out and touch

Cell/cell junctions come in a variety of shapes and sizes.

When cells come together—and stay together—they do more than just touch. Contiguous cells actually form highly specialized, highly organized interconnective structures. And with the development of a technique called freeze fracture, a method for separating the inner face of a membrane from its outer face, scientists have managed to extract many details about the micro-architecture of these fine intercellular contacts or junctions.

One of the earliest known of the intercellular junctions is the revet- or weld-like desmosome. Adjacent cell membranes linked by way of desmosomes don't actually touch; the neighboring surfaces are separated by a space of about 200 angstroms. (One angstrom equals one hundred-millionth—10^{-3}—of a centimeter.) And the space is packed with dense, extracellular material, which serves as binder.

Desmosomes have been found primarily between the cells of sheetlike epithelial tissue, which covers the surface of the body and the internal surfaces of most organs. A variation of the desmosome, the half-desmosome, does not link cells to other cells but anchors individual cells to the underlying mesh of connective tissue, between the epithelium and the underlying organ. According to Norton B. Gilula of Rockefeller University, an authority on intercellular junctions, the function of desmosomes is now thought to be strictly that of adhesion—to keep tissues and organs intact despite mechanical stress. That belief is prompted in part by the fact that cells normally subjected to severe, repeated stress—heart muscle, for instance, or the cervix and uterine walls—possess unusually large numbers of desmosomes.

Tight junctions

Epithelial tissue is especially rich in another kind of distinctive intercellular connection called the tight junction, clearly characterized by Marilyn Farquhar and George Palade of Rockefeller Institute (now Rockefeller University) in 1967. Unlike the desmosome, with its 200-angstrom separation between cell surfaces, the tight junction, as the name implies, seems to represent a true contact between cells. No intervening space is evident. It's as if the outer surfaces of the two bilayers of neighboring cell membranes had melted into each other.

Found at the apex, or shoulder, of epithelial cells, tight junctions act as sealants, occlusions, or barriers that enable an organism or an organ to maintain an internal environment that is chemically distinct from its surroundings, explains the California Institute of Technology's Jean-Paul Revel, another leading figure in intercellular junction research. Tight junctions he says, abound in places where a sharp physical separation between two compartments—or from the outside world—is essential. It's no surprise, then, that tight junctions have been discovered in embryonic cells at the very early stages of development.

For all that, tight junctions may not be totally impermeable. Freeze-fracture studies by Philippa Claude and Daniel A. Goodenough of Harvard Medical School show that different kinds of tissue have structural differences in their tight junction that appear to correlate with varying degrees of leakiness: the bigger the junction, the less the leakage.

[Patrusky, Ben, "The Social Cell," MOSAIC. 12:4, July/August 1981, National Science Foundation, p. 9] fairly even balance between given information and new information. This is what you should strive for in your own writing. This means that you must have some idea of who your readers are and what sort of background knowledge they have; as illustrated by the rockclimbing and Pope examples, "givenness" and newness are partly functions of the knowledge a reader brings to a test—and this can vary from reader to reader and text to text. To a rockclimber, the sentence about big-wall Grade VI's would be perfectly comprehensible; to anyone else, it wouldn't be. Thus, if for some reason you had to communicate that kind of technical information to a nonspecialist reader, you would have to insert some background information more familiar to the reader to provide a proper framework for interpreting the new information. In so doing, you would be creating a better balance between given and new.

In technical writing, it frequently happens that the writer feels it necessary to introduce key concepts that may be unfamiliar to the reader. Sometimes these key concepts even occupy topic roles: topics of paragraphs, topics of sections, perhaps even the topic of the entire text. In general, it's important to define such concepts, not necessarily with a formal definition but rather with some kind of *illustration*. How is the concept used? What is it similar to? What does it look like? If technical terminology is used, what is a nontechnical way of saying more or less the same thing? Not only will answering such questions with the reader's needs in mind help the reader understand that particular concept but, more important—especially if the concept is a topical one—it will enrich and sharpen the reader's interpretation of the text as a whole. It will provide some of the given information that a specialist reader would automatically and implicitly associate with that particular concept but which a nonspecialist reader would not.

There are several ways to illustrate and explain unfamiliar concepts for the nonspecialist reader. *Visual aids,* of course, should be used whenever the concept is suited to visual representation (Chapter 7). Often, however, a concept is too abstract to be represented visually. In such cases, specific *examples* of the concept are usually the most powerful means you can use to help the nonspecialist reader. Research by cognitive psychologists indicates that readers confronted with an unfamiliar abstract concept will often try to construct a concrete example they can relate to; in some cases, they will actually build a scenario with themselves as the principal actor acting out the concept. As a writer, you can sometimes save the reader this effort by providing an example, or even a scenario, yourself. This will also prevent the reader from constructing a misleading example or scenario. *Analogies* help explain an unfamiliar concept by showing that it is similar in certain ways to a familiar concept; they are useful in situations where the concept is so unfamiliar that you simply cannot think of any ordinary examples of it. *Paraphrases,* on the other hand, are useful in precisely the opposite situation:

where the concept is familiar to the reader but only if restated in more recognizable terms. Paraphrases have a distinct advantage over examples and analogies in that they usually take up less space; sometimes even a one-word paraphrase will accomplish the purpose. *Definitions,* of course, are a familiar way of explicating new concepts. They can be combined with some of the techniques mentioned above to form *extended definitions.*

Here is an example of an extended definition, explaining what the technical term *Remrak coefficient* means:

THE REMRAK COEFFICIENT

	In the production of powered detergents, spray drying is the technique used to evaporate the solvent from the liquid reaction mixture and physically form the finished powder product. In spray drying, the liquid is sprayed into the top of a tall tower and allowed to fall freely to the bottom of the tower, where it is removed as a dry powder. The solvent evaporates during the course of the fall. Particles dried in this
Analogy	fashion have an unusual shape, *like that of a saddle (or a Pringle's potato chip),* and consequently fall through the air in an unusual manner. Rather than falling in a vertical path, the
Paraphrase	particles fall in a helical *(spiral)* path. The shape of the helical path is described by the Remrak coefficient, *which is the ratio*
Definition	*of the diameter of the helix to the height required for one passage of the particle around the perimeter of the helix.* The coefficient, which is a function of drying conditions, is sought to be maximized, *so that the length of flight of the particle is*
Paraphrase	*made much greater than the actual height of the spray-drying tower.* [italics added]

The writer of this passage has obviously gone to considerable lengths to help us understand what the Remrak coefficient is. Among other things, most of the standard forms of verbal illustration have been used: analogy, paraphrase, simple definition, and extended definition. Has it been worth the trouble? Yes, if we didn't already know the concept and were interested in finding out; no, if we already knew or weren't interested in knowing. In general, verbal and visual illustrations are powerful devices, but they work only under the following conditions:

1   The concept is not already familiar to the reader.

2   The information used to illustrate the concept *is* familiar to the reader.

3   The concept being illustrated is an important one in that particular context.

4   The information used to illustrate the concept focuses on features of that concept that are relevant to that particular context.

Do take advantage of the power of illustrations in explaining unfamiliar technical terms to nonspecialist readers. When used correctly, illustrations can clarify things in an instant.

13.4
USE STANDARD TERMINOLOGY WHEN WRITING FOR SPECIALIST READERS

When writing for specialists, on the other hand, do not overexplain. That is, do not exemplify, define, illustrate, paraphrase, or otherwise explain concepts the reader is likely to already be familiar with. Instead, simply refer to such concepts with the standard terminology of the field. Part of what it means to be a specialist in a given field is to know the standard technical terminology of that field. Technical terms permit efficient and precise communication between specialists who know the concepts that such terms refer to. They should be used for that purpose, and used freely, even if they appear to be incomprehensible jargon to an outsider. When used among specialists, standard technical terms are not only comprehensible but are often "information-rich" in the sense that they may trigger a host of associated concepts in the reader's memory. These associated concepts then become part of the "given information" in the message. Thus, in the jargon-laden writing of specialists for other specialists, there is usually more than enough given information. Adding more given information in the form of examples, analogies, etc., would only produce a disproportionate and inefficient given/new ratio for that type of reader.

What do you do, though, if you are writing to a *mixed* audience of specialists and nonspecialists? This is always a very challenging—sometimes impossible!—situation, but there are a few things you can do. First, you might divide and conquer: produce two separate pieces of writing, or a single piece with two parts to it, so that each group of readers can be addressed with appropriate terminology. Alternatively, you might stick to a single text but briefly define the technical terms as you go along. The least objectionable way of doing this, usually, is to insert a short familiar paraphrase immediately after each technical term; in the Remrak coefficient example, for instance, notice how the writer has inserted the paraphrase *(spiral)* after the less familiar term *helical*.

13.5
STRUCTURE YOUR TEXT TO EMPHASIZE IMPORTANT INFORMATION

Structure the different parts of the text so as to give greatest prominence to the information you expect the reader to pay most attention to. For main ideas, use a hierarchical (general-to-particular) structure; for details, use a listing (coordinate) structure. As discussed above, a hierarchical text structure allows the reader to move quickly through a text, seeing what the main ideas are, how they're linked together, and what kind of detailed support they have. Many readers, expecially busy decision makers, habitually read this way. Thus, if you are writing for that type of reader, you should try to organize and present your information in a highly hierarchical pattern, with many levels of subordination.

On the other hand, if you are writing for a reader who will be focusing more on details, try to use a more coordinate structure—that is, with the details arranged in lists. A listlike structure, whether it's formatted as a list or not, draws the reader's attention to all of the items making up the list. Instead of one statement being subordinated to another, as in a hierarchical structure, the statements in a list are all on the same level and thus share equal prominence. Perhaps the most obvious examples of this phenomenon are lists of instructions, which are expected to be read and followed step by step. The same phenomenon can also be seen in carefully reasoned arguments and explanations, which are often cast in the form of a listlike sequence of cause-and-effect statements. Chronological sequences, too, as found in descriptions of test procedures or in progress reports, are often presented as lists. (As Chapter 3 emphasizes, remember to use parallelism for all lists, formatted or not.)

13.6

CONSTRUCT WELL-DESIGNED PARAGRAPHS

Make sure that each paragraph has a good topic statement and a clear pattern of organization. The paragraph is a basic and highly functional unit of discourse in scientific and technical writing. By definition, a paragraph is a group of sentences focusing on one main idea. If you use a topic statement to capture the main idea and a clear pattern of organization to develop it, you make it easy for the reader to either read the paragraph in detail or read it selectively. The topic statement, of course, should be presented within the first two sentences of the paragraph, and it should contain one or more key words for readers to focus their attention on. The pattern of organization you select for the remaining sentences in the paragraph should (1) be consistent with expectations likely to be raised by the topic statement, (2) be appropriate to the subject matter, and, most important, (3) be appropriate to the anticipated use of the paragraph by the reader (see, for example, Section 13.5). If you adhere to these principles with all your paragraphs, you will greatly enhance the overall readability of your writing.

(For a thorough discussion and review of the principles of paragraph writing, see Chapter 2.)

13.7

FIELD TEST YOUR WRITING

Field test your manuscript with its intended users or with representative substitutes. Up to this point, you've had to make guesses about whether or not you're providing your readers with a proper mix of given information and new information for their purposes. Your decisions about what kind of terminology to use, what kind of structure to use, when to use verbal or visual illustrations, and so on, have been made on the basis of guesswork about your readers' background knowledge and the reasons they will have for reading your writing. Educated guesswork, perhaps, but guesswork nonetheless.

This is why field testing with actual users (or representative users) is an essential part of making any manuscript maximally useful. Field testing allows you to see whether the assumptions you have made about your readers are accurate or not. This is so important that you should not put it off until the final editing stage (though you might want to do a second round of field testing at that time). As soon as you've finished writing a good first or second draft, try it out on a few intended users. Have them read it as if it were the final draft submitted for actual use. Tell them to mark it up, raise questions about it, criticize it. Talk to them about it, ask them for their comments. Does it leave anything out? Does it mislead them? Does it raise unanswered questions? If they're using it for reference purposes, can they easily find what they need? If they're skimming it for main points, can they easily locate and understand them?

If your manuscript contains any instructions or any other material designed to be acted on, have your readers read and try to follow these instructions while you observe. Take note of places where they have to pause and ponder. Take note also of places where they go astray. Do not interfere with their efforts—let the manuscript speak for itself. Later, do a postmortem with your readers. Ask them where they had trouble. Did something confuse them? Did something seem to be missing? Would they prefer to have the material written in some other way?

This is also a good stage at which to consult experts, to make sure that nothing you have written is substantively wrong. If you're writing a research proposal or article, for example, you might want to show your draft to other researchers in that area, so as to guard against the possibility that you've overlooked something important or misrepresented someone else's research. If you're writing a progress report for a group project, this would be a good time to show it to other members of the team.

In all of this testing, both with intended users and with experts, take note of *all* trouble spots—even those that do not seem to pertain to inadequate selection or structuring of information. Some trouble spots will probably be due to sentence-level problems: problems of phrasing, wordiness, diction, sentence structure, lack of emphasis, etc. These will be addressed in Chapters 14 and 15.

EXERCISE 13-1

A Take a highly technical term or concept from your field and define it in standard terminology so that another specialist in the same field can quickly and easily understand it. Then check your definition against that found in a technical dictionary.

B Take the same term or concept you chose for part A and write an extended definition of it so that a nonspecialist can understand it. Then test your definition by having a nonspecialist read it; if he or she has any trouble understanding it, make appropriate modifications.

C The following paragraph is taken from a student report on two prototype electric-powered cars, an AMC Pacer and a Ford Pinto. Reorganize and rewrite it in two different versions: (1) one that emphasizes a single main point and (2) one that emphasizes details.

The fuel system integrity test is concerned with fuel leakage from the fuel lines in crash and rollover tests. The rollover test involves turning the car on its side at a 90 degree angle to the ground. Fuel leakage in electric cars is primarily concerned with the gas used in the heating system. The Ford Pinto had no problem in either test. The AMC Pacer had no leakage in the crash test, but it had excessive leakage in exceeding the standard in the rollover test. In the rollover test, the problem area was where the gas lines connected to the holding tank. Redesign of this setup could alleviate this problem. The fuel system integrity test can be met easily for electric cars.

D Take a full-length sample of your own writing and field test it with one or more intended users (or representative substitutes). Then revise it according to the guidelines laid down in this chapter.

REFERENCES

1 Julie Wei, "Pure Mathematics: Problems and Prospects in Number Theory," *Research News* **30**(3):20 (1979).

ADDITIONAL READING

R. C. Anderson, R. E. Reynolds, D. L. Schallert, and E. T. Goetz, "Frameworks for Comprehending Discourse," *American Educational Research Journal* **14**:367–381 (1977).

Marshall Atlas, "The User Edit: Making Manuals Easier To Use," *IEEE Transactions on Professional Communication* **PC-24** (1):28–29 (1981).

F. C. Bartlett, *Remembering*, Cambridge University Press, London, 1932.

Walter Kintsch and Douglas Vipond, "Reading Comprehension and Readability in Educational Practice and Psychological Theory," in *Perspectives in Memory Research*, L. G. Nilsson (ed.), Lawrence Erlbaum, Hillsdale, N.J., 1979, pp. 329–365.

James R. Miller and Walter Kintsch, "Knowledge-Based Aspects of Prose Comprehension and Readability," *Text* **1**(3):215–232 (1981).

J. M. Royer and G. W. Cable, "Illustrations, Analogies, and Facilitative Transfer in Prose Learning," *Journal of Educational Psychology* **68**:205–209 (1976).

D. E. Rumelhart and A. Ortony, "The Representation of Knowledge in Memory," in *Schooling and the Acquisition of Knowledge*, Lawrence Erlbaum, Hillsdale, N.J., 1977.

D. Schallert, "Improving Memory for Prose: The Relationship between Depth of Processing and Context," *Journal of Verbal Learning and Verbal Behavior* **15**:621–632 (1976).

M.N.K. Schwarz and A. Flammer, "Text Structure and Title: Effects on Comprehension and Recall," *Journal of Verbal Learning and Verbal Behavior* **20**:61–66 (1981).

G. J. Spilich, G. T. Vesonder, H. L. Chiesi, and J. F. Voss, "Text Processing of Domain-Related Information for Individuals with High and Low Domain Knowledge," *Journal of Verbal Learning and Verbal Behavior* **18**:275–290 (1979).

A. B. Tenenbaum, "Task-Dependent Effects of Organization and Context upon Comprehension of Prose," *Journal of Educational Psychology* **69**:528–536 (1977).

Perry Thorndyke and Barbara Hayes-Roth, "The Use of Schemata in the Acquisition and Transfer of Knowledge," *Cognitive Psychology* **11**:82–106 (1979).

14

MAKING YOUR
WRITING READABLE:
INFORMATION ORDERING

One of the most important parts of speech in scientific and technical writing is the *noun phrase* (NP), which can be defined as any noun or noun-plus–modifier combination (or any pronoun) that can function as the subject or object of a sentence. Some examples are *tables, water, we, a potential buyer, the growing demand for asphalt,* and *strict limitations on the size of plates that can be handled.*

Notice that each of these NPs can serve as the subject of a sentence:

Tables usually have four legs.

Water can be dangerous.

We have an emergency.

A potential buyer has arrived.

The growing demand for asphalt is obvious.

Strict limitations on the size of plates that can be handled have been established.

By contrast, a singular countable noun, such as *table,* is not an NP because it cannot function by itself as the subject or object of a sentence. We cannot say

Table usually has four legs.

Instead, we would have to say

A table usually has four legs.

or

The table has four legs.

or

John's table has four legs.

NPs are important on the grammatical level because they account for most of the principal parts of a sentence: subject, indirect object, direct object, and object of a preposition. Not all of these parts are found in all sentences, of course, but here is an example that does have all four:

<u>*We*</u>   are sending <u>*your company*</u> <u>*the new TK-I40 model*</u>
SUBJECT   INDIRECT OBJECT   DIRECT OBJECT

on <u>*the assumption that you need an all-purpose sorter.*</u>
            OBJECT OF PREPOSITION

NPs are important on the functional, or communicative, level because they carry the main information of the sentence. In the example just cited, for instance, notice how much information is contained in the NPs (italicized)—almost all of it.

14.1

OPTIMAL ORDERING OF NOUN PHRASES

In English, NPs are expected to occur in certain orderings according to grammatical and functional criteria. These will be discussed in order of importance, beginning with the most important.

Put Given Information Before New Information

As with all languages, English sentences typically contain a mixture of given information and new information. That is, some NPs in a sentence refer to concepts or objects that have already been discussed or that are presumed to be understood from the context; this is *given information.* Other NPs refer to concepts or objects that have not yet been discussed and are not presumed to be understood from the context; this is *new information.* The optimal ordering of NPs within a sentence according to this view is *given* before *new* because the given information can serve as a frame within which the new information can be understood. Let us consider a specific example of this ordering:

> *The 5-year plan* does not indicate a clearly defined commitment to *long-range environmental research.* For instance, where *the plan* does address *long-range research, it* discusses the development of techniques rather than the identification of important long-range issues.

The key NPs in both sentences are in italics. By the time the first sentence has been read and understood, the phrases *the 5-year plan* and *long-range environmental research* have been mentioned and are part of the given information possessed by the reader. Notice that these words—the given information—come at the beginning of the second sentence and that the new noun phrases—the new information—come at the end of the second sentence. This ordering of given before new is desirable because the given information of the second sentence serves as a kind of glue between the information presented in the first sentence and the new information presented in the second sentence. Such an ordering allows a reader to more easily fit the new information into a meaningful context and to see the connection between the two sentences.

To see how much easier it is to understand sentences which put given before new information, compare the passage just discussed with the following one (which has the opposite ordering):

> The 5-year plan does not indicate a clearly defined commitment to long-range environmental research. For instance, the development of techniques rather than the identification and definition of important long-range issues is the subject of the plan where it does address long-range research.

Notice how hard it is to place the new information of the second sentence in an appropriate context.

Put Topical Information in Subject Position

Often, more than one NP in a sentence carries given information. In that case, which of these NPs should be promoted to subject position? Ideally (all other things being equal), the NP that carries information most closely related to the paragraph topic—call it "topical information"—should go there. Consider this example:

> Not all investors will benefit from All-Savers Certificates. Investors exceeding a deposit of $7931 ($15,861 joint return) would have an after-tax yield far lower than with alternative investments, such as money market funds or Treasury bills. Alternative investments would also yield better after-tax yields and no penalty if the certificate was redeemed within the one-year maturity period.

The last sentence in this paragraph has three overt NPs which contain given information: *Alternative investments, after-tax yields,* and *the certificate.* Of these, the last seems to come closest to being thought of as topical information; the word *Certificate,* after all, does appear in the topic statement. But what is the *real* topic of this paragraph? Isn't it *different kinds of investors?* Notice, for example, that the word *investors* appears not only in the topic statement but in the subject position of the next sentence. Notice also that investors are referred to by implication as the delegated agent of the passive main verb: *was redeemed (by investors).* . . . Ideally, then, we should try to insert the word *investors* in the subject position of the third sentence, too, if it is at all possible. Indeed it is:

Not all investors will benefit from All-Savers Certificates. *Investors* exceeding a deposit of $7931 ($15,861 joint return) would have an after-tax yield far lower than with alternative investments, such as money market funds or Treasury bills. *Investors* redeeming their certificates within the one-year maturity period would also have a lower after-tax yield and would pay a penalty besides.

Not only does this rewritten version keep the focus on the topic of the paragraph—and thus contribute to paragraph unity—it also establishes parallelism between the second and third sentences, thus making it much clearer to the reader that we are talking about two different classes of investors: those who exceed a deposit of $7931 ($15,861 joint return) and those who redeem their certificates early. These two classes do not necessarily intersect. The original version, by contrast, links the third sentence to the second through the repeated use of the term *alternative investments,* with the second use appearing in subject position. The third sentence might therefore easily be misunderstood as simply modifying or adding detail to the second, and the paragraph as a whole might be misinterpreted as having a general-to-particular (not listing) pattern.

Put "Light" NPs Before "Heavy" NPs

As can be seen in the examples at the beginning of this chapter, NPs vary considerably in length, complexity, preciseness, etc. If we use the word *heavy* to describe NPs which are long and complex and the word *light* for NPs which are short and simple, the preferred stylistic ordering is light NPs before heavy NPs. For instance, consider the following passage:

We have received and acted upon requests for equipment from several branch offices. We have sent the research, development, and testing office in Chicago a gas analyzer.

The second sentence of this passage is awkward and difficult to read. It has a very heavy indirect object—*the research, development, and testing office in Chicago*—and a very light direct object—*a gas analyzer.* Thus, the ordering of NPs in this sentence, as it stands, is heavy . . . light. A more readable version of the second sentence and thus a better version would order the NPs light . . . heavy as follows:

We have sent *a gas analyzer*
                DIRECT OBJECT

to *the research, development, and testing office in Chicago.*
        OBJECT OF PREPOSITION

Notice that in moving the heavy NP to the end, we have to insert the preposition *to.*

Sentences are more readable if they are ordered light . . . heavy because of important limitations on the capacity of the mind to process information. These limitations have been described by George Miller, a famous psychologist, who demonstrated that the mind can usually process a maximum of nine separate items of information at a time. This means that as the mind reads through a sentence, it must try to make sense out of the sentence as quickly as possible. Ideally, the mind would see the entire structure of the sentence—subject, verb, indirect object, direct object—all within the nine-word processing limit. Unfortunately, this often does not happen since the main constituents of a sentence often span more than nine words.

However, even if you cannot put *all* of the main sentence constituents into a span of nine words or less, you can try to put *most* of them into such a span. One way to do this is to put light units before heavy ones. Let us consider our heavy . . . light sample sentence again:

```
 1     2    3
We    have sent
```
SUBJECT   VERB

```
 4      5           6            7       8       9    10    11
the research, development, and testing office in Chicago
```
INDIRECT OBJECT

```
12 13      14
a  gas analyzer.
```
DIRECT OBJECT

In this version of the sentence, the reader sees only the subject, the verb, and part of the indirect object before reaching the nine-word limit. However, in the light . . . heavy version shown below, the reader sees the subject, verb, direct object, and part of the indirect object (now the object of the preposition *to*) before reaching the nine-word limit:

```
 1      2    3    4 5       6
We    have sent a gas analyzer
```
SUBJECT   VERB   DIRECT OBJECT

```
 7  8      9           10          11    12     13   14      15
to the research, development, and testing office in Chicago.
```
OBJECT OF PREPOSITION

In the light . . . heavy version, then, by the ninth word the reader has a much fuller sense of the structure of the whole sentence and thus has a better chance of understanding the sentence with a minimum amount of effort.

From this discussion, it should be obvious that a heavy NP in the subject position would cause even more trouble for a reader than a heavy NP in the indirect object position. In a sentence with a heavy subject NP, the reader may not even get to the main verb before reaching the nine-word memory limit. Consider the following:

> The idea of designing and producing an economical AM/FM receiver that is both affordable for the average consumer and profitable for the company was introduced.

Here, all of the sentence before *was introduced* is the subject. This is a 23-word subject and poses serious comprehension problems for the reader, especially since the head noun, *idea,* must be held in memory while the reader processes 21 words before reaching the main verb, *was introduced.*

Note how much easier the sentence is to read if it is rewritten as follows:

> It was suggested that we design and produce an economical AM/FM receiver that is both affordable for the average consumer and profitable for the company.

In this rewritten version the reader has a much fuller sense of the sentence's structure earlier on and thus a better chance of understanding the sentence with a minimum amount of effort. This greater chance of easy understanding is the main reason writers try to order their NPs light . . . heavy.

One other situation should be considered before leaving this discussion of light . . . heavy ordering. That situation is the processing and "counting" of long introductory units, as in the following sentence:

> At the September and October meetings of all company engineers, it was suggested that we design and produce an economical AM/FM receiver that was both affordable for the average consumer and profitable for the company.

From what has been said so far, we would expect this example to be harder to read than the previous example because it begins with a 10-word phrase, *At the September and October meetings of all company engineers,* which comes before the subject and verb in the main clause. However, it doesn't seem to be any harder to read. This is because when readers process a sentence, they first process introductory clauses or phrases as a unit and then start over again with the main clause. Thus, in ordering units light . . . heavy, a writer should first order the units within the introductory clause or phrase and then order the units within the main clause:

> Light . . . heavy, light . . . heavy.

For ease of reading, the writer should also try to see that the introductory clause or phrase is no longer or not much longer than the nine words we have defined as the limit for easy remembering.

Sometimes a long subject is followed by a very short unit after the verb:

<u>*That we are in the midst of an energy crisis* is</u>   <u>   *obvious.*   </u>
    SUBJECT   PREDICATE ADJECTIVE

Such a construction sounds strange to most people because it violates the criterion of light before heavy. In such cases, a writer might shift the heavy subject NP to the end of the sentence and replace it with the pronoun *it* (this last step satisfies the requirement that all English sentences have a subject NP):

It is obvious that we are in the midst of an energy crisis.

Of course, if the unit after the verb is quite long and complex itself, then the need to shift the subject is not so strong:

<u>*That we are in the midst of an energy crisis* is</u>
                    SUBJECT

obvious to everyone in the country except the big politicians.

This sentence is more balanced than the first version, does not violate our light . . . heavy rule, and sounds better to most people.

The three situations we have discussed are (where S stands for *subject* and V for *verb*):

S __ V _____.   preferred

S _____ V _____.   acceptable if subject not too long

S _____ V __.   not preferred

14.2

WAYS OF SATISFYING THESE CRITERIA

In most cases, sentences can be constructed so that at least the first two criteria, if not all three, can be satisfied. Sometimes this procedure is straightforward. At other times, however, writers may find themselves boxed in, having written a sentence that satisfies the light-heavy criterion, say, but not the other two. The purpose of this section is to describe several of the most common ways that good writers get themselves out of such a situation.

Passive-Active Alternation

We often have a sentence such as the following, consisting of a subject NP, an active verb, and an object NP:

<u>Christopher Sholes</u> invented *the typewriter.*
        SUBJECT   DIRECT OBJECT

We can create a passive form of such a sentence by (1) interchanging the two NPs, (2) inserting an appropriate form of the verb *to be,* and (3) inserting the preposition *by:*

The typewriter was invented by <u>Christopher Sholes.</u>
        SUBJECT   OBJECT OF PREPOSITION

Although these two sentences have the same meaning, they differ in word order and thus should be used in different contexts. For example, suppose the topic of discussion is *typewriters* and no mention has been made of *Christopher Sholes:*

> A typewriter is a machine that prints alphabetic characters, numbers, and other symbols. It [the typewriter] was invented by Christopher Sholes.

In this case, the passive form used in the second sentence is appropriate because the NP representing given information *(the typewriter)* comes before the NP representing new information *(Christopher Sholes).* On the other hand, if the topic of discussion is *Christopher Sholes,* as in the passage below, then reference to him would constitute given information and the active form of the second sentence would be preferred.

> Christopher Sholes was a nineteenth-century inventor who made the life of the secretary much easier. He [Christopher Sholes] invented the typewriter.

Equative Shift

Equative sentences consist essentially of two NPs connected by a linking verb (usually some form of the verb *to be*). Furthermore, as the name implies, the NPs are interchangeable without any change of meaning:

> Air pollution is one major form of pollution.
> One major form of pollution is air pollution.

Here again the choice of which form to use depends mainly on given-new considerations. If *air pollution* is given information and we want to add new information about it (namely, that it's a major form of pollution), then the first version is preferred:

There are several things the average person should know about air pollution. *Air pollution is one major form of pollution.* It exists in all large cities and most small ones. And it is a secondary cause of death among the sick and elderly.

If we have already been talking about different forms of pollution, however, and now want to introduce air pollution as one such form, as in the next example, then *air pollution* represents new information and should come after the verb:

There are several types of pollution with which the average person should be familiar. *One major form of pollution is air pollution.* Other major forms of pollution are water pollution and noise pollution.

Caution! Do not mistake a nonequative sentence for an equative one! Equative shift, which we have just been describing, cannot be applied to nonequative sentences without changing the meaning! For example, consider this sentence:

Football players are good athletes.

This implies that all football players are good athletes, which is true. However, if the NPs are interchanged, we have

Good athletes are football players.

This implies that all good athletes are football players, which is not true. Clearly, these two sentences do not mean the same thing. Thus, they cannot undergo equative shift.

Indirect Object Shift
As was mentioned earlier, the relatively fixed word order of English sentences requires indirect objects to precede direct objects. But then what do you do if, in a given situation, the indirect object NP is either heavier or newer (or both) than the direct object NP? The answer is that you shift the indirect object and convert it into the object of a preposition, following the direct object NP. Consider this sentence, for example:

We are sending <u>*your branch office in Longview*</u>   <u>*a copy.*</u>
                            INDIRECT OBJECT   DIRECT OBJECT

Suppose this sentence were to occur in a context where a report is being discussed:

We have finished our final report on the new product line. We are sending your branch office in Longview a copy.

In this case, *a copy* (of the report) would not constitute much new information since reports are customarily made in copies, but the location where one of the copies is to be sent probably does constitute new information. Furthermore, the location is a longer, more complex (i.e., heavier) element in this case. Given these considerations, a preferred version of the sentence would shift the heavier and newer information to the end of the sentence:

We have finished our final report on the new product line.

We are sending ____*a copy*____
                 DIRECT OBJECT

to *your branch office in Longview.*
   OBJECT OF PREPOSITION

14.3

A PROCEDURE FOR PRODUCING MORE READABLE SENTENCES AND PARAGRAPHS

The purpose of this section is to summarize the material presented in this chapter and to outline a procedure for applying it. The chapter provides a number of criteria and suggestions for producing functional and readable sentences. By applying the criteria given in this chapter—in the order they are given—to each sentence as it appears in a paragraph or text, a writer should consistently produce more readable prose.

A summary of these criteria and a procedure for applying them are given in the flow chart in Figure 14-1. To illustrate the use of this chart, let us consider this paragraph:

> The sound reproduction from a radio can be quite inferior to that from a record or tape. One of the causes of radio's inferior sound reproduction, and in some environments perhaps the major cause, is multipath distortion. This type of distortion occurs when there are tall structures or buildings in the receiver's environment. Most of the radio signal is reflected and diffracted by these tall structures before it reaches the radio's antenna. When the signal is received by the radio, the radio interprets it as distorted and can only produce an inferior sound from it. In contrast, the sound on a record or tape suffers none of this distortion.

To evaluate this paragraph and revise it where necessary, let us follow the procedure outlined in Figure 14-1. Step 1 asks us to read the paragraph, and then step 2 asks us to decide if the topic sentence (the first sentence of the passage) is adequate. The answer is "yes" here because the topic sen-

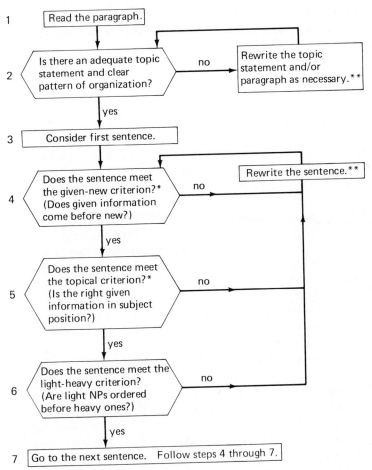

1 Read the paragraph.

2 Is there an adequate topic statement and clear pattern of organization? — no → Rewrite the topic statement and/or paragraph as necessary.**

yes

3 Consider first sentence.

4 Does the sentence meet the given-new criterion?* (Does given information come before new?) — no → Rewrite the sentence.**

yes

5 Does the sentence meet the topical criterion?* (Is the right given information in subject position?) — no

yes

6 Does the sentence meet the light-heavy criterion? (Are light NPs ordered before heavy ones?) — no

yes

7 Go to the next sentence. Follow steps 4 through 7.

*If the sentence being analyzed is the first sentence of the paragraph, this question may not be relevant.

**Sometimes it will not be possible to rewrite a sentence to meet all of the desired criteria.

FIGURE 14-1 Flow chart for editing sentences in paragraphs.

tence introduces the main idea, that *sound reproduction from a radio can be quite inferior to that from a record or tape,* and the rest of the paragraph goes on to describe why this is so.

SENTENCE 1 We then consider whether or not the first sentence meets the criteria outlined in steps 4 to 6.

STEPS 4 AND 5

It does not violate given-new or topical criteria since there is no previous given or topical information to consider here.

STEP 6

It meets light-heavy criteria because the subject NP *(the sound repro-duction from a radio)* is not significantly heavier that the object NP *(that from a record or tape).*

SENTENCE 2 Since the first sentence does not require any revision, we move on to step 7 and decide whether the second sentence meets the criteria outlined in steps 4 to 6.

STEPS 4 AND 5

It does not really violate given-new or topical criteria, although it would be desirable to get the given and topical information *(radio's inferior sound reproduction)* closer to the beginning of the sentence. We could rewrite the sentence to read

One cause of radio's inferior sound reproduction, and in some environments perhaps the major cause, is multipath distortion.

This sentence meets given-new and topical criteria.

STEP 6

Revised sentence 2 does not meet light-heavy criteria at all. It has a 15-word subject NP (everything before *is*) and only two words after *is.* Thus, to meet light-heavy criteria, we must rewrite the sentence again. Three possible revisions which meet all of the criteria in steps 4 to 6 are:

1   One cause of this inferior reproduction is multipath distortion; in some environments, this is perhaps the major cause.

2   Radio's inferior reproduction is partly caused by multipath distortion; in some environments, this is perhaps the major cause.

3   This inferior reproduction is partly caused by multipath distortion; in some environments, this is perhaps the major cause.

Since all of these revisions are quite similar, we shall choose the first for our revision.

SENTENCE 3 Since the second sentence does not require any more re-vision, we move on to step 7 and consider whether the third sentence meets the criteria outlined in steps 4 to 6.

STEPS 4 AND 5

It meets given-new and topical criteria because the first NP *(This type of distortion)* is given and topical information. However, since we have moved *multipath distortion* away from the end of sentence 2, we may wish to repeat *multipath distortion* at the beginning of sentence 3:

Multipath distortion occurs when there are tall structures or buildings in the receiver's environment.

This new version meets given-new and topical criteria.

STEP 6

It also meets light-heavy criteria because the subject NP (now *multipath distortion*) is lighter than the other NP *(tall structures or buildings in the receiver's environment)*.

SENTENCE 4   Since sentence 3 does not require any more revision, we move on to decide whether sentence 4 meets the criteria outlined in steps 4 to 6.

STEPS 4 AND 5

It violates given-new and topical criteria since the given and topical information *(these tall structures)* occurs after significant new information *(Most of the radio signal is reflected and diffracted)*. We could revise the sentence to meet given-new and topical criteria as follows:

These tall structures reflect and diffract most of the radio signal before it reaches the radio's antenna.

This meets given-new and topical criteria.

STEP 6

The rewritten sentence also meets light-heavy criteria.

SENTENCE 5   Since sentence 4 does not need any more revision, we move on to consider sentence 5.

STEP 4

It violates given-new criteria in the main clause because the given information *(it,* meaning *the signal)* follows the new information *(the radio)*. We could revise the sentence to meet given-new criteria as follows:

When the signal is received by the radio, it is seen as distorted and produces an inferior sound.

STEP 5

The revised sentence also meets topical criteria since the previous sentence talks about the *radio signal:*

These tall structures reflect and diffract most of *the radio signal* before *it* reaches the radio's antenna. When *the signal* is received by the radio, *it* is seen as distorted and produces an inferior sound.

If we rewrite sentence 5 with *the radio* as the subject, it would violate topical criteria:

These tall structures reflect and diffract most of *the radio signal* before *it* reaches the radio's antenna. When *the radio* receives the signal, *the radio* sees it as distorted and produces an inferior sound from it.

While *the radio* is mentioned in sentence 1 and thus is given information, sentence 4 introduces the topic of *the radio signal,* not *the radio.* Thus, to satisfy the topical criterion set up in sentence 4, sentence 5 must place *the radio signal* (or *the signal*) in the subject position:

These tall structures reflect and diffract most of *the radio signal* before *it* reaches the radio's antenna. When *the signal* is received by the radio, *it* is seen as distorted and produces an inferior sound.

STEP 6

Sentence 5 now also meets light-heavy criteria.

SENTENCE 6 Since sentence 5 does not require any more revision, we move on to consider sentence 6. It meets all of the criteria outlined in steps 4 to 6. The revised paragraph follows.

The sound reproduction from a radio can be quite inferior to that from a record or tape. One cause of this inferior reproduction is multipath distortion; in some environments, this is perhaps the major cause. Multipath distortion occurs when there are tall structures or buildings in the receiver's environment. These tall structures reflect and diffract most of the radio signal before it reaches the radio's antenna. When the signal is received by the radio, it is seen as distorted and produces an inferior sound. In contrast, the sound on a record or tape suffers none of this distortion.

Once we have arrived at this reordered version, we can combine and cut the individual sentences to produce a more concise version of the paragraph:

The sound reproduction from a radio can be quite inferior to that from a record or tape. One cause of this is multipath distortion; in some environments it is perhaps the major cause. This distortion occurs when there are tall structures in the receiver's environment which reflect and diffract the signal before it reaches the radio's antenna. The signal is thus distorted and can be used only to produce an inferior sound. In contrast, the sound on a record or tape suffers none of this distortion.

Procedures for effectively combining and cutting sentences are discussed in Chapter 15.

EXERCISE 14-1 Where necessary, reorder the noun phrases in the following passages to create more readable sentences.

A   The labor movement has existed in the United States for about 180 years. In the past 40 years, labor unions have become an integral part of the workplace. Over 19 million workers belong to labor unions. The wage rates, working hours, and working conditions of both unionized and nonunionized industries are influenced and determined by organized labor. Clearly, unions are permanent and powerful institutions in America.

B   *Advanced Component Technology for the Ford Pinto*   A number of systems, devices, and technologies to work with in meeting your specified requirements are available. One such device is the aneroid control. This device is used to limit the fuel while the turbocharger is building up manifold pressure during acceleration. The pressure-reducing valve is another such device. It reduces the pressure of fuel entering the carburetor. We have also developed other devices that we think will be useful in the future.

C   Several recent developments have encouraged a re-evaluation of wind-powered vessels for the American merchant fleet. The sharp rise in the cost of energy required to drive a fuel-powered ship is the most significant of these developments. A second development is the improvement of technology associated with wind-powered ships.

D   *General Appraisal of the EPA Plan*   The plan prepared by the Environmental Protection Agency (EPA) lacks several essential characteristics. It does not clearly outline program priorities nor does it relate priorities to overall program goals. The planning process is vague, and no guidelines are offered for future updates of the plan. It is difficult to see a rationale for the major thrusts suggested in the budget. For example, the plan offers no basis for the dominant expenditure on developing control technology over the 5-year period covered by the plan.

E   *Accomplishments of the Massachusetts Institute of Technology's Sea Grant Program*   The Sea Grant Program estimated the environmental effects which would be associated with oil exploration on the Georges Bank, off the New England coast. Among other matters, we reviewed what is now known about the biological effect of oil, particularly its toxicity. We analyzed how long a Georges Bank oil spill would be likely to stay on the Bank and where it would be likely to go. Available statistics on oil spills were reviewed, and the amounts of oil which might be spilled in New England under a number of developmental hypotheses were estimated. Also the effect of offshore drilling and of possible oil spills on the fishing industry were estimated. Finally, we made a preliminary analysis of the loss in regional income which would be associated with oil spills as a result of cleanup costs.

F   [From a report written by a geologist to a township council]

FOREWORD

Within this township, the greatest industrial and residential development has occurred in areas with limited groundwater resources. This development has already created water shortages and a need to deepen existing wells; more seriously, the development of new industry has been greatly handicapped or prevented. Thus, the township council authorized Moody and Associates to locate, if possible, several high-yield water wells to provide a sufficient municipal water supply. The purpose of this report is to describe the successful location of sand and gravel materials with excellent groundwater yielding capability.

SUMMARY

We conducted studies of six test drillings to locate an area in which a reliable, high-quality, municipal water well field could be developed. We have located such a site.

Results: Four of the six wells drilled were productive and penetrated sand and gravel suitable for development of municipal wells. Each of the four productive wells showed very good water-producing capability, yielding 300 to 500 gallons per minute. Further, generally excellent-quality water, though with a potentially high but treatable concentration of manganese, was produced. One test well showed manganese concentration now high enough to require treatment, and there is a concern that manganese levels in all the wells might increase to such a level. However, this would not prevent development of the well field, and additional study will be required to determine if manganese removal is necessary at all.

G   *The Future of Solar Energy*   Various studies have shown that solar energy could supply between 10 and 35 percent of U.S. energy by the turn of the century. The 20 percent estimate is the most given figure, which is approximately equal to current imported oil levels. Much of the 20 percent supplied would be used for space heating and for heating of hot water. Shown below is a graph which gives the results of a Stanford Energy Institute Research Project. It states three different scenarios and shows the effect of each on the percentage of solar energy in the United States.

H   In comparing the costs of electric cars to the gas car, the major factors to be considered are the price of the car and the total fuel cost. Maintenance costs do not have to be considered because they are about the same for both cars. The price of the cars considered here is based on a standard model with minimum options. The total fuel cost is composed of battery replacement and electricity consumption for the electric car, whereas the cost of gas is the total fuel cost for the gas car. The following results were used for comparison.

I   Industry will be needed to produce the electric car. The present U.S. auto companies will be the ones who will have the capabilities to produce an electric car. All four U.S. auto companies (General Motors, Ford, Chrysler, and American Motors) are encouraged about the development and production of electric cars. These companies are all doing research in the area of electric cars, with General Motors leading the way.

ADDITIONAL READING

Herbert Clark and Susan Haviland, "Comprehension and the Given-New Contract," *Discourse Production and Comprehension*, R. Freedle (ed.), Ablex, Norwood, N.J., 1977.

Barbara Hayes-Roth and Perry W. Thorndyke, "Integration of Knowledge from Text," *Journal of Verbal Learning and Verbal Behavior* **18**:91–108 (1979).

Michael P. Jordan, "Some Associated Nominals in Technical Writing," *Journal of Technical Writing and Communication* **11**(3):251–264 (1981).

Ronald Langacker, "Movement Rules in Functional Perspective," *Language* **50**:629–664 (1974).

Charles Li (ed.), *Subject and Topic*, Academic Press, New York, 1976.

C. A. Perfetti and S. R. Goldman, "Thematicization and Sentence Retrieval," *Journal of Verbal Learning and Verbal Behavior* **13**:7–79 (1974).

15

MAKING YOUR WRITING READABLE: EDITING FOR EMPHASIS

Although some readers may prefer to skim-read, others have to read more closely and thoroughly, concentrating on details. For these readers, there is a danger of getting *lost* in the details, of overlooking main points and "not seeing the forest for the trees," so to speak. Consequently, the details themselves begin to lose significance: the reader cannot see exactly how they fit into the larger picture and thus cannot evaluate their importance. The reading process as a whole bogs down at this point, and the reader is forced to stop and start over.

When readers get bogged down in detail like this, it's often the writer's fault. Many writers make little effort to organize details in a coherent, unified way, preferring instead to have the reader do all the work. But this invites the kind of failure just described. Readers are often pressed for time, or tired, or have other things on their mind. Many readers lack the kind of background knowledge the writer has. Still others have poor reading techniques and are unable to decipher poor writing, no matter how hard they work at it. In general, readers are at the mercy of the writer; they depend on the writer to present details in such a way that the role of these details in support of main points is readily apparent. If the writer fails to do this, there is little the reader can do except try to figure things out.

It thus falls on the writer to mold the details of a text so that they reinforce main points in unified fashion. This is somewhat similar, actually, to the demands made on a speaker engaged in serious conversation. Face-to-face conversation is an intensive form of communication in which the speaker is acutely aware of the listener and vice versa. Because of this close speaker-listener relationship, conversations are governed by certain unwritten rules: say what you mean, don't beat around the bush, get to the point, be honest, etc. If the speaker violates any of these rules, the conversation will begin to break down unless the listener rescues it with a corrective comment, such as "I don't see what you're driving at" or "What's your point?" The possibility of such immediate feedback from the listener forces the speaker to make every detail relevant to the conversation; most listeners are simply intolerant of irrelevant details and will either intervene or break the conversation off if the speaker strays too far from the topic of discussion.

Good conversationalists, of course, are aware of such constraints and employ various techniques to make it clear to the listener that they are observing the rules. For one thing, they use emphatic intonation, physical gestures, inverted sentence structure, intensifiers, and other devices to signal important words—key words, topical words, words carrying new information. Conversely, they use none of these devices for the less important words—those that carry given information or redundant information. As for empty, meaningless words that serve no communicative purpose at all, they are simply omitted. In general, both by giving prominence to important words and by subordinating or omitting unimportant ones, good conversationalists emphasize those aspects of a detailed discussion that link the discussion to the main point or purpose of the conversation. As a result, the listener not only absorbs those details but also sees just how they support the main point.

Writers should do the same kinds of things as good conversationalists. They may not be in as close touch with their audience as speakers are and so they may not have such immediate demands placed on them, and they cannot, of course, use intonation and gestures in their writing. But writers do have an audience, and this audience needs to know, just as listeners do, how the details of a discussion are related to the main points. Furthermore, writers have as many devices as speakers do for helping the reader see how details support main points. In short, the use of emphasis is as appropriate and indeed necessary to good writing as it is to good conversation.

This chapter describes the most common and useful devices used by good writers to create emphasis within individual sentences. These fall into three categories: devices used to highlight important words and phrases, devices used to subordinate relatively unimportant words and phrases, and devices used to eliminate unnecessary words and phrases. This chapter thus complements several earlier chapters (2, 13, and 14), which also suggest ways of emphasizing details, but with more of a paragraph orientation.

15.1

COMBINE CLOSELY RELATED SENTENCES

Combine closely related sentences unless there is a compelling reason not to (such as maintaining independent steps in a list of instructions or avoiding extreme sentence length); put main ideas in main clauses. Many inexperienced writers have a tendency to use nothing but short, simple sentences, producing a very choppy style of writing which irritates the reader with its sing-song rhythm and, worse, fails to put emphasis on important ideas. This tendency derives, probably, from two principal sources: (1) an overemphasis in many quarters on the need to avoid dangling modifiers, comma splices, and other problems associated with complex sentence structures and (2) an erroneous belief, promoted by readability formulas, that short sentences make reading easier. Dangling modifiers, comma splices, and other errors of sentence structure and punctuation should, of course, be avoided—but not at the expense of emphasis, unity, and coherence. And although a short sentence by itself may be easier to read than a long sentence, the repeated use of short sentences may have just the opposite effect.

The best approach to take regarding sentence length is to *let the form reflect the content.* If an idea is complex enough to require qualification, the best way to qualify it may be with a relative clause, an adverbial phrase, or some other complex modifier. On the other hand, if an idea is simple and straightforward, a simple sentence may be the best way to represent it. Often, these choices can be made properly only within the context of an entire paragraph. For example, consider the following paragraph from a student report:

ORIGINAL VERSION

At the present time electric car utilization is not possible. The problems holding it back are satisfactory performance and costs. Performance problems of lack of speed, short mileage range, and lack of acceleration are present. Cost problems are the price of battery replacement and the base price of the electric car. It is possible though, with research and development, that these problems can be solved in the future.

Each of the first two sentences, taken in isolation, is grammatically correct and easy to read. When you look at them together, however, you notice that there is excessive overlap between them: sentence 2, in other words, contains too much given information *(The problems holding it back).* This unnecessary redundancy can be eliminated by combining these sentences:

At the present time electric car utilization is not possible because of performance and cost problems.

Not only does this move reduce the wordiness of the first two sentences, it also creates a better topic statement: it's more unified and emphatic, and it introduces the key terms *performance problems* and *cost problems.* (Notice how these terms are the subjects of the next two sentences.) If we also change sentence 3 to satisfy given-new and light-heavy criteria, we can reduce the wordiness of the paragraph and increase its readability still further. The overall result is this:

FIRST REWRITE

At the present time electric car utilization is not possible because of performance and cost problems. The performance problems are lack of speed, short mileage range, and lack of acceleration. The cost problems are the price of battery replacement and the base price of the car. It is possible, though, with research and development, that these problems can be solved in the future.

This is a significant improvement, but we have other options that might improve it even more. For example, now that we've converted the original sentence 2 into a prepositional phrase, we can shift it into presubject position in place of the time adverbial originally there:

> Because of performance and cost problems, electric car utilization is not possible at the present time.

This puts more focus on the key terms *performance problems* and *cost problems* and less focus on the less important time adverbial. Another change we could make, though not as compelling a one as those just described, would be to combine the two sentences in the middle with a semicolon. These two sentences are closely related in function; linking them formally would reflect this relatedness.

FINAL VERSION

Because of performance and cost problems, electric car utilization is not possible at the present time. The performance problems are lack of speed, short mileage range, and lack of acceleration; the cost problems are the price of battery replacement and the base price of the car. It is possible, though, with research and development, that these problems can be solved in the future.

We might entertain (though ultimately reject) one other change: substituting *its* for *the* in sentence 2 so as to have *Its performance problems* and *its cost problems*. This might be desirable from the point of view of reiterating the paragraph topic *(electric car utilization).* Unfortunately, it would create excessive repetition of the word *it* in the subject position: notice that the last sentence of the paragraph also begins with this word. Using the same pronoun to lead off all three sentences would constitute misleading parallelism, since *it* does not refer to the same thing in all three cases.

One must often make trade-offs like this in deciding how best to structure sentences and express ideas. This is simply inevitable. Language is highly complex and can trigger highly complex perceptual processes in the reader's mind. Expressing an idea in a certain way may have not only some benefits but also some undesirable side effects. Expressing the same idea in a different way may have fewer (or more) benefits but also fewer (or more) undesirable side effects. In such cases, the writer should strive for optimization, not perfection.

In the example just discussed, for instance, we could retain the two *its* in sentence 2 and avoid the misleading parallelism by simply changing the form of the final sentence into something like *Research and development, however, may be able to solve these problems in the future.* This would give us the benefits of repeated reference to the paragraph topic without the undesirable side effect. However, in doing so we would be sacrificing a number of benefits associated with the original form of the final sentence. For one thing, the original form begins with *It is possible,* which contrasts nicely with *is not possible* in the topic statement; this emphasizes the fact that these two sentences have contrasting functions. The revised alternative does not contain this phrase. Second, the original form further emphasizes this contrast with the topic statement by having two "possibility phrases": *is possible* and *can be;* the revised alternative has only one *(may be able)*. Finally, the original form gives more prominence to the key word *problems* by putting it in subject position in the *that* clause; the alternative form, by contrast, has it in object position.

There are, of course, many other alternatives that could be considered (though we might not want to, given the usual time constraints!). Let us assume, however, that the second rewrite is the best version we can devise, and now let us take stock of the benefits we've gained. First and foremost, we've produced a paragraph whose form reflects its content:

Because of performance and cost problems, electric car utilization is not possible at the present time.	MAIN IDEA
The performance problems are lack of speed, short mileage range, and lack of acceleration; the cost problems are the price of the battery replacement and the base price of the car. It is possible, though, with research and development, that these problems can be solved in the future.	DETAILS OF PROBLEM POSSIBLE SOLUTION

The ideas in this rewritten version are grouped so that there is a one-to-one correspondence between form and content: each sentence has a particular, identifiable function. Most important, we now have an effective topic statement. Second, we've put more prominence on the key words and phrases of the paragraph by placing them at the beginning of the topic statement. They are then reinforced in the following sentences, where they appear in subject and initial positions. Finally, we've eliminated a little wordiness (64 words instead of 68).

In general, combining sentences is often a good way to create emphasis in your writing. By making it easy for your readers to see the relatedness of ideas, you make it easier for them to absorb these ideas. You can also show explicitly that one idea is logically subordinate to another by putting the more important idea in the main clause of the sentence and the less important idea in a subordinate clause. For example, suppose you wanted to combine the two sentences in italics in the following paragraph:

NEGATIVE EXAMPLE
Electric cars must be able to meet the same safety standards that gasoline cars must meet as set up by the Department of Transportation. *These standards are derived from an established crash test. In the crash test, the car is propelled against a solid wall at 30 mph.* The data obtained from the crash test are analyzed for fuel spillage, fuel system integrity, windshield retention, and zone intrusion.

In combining the two italicized sentences, we could subordinate the more detailed second sentence to the more general first one:

These standards are derived from an established crash test, in which the car is propelled against a solid wall at 30 mph.

Alternatively, we could maintain prominence on the details and subordinate instead the idea that the crash test is an established one:

These standards are derived from propelling the car against a solid wall at 30 mph, which is an established crash test.

Clearly, the first option is the more appropriate one in this context: the fact that the crash test is an established one underscores the main idea of the paragraph, as stated in the topic sentence.

REVISED VERSION

Electric cars must be able to meet the same safety standards that gasoline cars must meet as set up by the Department of Transportation. These standards are derived from an established crash test, in which the car is propelled against a solid wall at 30 mph. The data obtained from the crash test are analyzed for fuel spillage, fuel system integrity, windshield retention, and zone intrusion.

There are times when it is best *not* to combine sentences. For example, if you are giving a list of instructions and want to emphasize independent steps in accordance with how the user might carry out the instructions, you might want to state these steps in independent sentences. To see how this might apply in a specific case, consider the following set of instructions for replacing a brake line in an automobile:

1 Disconnect the union nuts at both ends.

2 Unclip the line from the chassis.

3 Pull the line out.

4 Install the new line in the chassis clips.

5 Moisten the ends in brake fluid.

6 Tighten the union nuts.

You could leave this set of instructions as is, in the form of a formatted list. Or you could combine some of the steps (2 with 3, 5 with 6) to create a more realistic four-step sequence of disconnect-remove-install-reconnect, as is done in this excerpt from a repair manual:

> To replace a brake line, disconnect the union nuts at both ends. Unclip the line from the chassis and pull it out. Install the new line in the chassis clips. Moisten the ends in brake fluid, then tighten the union nuts.[1]

To combine sentences beyond this, however, would be a mistake because it would destroy the emphasis we want to maintain on certain individual steps. For example, if we were to combine sentences 2 and 3 in the repair manual version, this would be the result:

NEGATIVE EXAMPLE

> To replace a brake line, disconnect the union nuts at both ends. Unclip the line from the chassis, pull it out, and install the new line in the chassis clips. Moisten the ends in brake fluid, then tighten the union nuts.

By lumping together the remove and install steps like this *(Unclip the line from the chassis, pull it out, and install the new line in the chassis clips),* we would be creating an imbalance in the sequence: no mechanic would consider this to be a single step, as the form of the description implies.

It's also best not to combine sentences when the result would be too long a sentence. Suppose, for example, you have been writing a proposal for a computer-aided design system and have included this paragraph in your summary:

> The proposed system is required to alleviate the increase in demand. The system will do that by removing the burden of data entry from the present system, CADDS. This is accomplished by utilizing the microcomputer as a stand-alone data entry system. The microcomputer has all of the graphics and software capabilities required to implement this concept.

As it stands, this paragraph is a nicely written one, with an adequate topic statement, a clear general-to-specific pattern of development, and properly constructed sentences satisfying the given-new, light-heavy, and topical criteria discussed in Chapter 14. The result is a highly readable paragraph with appropriate emphasis on the main ideas and key words. If you were to combine the sentences into one, on the other hand, much of this emphasis would be destroyed:

NEGATIVE EXAMPLE

> The proposed system is required to alleviate the increase in demand by utilizing the microcomputer as a stand-alone entry system with all the necessary graphics and software capabilities to remove the burden of data entry from the present system, CADDS.

This is a more economical version, no doubt, insofar as it contains 16 fewer words than the original. But is it more readable? Absolutely not! In fact, it's a perfect example of the kind of incomprehensible gobbledygook that so many readers of technical writing complain about. The lesson to be learned from this example, then, is this: do not combine sentences just for the sake of doing so; do it only when it serves a purpose.

EXERCISE 15-1 Each of the following paragraphs lacks flow and emphasis because the sentences are all short and choppy. Break up this choppiness by combining those sentences that are most closely related. Be sure to put main ideas in main clauses.

A The Cleveland Engine Plant will begin making diesel engines in 1982. The diesel engines will be produced along with the gas engines already in production. However, there is only one area for machining pistons. Both gas pistons and diesel pistons must be machined there.

B I recommend using the cart system to reduce the costly existing production layout for machining gas pistons and diesel pistons. The cart system is the least expensive of the three systems. Also, the cart system meets the desired production rate for all probable product mixes. The cart system is the most compact of the three systems and fits in the existing piston machining area. The cart system layout should be used in the piston machining area.

C The maximum speeds of the electric cars are much lower than those of the gas cars. The maximum speeds of the electric cars are barely fast enough to maintain freeway speeds. This causes problems with utilization of electric cars on freeways, and it almost limits their use to residential streets. The maximum speed must be increased to give electric cars a better chance for utilization.

D Operation of the ammonia process has been only marginally profitable. Deterioration of the hydrogen recovery unit to unacceptably low efficiency has contributed to the high operating costs of this process. Process Engineering is preparing recommendations for replacing the hydrogen process unit with a similar unit. However, an alternative process is available. The alternative is to recover argon from the synthesis loop purge gas. On my own initiative, I have investigated the applicability of this process to the Des Plaines River Ammonia Plant. In this report, I present the results of my study of product outputs, utility requirements, and capital cost. I conclude that this process could improve considerably the ammonia process profitability. I seek approval from management to investigate this option in more detail.

15.2

USE INTENSIFIERS AND SIGNAL WORDS

All main points of a discussion should be supported, of course, by details; conversely, all details used should support main points. When this relationship between main points and details is established in an obvious way, emphasis is created. On the other hand, when it is left up to the reader to figure out how the details and main points are related, emphasis is lost. This means that you should take pains to show how details and main points are related. Make it obvious! If this means taking sides in an issue, so be it! In most situations, readers won't expect you to be totally neutral; indeed, they will want you to cast information in a certain light so as to more easily give it a single, unambiguous interpretation. This does not mean that you should forsake your objectivity. Not at all. It means simply that once you have used your objectivity to reach a certain conclusion, you should then make every effort to orient your writing toward that conclusion. Don't make the mistake of doing what many inexperienced writers do, just presenting detail after detail in a flat, colorless, pseudoobjective way and forcing the reader to see what purpose these details are supposed to have; you will only be wasting the reader's time and risking misinterpretation as well. Instead, take care to present details in such a way that the reader can easily see how they support your main points.

How can you do this? In addition to a number of techniques already discussed—such as using good topic statements, putting topical words in the subject position, and combining sentences—you can use appropriate signal words and phrases at the beginning of sentences: conjunctive adverbs like *specifically, in general, conversely, furthermore, on the other hand, however, therefore* and subordinating conjunctions like *since, because, although, despite.* (Many of these have already been discussed in connection with the paragraph patterns described in Chapter 2.) You can also use short adverbial phrases like *for this purpose, with this in mind, before taking any further steps, in order to determine the market for this product* at the beginning of sentences.

To see the effect of using signal words and phrases, consider first this negative example:

NEGATIVE EXAMPLE

The more-complete-expansion cycle is more efficient than the conventional diesel cycle. It uses the same four strokes but a relatively smaller combustion chamber. The intake valve closes when the piston is at some point A between TDC and BDC in its intake stroke. This draws in less air-fuel mixture. The compression, power, and exhaust strokes are identical to those of the diesel cycle. The increase in volume of the air-fuel mixture due to the piston moving to BDC is the same. The temperature and pressure of the gases after the power stroke is over are lower. More of their energy has been utilized. This increases the efficiency of the cycle. The smaller combustion chamber is necessary. The temperature and pressure at TDC in the compression stroke must be high enough to cause combustion of the mixture. The mixture must be squeezed into a smaller volume to get the same temperature and pressure.

It's pretty hard to make sense of this, isn't it? Certainly there's little emphasis on main points. Notice what happens, however, when we insert a few conjunctive adverbs, subordinating conjunctions, and adverbial phrases:

REVISED VERSION

The more-complete-expansion cycle is more efficient than the conventional diesel cycle. It uses the same four strokes but a relatively smaller combustion chamber. *In its intake stroke,* the intake valve closes when the piston is at some point A between TDC and BDC, *thus* drawing in less air-fuel mixture. The compression, power, and exhaust strokes are identical to those of the diesel cycle. *However, since there was less mixture to start with and since* the increase in volume due to the piston moving to BDC is the same, the temperature and pressure of the gases after the power stroke is over are lower. More of their energy has been utilized, *thus* increasing the efficiency of the cycle. The smaller combustion chamber is necessary *because* the temperature and pressure at TDC in the compression stroke must be high enough to cause combustion of the mixture. *Since there is less mixture in the chamber,* the mixture must be squeezed into a smaller volume to get the same temperature and pressure.

Clearly, this rewritten version is far more comprehensible than the original. The signal words and phrases help us see the logic of the paragraph—how the sentences work together to explain how the more-complete-expansion cycle is more efficient than the diesel cycle.

Another way you can highlight details and support main points is to use intensifiers, that is, words that amplify the meaning or connotation of other words. For example, if you say that the loss of certain equipment would impede progress on a certain construction project, that's one thing; if you say that it would *greatly* impede progress, that's quite another. The word *greatly* is an intensifier: it intensifies the meaning of an accompanying verb (in this case, *impede*) and thus emphasizes that meaning. Some other commonly used intensifiers in scientific and technical writing are *significant, significantly, considerable, considerably, obvious, obviously, easy, easily, much, many, complete, completely,* and *very.*

The use of intensifiers is particularly important when you are presenting numerical quantities whose significance is not immediately apparent to the reader. Don't just say "The present worth of the hydrogen process is $4.14 million" and expect the reader to know how to evaluate that figure. Is $4.14 million a lot for such a process, in the circumstances given, or isn't it? Many readers wouldn't know. If you insert the word *only,* however, you can give the reader a big assist: "The present worth of the hydrogen process is *only* $4.14 million." This immediately indicates that $4.14 million is a relatively small figure for that set of circumstances.

Intensifiers should be used even when the significance of a numerical quantity is fairly apparent—simply for emphasis. For example, instead of writing,

> The present worth of the argon process is $10.25 million. The present worth of the hydrogen process is $4.14 million.

it would be more effective to write

> The present worth of the argon process is $10.25 million whereas that of the hydrogen process is *only* $4.14 million.

This use of *only* may not be necessary for clarity, but it provides emphasis and thus underscores the fact that the argon process is more valuable than the hydrogen process. It is often bits and pieces of emphasis like this that add up and have a significant cumulative impact on the readability of a passage as a whole.

Some other intensifiers that can be used to emphasize numerical values are *more than, at least, almost, less than, not even, barely more than, well under,* and *at most.* These usually put a positive or negative slant on values and so should be used judiciously, that is, only in cases where the reader expects you to take more of an advocate's role, as when making a proposal or making recommendations. Here is an example, taken from a report making recommendations to meet EPA clean air requirements:

> Repairing the flyash returns is the optimal solution because of economic and downtime considerations. The three return units can *easily* be repaired by three of our employees with a downtime of *no more than* two days and a cost of *less than* $1,375, including labor. To have the units replaced, *on the other hand,* would take *at least* two months' downtime and, because of our limited staff, would have to be done by outside contractors at a cost of $70,000 *or more.*

Leaving these intensifiers out would not change the meaning of the paragraph significantly but would definitely reduce its effectiveness:

> Repairing the flyash returns is the optimal solution because of economic and downtime considerations. The three return units can be repaired by three of our employees with a downtime of two days and a cost of $1,375, including labor. To have the units replaced would take two months' downtime and, because of our limited staff, would have to be done by outside contractors at a cost of $70,000.

There are many situations, of course, where facts and figures are expected to be given in a neutral, unbiased fashion, e.g., in reporting test results to other specialists. In such cases, it is presumed that the readers are very knowledgeable about the subject matter and can easily interpret the data themselves. There are many other situations, however, where the readers are not so knowledgeable and need a helping hand. Although you should never use intensifiers to create a false impression, you should use them to help readers gain a true impression.

EXERCISE 15-2 Pretend that you are the fleet manager of a taxicab company. Your profit margin was much lower than expected last year because your costs soared while revenues remained about the same. In light of this problem, the company president several months ago had you purchase and field test several new front-wheel-drive automobiles. You kept careful records of each car's performance, as summarized in the following table:

CAR	FUEL COST PER MILE	TURNING CIRCLE (FEET)	EPA VOLUME INDEX*	MAINT. COST PER MILE	PURCHASE PRICE
Old fleet average	10.0¢	41.6	150	3.6¢	$6800
Buick Skylark	6.8	33.5	135	3.3	7600
Plymouth Horizon	6.4	32.8	121	3.8	7200
Volkswagen Dasher	5.9	33.2	129	3.3	9100

*The EPA volume index is a measure of relative roominess and cargo capacity.

Write the Summary section of a short informal report to the company president describing these test results and recommending that the company replace its present fleet with one of these new models. Be sure to use signal words and intensifiers to emphasize your point of view.

EXERCISE 15-3 The following samples of technical writing fail to give proper emphasis to important points. Correct this deficiency by providing topic statements or summarizing statements, by combining sentences, by inserting intensifiers or signal words, by eliminating unnecessary words, and/or by making other appropriate changes.

A   The fuel cost for the electric car is much lower than that for the gas car. When the price of battery replacement is added, the fuel cost rises considerably. For an average distance of 10,000 miles/yr, the total fuel cost of the electric car is about 33% greater than that of the gas car. This amount may seem to be significant, but gas prices are rising faster than electricity prices. Also, the price of battery replacement should go down slightly as batteries improve. The total fuel cost poses a problem for utilization at the present time, but it could be alleviated in the near future.

B   *Labor Cost*   The same number of workers is not used for each layout, but each layout uses six machine operators. Four extra workers are needed to operate the forklifts in the process layout. The workers make $10 per hour. Based on a 40-hour work week, 50 weeks a year, the extra workers cost $20,000 per worker per year. No extra workers are needed in the product layout, but its production rate is well below that needed to meet forecasted demand. More worker-hours will be needed to meet production. This would require a second shift or overtime to be used. Overtime is less costly than running a second shift to meet required demand (see Appendix B for the calculations). There is no extra labor cost in the cart system since it is fully automated. The labor costs associated with each layout are given below. All three include the labor cost of the six machine operators.

LAYOUT	LABOR COST
Product Layout	$228,000
Cart System	120,000
Process Layout	200,000

15.3

BE CONCISE

Be concise. While the more important words and phrases of a text should be highlighted, the less important ones should be subordinated—or perhaps even eliminated altogether. Unnecessary words and phrases will only detract from the emphasis you've carefully tried to build up through the use of combined sentences, signal words, and intensifiers. A bloated, wordy style can submerge your readers in a sea of empty terms, making it next to impossible for them to follow your main points and be persuaded to your point of view. In fact, foggy language is more likely than not to turn readers against you. Recall the MacIntosh survey cited in Chapter 1: of the 182 company executives, senior scientists, project leaders, and other prominent scientific and technical people who were asked to evaluate the writing they read, all 182 complained about "generally foggy language." More specific complaints were "failure to connect information to the point at issue" (169 respondents), "wordiness" (164), and "lack of stressing important points" (163).

Inexperienced writers sometimes think that they *must* use a wordy, bloated style of writing in order to create a certain professional image. They seem to believe that by using pretentious language they will enhance their image as experts in their field. Actually, what evidence there is suggests just the opposite: pretentious, wordy language is less likely to promote one's credibility as an expert than is concise, direct, simple language. For example, in a recent survey[2] taken at a conference of the British Ecological Society, 74 scientists were asked to read two versions of the same information, one version (attributed to a person named "Brown") being noticeably wordier than the other (by a person named "Smith"):

BROWN'S VERSION

In the first experiment of the series using mice it was discovered that total removal of the adrenal glands effects reduction of aggressiveness and that aggressiveness in adrenalectomised mice is restorable to the level of intact mice by treatment with corticosterone. These results point to the indispensability of the adrenals for the full expression of aggression. Nevertheless, since adrenalectomy is followed by an increase in the release of adrenocorticotrophic hormone (ACTH), and since ACTH has been reported (P. Brain, 1972) to decrease the aggressiveness of intact mice, it is possible that the effects of adrenalectomy on aggressiveness are a function of the concurrent increased levels of ACTH. However, high levels of ACTH, in addition to causing increases in glucocorticoids (which possibly accounts for the depression of aggression in intact mice by ACTH), also result in decreased androgen levels. In view of the fact that animals with low androgen levels are characterized by decreased aggressiveness the possibility exists that adrenalectomy, rather than affecting aggression directly, has the effect of reducing aggressiveness by producing an ACTH-mediated condition of decreased androgen levels.[2]

SMITH'S VERSION

The first experiment in our series with mice showed that total removal of the adrenal glands reduces aggressiveness. Moreover, when treated with corticosterone, mice that had their adrenals taken out became as aggressive as intact animals again. These findings suggest that the adrenals are necessary for animals to show full aggressiveness.

But removal of the adrenals raises the levels of adrenocorticotrophic hormone (ACTH), and P. Brain found that ACTH lowers the aggressiveness of intact mice. Thus the reduction of aggressiveness after this operation might be due to the higher levels of ACTH which accompany it.

However, high levels of ACTH have two effects. First, the levels of glucocorticoids rise, which might account for P. Brain's results. Second, the levels of androgen fall. Since animals with low levels of androgen are less aggressive, it is possible that removal of the adrenals reduces aggressiveness only indirectly: by raising the levels of ACTH it causes androgen levels to drop.[2]

Not surprisingly, the vast majority of these scientists (88 percent) found Smith's version easier to read. They also found Smith's style "more appropriate" (58 percent versus 8 percent for Brown, with 33 percent finding both styles equally appropriate). What is most significant, however, is that those who were willing to venture an opinion about the authors' probable scientific competence on the basis of their writing styles clearly favored Smith, as can be seen in Table 15-1. A larger sample, with 338 other respondents, was taken elsewhere, with similar results.

TABLE 15-1 RESPONSES FROM 74 SCIENTISTS JUDGING
THE SMITH AND BROWN ABSTRACTS

	YES, SMITH (%)	YES, BROWN (%)	NO DIFFERENCE (%)
Does one author give the impression of being more competent?	31	12	57
Does one author seem to have a better organized mind?	73	3	23
Does one author seem to be more objective?	24	8	68
Does one author inspire more confidence in what is being said?	44	15	41
Does one author seem to have a more dynamic personality?	46	7	47

SOURCE: Turk, Reference 2.

The more concise style of Smith's abstract (155 words versus 179 for Brown's) was definitely preferred by these scientists, to the point even of crediting Smith with probably being a better scientist. What is it, aside from the better formatting, that makes Smith's style so appealing? For one thing, it's not so "noun-heavy": it *has a higher percentage of verbs and adjectives* than does Brown's style. For example, instead of saying *effects reduction of*, it simply says *reduces*. Instead of *point to the indispensability of the adrenals*, it has *suggest that the adrenals are necessary*. Instead of *producing . . . a condition of decreased androgen levels*, it has *causes androgen levels to drop*. Second, the Smith style *has simpler sentence structure*, with fewer and shorter adverbial phrases before the sentence subject. This means that the reader reaches the main verb of the sentence sooner, making it easier to process the sentence as a whole (see Chapter 14 for a detailed explanation). For example, in the Brown version, the third sentence (beginning with *Nevertheless*) has 32 words before the main verb, and the sentence immediately following that one has 26 before the verb! In fact, the average number of words preceding the main verb in the Brown version is 17.6, versus only 6.8 for the Smith abstract. Thirdly, the Smith style *avoids unnecessary technical terms* in favor of more commonplace equivalents, even when it requires more words to make the substitution. In place of *adrenalectomised mice*, for example, Smith has *mice that had their adrenals taken out;* instead of *are a function of*, there is *are due to*. Finally, Smith's style *uses more pronouns and demonstrative adjectives: their* in sentence 2, *these* in sentence 3, *this* and *it* in sentence 5, and *it* in the last part of sentence 9. By contrast, Brown's style has only the one demonstrative *These*, leading off sentence 2. Pronouns and demonstrative adjectives, in general, help make a text more cohesive—provided, of course, that it's clear to the reader what they refer to.

This last point deserves some discussion before we end this chapter. Scientists, engineers, and other technical people sometimes use full noun phrases repeatedly to avoid being "imprecise." They've heard of cases, perhaps, where a single misinterpretation of a pronoun by a single reader has led to some accident or mishap, which in turn has led to the writer's company being sued for damages. Therefore they tend to avoid pronouns and demonstratives altogether, preferring instead to repeat full noun phrases over and over. This strategy is certainly a safe one, and indeed it should be used in appropriate circumstances (such as when writing operating instructions for a potentially hazardous machine or when writing a legally binding contract). There are many circumstances, however, where such caution is uncalled for, and where in fact it simply disrupts the coherence of the text. Consider this example:

NEGATIVE EXAMPLE
In order to keep from delaying the construction phase of the Parkway Office Building, the Geotechnical Division needs to know the loads that will be placed upon the footings. I have investigated the proposed use of the structure and various flooring systems to determine the loads that will be placed upon the footings. This report gives the loads on the footings and explains how these loads were derived.

There is no reason to *describe* the loads every time they're *referred* to! Pronouns and demonstratives can be used instead without any real risk of misinterpretation, and the result will be a more coherent and more concise text:

REVISED VERSION
In order to keep from delaying the construction phase of the Parkway Office Building, the Geotechnical Division needs to know the loads that will be placed upon the footings. I have investigated the proposed use of the structure and various flooring systems to determine these loads. This report gives the loads and explains how they were derived.

In general, when you have to refer repeatedly to some object or concept that has first been introduced with a long noun phrase, you can usually use a shortened version of this noun phrase and a demonstrative adjective or definite article without much, if any, risk of ambiguity. Here is a typical example:

Damper position indicators shall be provided on all damper units. *These indicators* shall be located a minimum of 1'-0" from the breeching to allow for installation of thermal insulation.

It would also be safe to substitute *The indicators* or perhaps even the pronoun *They* in this context. (Note, however, that the demonstrative pronoun *These* by itself could be misinterpreted as referring to the damper units.) It would not be necessary to repeat the full noun phrase *damper position indicators*.

Sometimes, however, it *is* necessary to repeat full noun phrases to assure proper interpretation. Consider this example:

NEGATIVE EXAMPLE
The two dampers installed in the flues shall be tied with the inlet dampers of the fans located at the exit side of the electrostatic precipitators. *These dampers* will open automatically in the event of any failure in the electrostatic precipitator system. Contractor shall provide proper instrument and control for the actuation of *these dampers*.

Now, what does *these dampers* refer to? The flue dampers? The inlet dampers? Or both? In cases like this, the writer has little choice but to repeat the full NP. In fact, it is often important in such cases for the writer to repeat the full NP *with exactly the same wording,* especially if the paragraph is long and complex. Look at the following badly written example and see if you can keep track of the different orders involved:

NEGATIVE EXAMPLE
On July 8, 1981, I was given the assignment of determining a method to reduce the problems associated with *lost unit down orders* and *special orders*. The misplacement of *special orders* in the shipping and receiving area has caused many customers to become unsatisfied with our services. This situation has also caused our partsmen to spend many hours each day locating *lost orders*. These problems have arisen from the increase in the number of *special orders* awaiting to be picked up in our Shipping and Receiving Department. Currently, there are between 200 and 300 *orders* awaiting customer pickup. *These parts* are stored on shelves, on the Shipping and Receiving room floor, and outside of the building. The Shipping and Receiving Department has no records at this time concerning which *special orders* have arrived. To correct this situation a system is needed which reduces employee work. This system must also minimize the number of *lost orders* at a low cost.

Do you know for sure what the term *lost orders* refers to in the third sentence? (Is it just lost *special* orders or is it *all* lost orders?) How about the *orders* in sentence 5? Or *These parts* in sentence 6? We didn't when we first read it, and we still don't.

EXERCISE 15-4   The verb phrases in column A are representative of those that characterize a wordy, noun-heavy style of writing. Use column B to provide one-word equivalents. (The first two have been done for you.)

A	B
effect a reduction in	reduce
accomplish a modification of	modify
put emphasis on	_____
come to the conclusion that	_____
provide with information	_____
increase by a factor of two	_____
give an explanation of	_____
have a deleterious effect upon	_____
create an improvement in	_____
do an analysis of	_____
make a recommendation that	_____
conduct an investigation of	_____

EXERCISE 15-5 The writing samples given below can all be made much more concise. Rewrite them, eliminating all unnecessary words.

A   At this point in time we are engaged in a re-evaluation of the budget.

B   The design that is recommended for adoption would accomplish the removal of particulates with the desired degree of efficiency.

C   According to our estimates, the cost of the system would be on the order of approximately $1.2 million.

D   It is the purpose of this report to provide you with information about the current status of the proteolipid experiment.

E   In our judgment, the optimal course of action would be to maintain continued use of the double-alkali reactant.

F   My telephone number is 221-3600 if you would like to contact me about any questions you might have pertaining to the contents of this report.

G   At our meeting of 9/4/82 we had a discussion about the fact that people have been making numerous complaints concerning vertical movement in the Hudson wing. The problem at hand was the excessive wait time people have been experiencing while waiting for the Hudson elevator.

H   The causes for this change in consumption were then determined. The nature of the causes indicated the most probable recommendations. If they were easily corrected, it was suggested that the corrections be made. If, however, the problem was unavoidable by using the existing units, new units were designed.

I For electrical generation it is desirable to have a windmill that has a rotational speed that is higher than that of the wind. The ratio of the rotational speed of the windmill to the wind speed is called the speed-tip ratio. High speed-tip ratios are therefore desired for windmills that are to be used for electrical generation.

J Industry has shown interest in the electric car through research. This research has revealed that electric car use is possible. Industry will also be needed to produce the electric car. It is of necessity that industry, the present auto companies, be interested in producing electric cars because they are the ones who have the capabilities. With the auto companies showing interest, electric car usage is possible.

K In order for electric cars to become of practical use in the United States, the government must have a direct hand in encouraging electric car research and development. Making it attractive to industry to produce electric cars is of grave importance. Thus, the government and industry becoming involved would be a major step forward in the production of electric cars.

In order for electric cars to operate in the United States, they must be able to pass safety standards set up by the Department of Transportation. It is very important that these standards be met not only from the standpoint of safety but also for consumer acceptance. Consumers are not receptive to anything that will put them in danger. The electric car must be safe because consumer acceptance will be hard to obtain, and if the car is not safe, it will be impossible.

In order for electric cars to be used in the United States, it must be shown that they have satisfactory performance and costs as compared to a gasoline car. If the electric car cannot perform up to the standards of a consumer's present gasoline car, the consumer will have no use for the electric car. The costs for electric cars must be competitive with gas cars in order to fit the consumer's budget.

REFERENCES

1 *Volkswagen Service-Repair Handbook,* Clymer, Los Angeles, 1971, p. 235.

2 Christopher Turk, "Do You Write Impressively?" *Bulletin of the British Ecological Society* 9(3):5–10 (1978).

ADDITIONAL READING

Jack Selzer, "What Constitutes a 'Readable' Technical Style?" *New Essays in Technical and Scientific Communication: Theory, Research, and Criticism,* P. Anderson, C. Miller, and J. Brockmann (eds.), Baywood Press, Farmingdale, N.Y., 1982.

Joseph Williams, *Style: Ten Lessons in Clarity and Grace,* Scott, Foresman, Glenview, Ill., 1981.

16

CONNECTIVES

As we emphasized in earlier chapters, effective communication requires much more than well-chosen words and well-made sentences. The words and sentences of a piece of written or oral communication must all fit together into a coherent whole, so that the reader or listener can grasp the "big picture." Usually, the proper use of organizational patterns, parallelism, key words, and other devices described earlier will suffice to do this. However, these devices alone are sometimes not enough, and they should be supplemented by explicit connective words and phrases ("connectives" for short): *therefore, however, although, on the other hand, in theory,* etc. Such words help clarify the logic of your communication, enabling your reader or listener to see the flow of your reasoning from sentence to sentence.

The power of connectives is such that you should use them fairly often, even when you think your audience can follow your reasoning without them. This will enhance the impression you want to give of having used sound and logical thinking in arriving at your conclusions. Be sure, however, to use these connectives correctly! If you misuse connectives, they will only mislead your audience and, in effect, will do more harm than good.

The purpose of this chapter is to cover the more commonly used connectives in scientific and technical English. These are presented in groups of virtually synonymous equivalents, with the most commonly used words and phrases appearing in upper-case letters. Those connectives generally restricted to informal uses are so designated. We have also provided for each group an example of usage. These examples should be sufficiently general and sufficiently elaborate to give you an idea of how each connective is properly used.

Connectives as a whole are divided into two categories according to grammatical function: conjunctive adverbs and subordinating conjunctions. It is important not to confuse the two. Conjunctive adverbs (*thus, however, for example,* etc.) are normally used with independent clauses to show how that clause is logically related to some independent clause preceding it. Although they usually occur at the beginning of the clause, they can also be embedded within it. Subordinating conjunctions (*since, although, if,* etc.), on the other hand, are used only with a subordinate clause to express a logical relationship between that clause and an independent clause which usually follows it. Subordinating conjunctions cannot be used in sentences of only one clause.

UNGRAMMATICAL

Since dissolution of an organic compound into an organic solvent does not produce ions. Most solutions of organic compounds do not conduct electricity.

GRAMMATICAL

Since dissolution of an organic compound into an organic solvent does not produce ions, most solutions of organic compounds do not conduct electricity.

16.1

CONJUNCTIVE ADVERBS

THEREFORE, THUS, consequently, as a result, for this reason, so [informal]

Combustion catalysts consist of various shapes of basic material coated with a metallic compound. The variety of shapes and catalytic materials provide a multitude of catalysts for each application. *Therefore,* a good general rule to follow is to consult with a catalyst manufacturer on the most suitable catalytic equipment configuration.[1]

HOWEVER

At one time the atom was believed to be the ultimate unit in the subdivision of matter. Subsequently, *however,* it became known that the atom is composed of still smaller units.[2]

FOR EXAMPLE, for instance

Variation is a very important characteristic of data. *For example,* if we are manufacturing bolts, excessive variation in the bolt diameter would imply a high percentage of defective product.[3]

FIRST, . . . SECOND(ly), . . . THIRD(ly), . . . FINALLY

A further elaboration of these facts falls naturally under two major headings. *First,* what are the quantitative relations between irradiation dosage and genetic damage? *Second,* to what dosages are people being exposed?[4]

IN ADDITION, FURTHERMORE, moreover, besides [informal]

All organic compounds, in principle, can be ranked according to their relative ability to undergo photochemical reactions characteristic of smog. For many, however, the data are not available. *Furthermore,* the ranking of a number of organic compounds on the basis of their rates of disappearance during photolysis would not necessarily be the same ranking of those same compounds on the basis of their ability to produce eye irritation.[5]

ON THE OTHER HAND

The increasing price of oil may cause a shift to coal in many countries. Worldwide, coal should last for at least 300 years. Oil, *on the other hand,* will be virtually depleted within 60 years.[6]

IN OTHER WORDS, that is

The chromosomes of higher organisms, including man, occur in pairs. *That is,* the genetic information found in any one chromosome is duplicated in another chromosome ordinarily identical in size and shape.[7]

OF COURSE, naturally [informal]

There are two types of kettles used for cooking varnish: the open kettle, which is heated over an open flame, and the newer totally enclosed kettle, which is set over or within a totally enclosed source of heat. *Naturally,* the open kettle allows vaporized material to be emitted to the atmosphere unless hooding and ventilation systems are provided.[8]

EVEN SO, NEVERTHELESS, nonetheless, still [informal]

Experiments on humans are essential; they have been the cornerstone of medical progress and will continue to remain so. *Even so,* many tests cannot be done on humans because of practicality or propriety; certain tests would be outrageously costly or unethical.[9]

as mentioned above (earlier, before), as shown in Figure _____

Conventional ultrasonic testing has its disadvantages: strong signals from one of the surfaces can mask the signal of a discontinuity near one of the surfaces. . . .

[three paragraphs later] *As mentioned earlier,* conventional ultrasonic testing generally fails to clearly distinguish discontinuities close to the surface . . .[10]

IN GENERAL

The conversion of low-Btu gas to high-Btu gas entails the production of much CO_2, each molecule of which represents the waste of a carbon atom originally present in the coal. *In general,* for processes now under development, it appears that the production of low-Btu gas will have a thermal efficiency up to 20% greater than that of high-Btu gas.[11]

IN FACT, indeed, as a matter of fact

Physics is far from being a solitary pursuit. *In fact,* physicists today interact intensely with each other.[12]

IN PARTICULAR, SPECIFICALLY

It is the role of astronomy to ascertain the earth's probable future as determined by the action of cosmic forces. *Specifically,* astronomy must seek to discover what the prospects are for the earth's continuing as a suitable abode for life, and study those events which could end the existence of mankind.[13]

actually, in actuality

Pythagoras was born in the year 569 B.C.(?) on the island of Samos, off the west coast of Asia Minor. It is claimed that he traveled extensively throughout Egypt, Babylon, and the Orient, where he supposedly learned of their mathematics, religions, and mysticism. *Actually,* all we know for certain is that he had a good knowledge of the beliefs and mathematical concepts of the people of Asia Minor.[14]

IN CONCLUSION, in summary, to conclude, to summarize

[after a two-page discussion of the "Synthetic Aperture Focusing Technique" as a new method of ultrasonic testing of materials] *In conclusion,* the SAFT UT system represents a significant improvement in modern methods of nondestructive evaluation. If continued funding by the NRC allows further research to proceed. . . .[15]

in theory, in principle, theoretically

The problem with solar energy is that, although sunshine itself costs nothing, it takes ingenuity and money to capture it. *Theoretically,* 1 horsepower can be generated on about 1 square yard of land under the best conditions. But such are the unavoidable losses that not fewer than 4 and probably as many as 10 square yards are needed to generate 1 horsepower.[16]

in practice

If an experiment is repeated a large number of times, *N,* and event *A* is observed *n* times, the probability of *A* is

$$P(A) = \frac{n}{N} \cdot$$

In practice, the composition of the population is rarely known and hence the desired probabilities for various events are unknown. But we ignore this aspect of the problem, since our aim is only to create a mathematical model.[17]

INSTEAD, rather

The most striking feature of the Cray-1 supercomputer is its simplicity. It does not use any sophisticated devices such as magnetic bubble memory or supercooled conductors. *Instead,* it uses only three types of integrated circuits: memory, logic, and arithmetic.[18]

IN COMPARISON, BY CONTRAST

Nuclear medicine differs from existing angiography procedures in that nuclear imaging is an emission effect caused by the emission of gamma rays from the decaying isotope. *In comparison,* angiography is a transmission effect caused by the transmission of X-rays through the patient, to be imaged on the opposite side.[19]

if so, in that case

The age of the engineer as a specialist may have reached its zenith and may now be starting to recede. *If so,* the development and implementation of computer-aided engineering may reestablish the importance of the engineer as a generalist.[20]

in most cases, generally, usually, for the most part

Large volumes of hydrocarbon gases are produced in modern refineries and petrochemical plants. *Generally,* these gases are collected and used as fuel or as raw material for further processing. Sudden or unexpected upsets in process units and scheduled shutdowns, however, can produce gas in excess of the capacity of the gas-recovery system.[21]

16.2

SUBORDINATING CONJUNCTIONS

ALTHOUGH, though, while, even though, despite the fact that

Even though automatic computing is much faster and cheaper than it has ever been, there are still problems that cannot be computed owing to their size or complexity, at least not in a practical sense. *Though* a computer wades through a problem at 100 million steps per second, it may be in a losing race with the scientist or engineer applying his or her ingenuity to complex problems.[22]

SINCE, BECAUSE, owing to

Since natural carbon contains approximately 1 percent C^{13} along with 98.9 percent C^{12}, the average atomic weight of carbon is about 12.011 amu.[23]

IF

The U.S. government considers computers to be strategic commodities with potential military applications. *If* an American firm wishes to export a computer, it must first obtain a license from the State

	Department (*if* the computer is intended for military use) or from the Commerce Department (*if* it is intended for civilian use).[24]
in order for (that)	For combustion of organic vapors and liquids, the concentrations of vapor and air must be within the limits of flammability. *In order that* a flame be self-sustaining, the mixture of air and combustibles must provide enough heat to maintain the combustion temperature.[25]
rather than, instead of [-ing]	The "ideal system concept" is a simple idea that anyone can understand. *Rather than* use the models of the present trouble-ridden system as a guide for developing a recommended system, this concept produces a model of the best or most ideal system as a guide.[26]

16.3

COMMON MISTAKES TO AVOID

Connectives are sometimes misused by nonnative speakers of English because of miscategorization, overgeneralization, interference from the native language, or the improper blending of two forms. Some of the most common misuses are as follows.

Using *actually* to mean *currently, presently*. This mistake occurs most frequently with speakers of Spanish, French, and other Romance languages since similar forms in those languages (Sp. *actualmente*, Fr. *actuellement*, etc.) do mean "currently, presently." *Actually* is closer in meaning to *in fact*, the only real difference being that it is normally used to play down or correct an earlier statement whereas *in fact* is used to upgrade or magnify one.

Using *after all* to mean *finally, in conclusion*. This phrase is a reduced form of *after all is said and done* and means roughly "regardless of what might be said about it." It is used to emphasize the fundamental essence of something, not just signal the end of a sequence or the end of a discussion.

Using *although* and *but* in the same sentence. "*Although* I studied hard, *but* I only scored 56 on the exam" is incorrect. This mistake results perhaps from confusion with the correlative conjunction *not only . . . but* or from native-language interference. In any event, it is permissible to use either *although* or *but*, but not both together.

Using *it* in idiomatic *as* phrases. Phrases such as "as was mentioned above" and "as is shown in Figure 12" are idiomatic and should be used as fixed forms, without an *it* subject. "As *it* was mentioned above" is incorrect.

Using *comparing to (with)*. The correct connective forms are *compared to* and *in comparison with*. *Comparing* can be used as a gerundive participle, but only if it is followed immediately by a direct object, e.g., "Comparing these latest results to (with) our earlier ones, . . ."

Using *conclusively* to mean *in conclusion*. *Conclusively* is the simple adverbial form of the adjective *conclusive* (= "decisive, supported by solid evidence"). It is not a connective.

Using *on the contrary* to mean *on the other hand*. *On the contrary* is properly used to deny some earlier statement, whereas *on the other hand* is used simply to present an alternative to some earlier statement. *On the one hand* and *on the other hand* are often used together to present a pair of alternative statements.

Using *particularly* to mean *in particular*. This error results probably from a blending of the synonymous connectives *in particular* and *specifically*. In any event, *particularly* means "especially"; it is an ordinary adverb, not a connective.

EXERCISE 16-1 In each of the following passages, fill in the blanks with the most appropriate connective from among the choices given.

A   Studying how light (or other electromagnetic radiation) interacts with matter is an important and versatile tool of the chemist. _____
     _____ , much of our knowledge of chemical substances comes from
     (Still, Indeed, However)
     their specific absorption or transmission of light.
         Suppose you look at two solutions of the same substance, one a deeper color than the other. Your common sense tells you that the darker colored solution is the more concentrated one. _____
     _____ , as the color of the solution deepens, you infer
     (In other words, Consequently, In addition)
     that its concentration also increases. This is an underlying principle of spectrophotometry: the intensity of color is a measure of the amount of a material in solution.
         A second principle of spectrophotometry is that every substance absorbs or transmits certain wavelengths of radiant energy but not other wavelengths. _____ , chlorophyll always ab-
     (Therefore, In general, For example)
     sorbs red and violet light while transmitting the yellow, green and blue wavelengths. The transmitted and reflected wavelengths appear green—the color your eye "sees." The light energy absorbed or trans-

mitted must match exactly the energy required to cause an electronic transition in the substance under consideration. Only certain wavelength photons satisfy this energy condition. _____ ,
(Thus, Nevertheless, Actually)
the absorption or transmission of specific wavelengths is characteristic for a substance, and a spectral analysis serves as a "fingerprint" of the compound.

B Before we can discuss the basic problem of the origin of our universe, we must ask ourselves whether such a discussion is necessary. Could it not be true that the universe has existed since eternity, changing slightly in one way or another in its minor features, but always remaining essentially the same as we know it today? The best way to answer this question is by collecting information about the probable age of various basic parts and features that characterize the present state of our universe. _____ , we may ask a physicist or chem-
(In general, In practice, For example)
ist: "How old are the atoms that form the material from which the universe is built?" Only half a century ago such a question would not have made much sense. _____ , when the existence of
(Instead, However, If so)
natural radioactive elements was recognized, the situation became quite different. It became evident that if the atoms of the radioactive elements had been formed too far back in time, they would by now have decayed completely and disappeared. _____ , the observed relative
(Besides, Since, Thus)
abundances of various radioactive elements may give us some clue as to the time of their origin.
We notice _____ that thorium and the common
(first of all, therefore, in fact)
isotope of uranium (U^{238}) are not markedly less abundant than the other heavy elements, such as, _____ ,
(in comparison, furthermore, for example)
bismuth, mercury, or gold. _____ the half-life periods
(Since, Although, Of course)
of thorium and of common uranium are 14 billion and 4.5 billion years, respectively, we must conclude that these atoms were formed not more than a few billion years ago. _____ , the fis-
(So, On the other hand, However)
sionable isotope of uranium (U^{235}) is very rare, constituting only 0.7% of the main isotope. The half-life of U^{235} is considerably shorter than that of U^{238}, being only about 0.9 billion years. _____
(Since, Therefore, While)
the amount of fissionable uranium has been cut in half every 0.9 billion years, it must have taken about seven such periods, or about 6 billion

years, to bring it down to its present rarity, if both isotopes were originally present in comparable amounts. . . .[28]

EXERCISE 16-2 Complete each of the following sentences with an appropriate line of development. (Do not worry about the factual accuracy of your answer; just provide a continuation that fits the given connective.)

A   Nuclear power plants do not produce chemical pollutants such as sulfur dioxide or PCBs. However, they _____

_____

_____

B   Synthetic fuels are difficult to manufacture. In fact, _____

_____

_____

C   White light appears to be devoid of any color. In actuality, though, _____

_____

_____

D   More and more automobile engines are being built today that use fuel injection devices instead of carburetors. Nevertheless, _____

_____

_____

REFERENCES

1   "Control Techniques for Hydrocarbon and Organic Solvent Emissions from Stationary Sources," US DHEW. PHS. EHS. National Air Pollution Control Administration. Washington, D.C. Publication No. AP-68. March 1970, p. 3-5.

2   Lawrence Van Vlack, *Elements of Materials Science and Engineering,* 3d ed., Addison-Wesley, Reading, Mass., 1975, p. 29.

3   William Mendenhall, *Introduction to Probability and Statistics,* 2d ed., Wadsworth, Belmont, Calif., 1967, p. 29.

4   A. H. Sturtevant, "The Genetic Effects of High Energy Irradiation of Human Populations," in Edward Hutchings Jr. (ed.), *Frontiers in Science,* Basic Books, New York, 1958, p. 71.

5   "Control Techniques . . . ," p. 3-23.

6   "Coal Technology for Energy Goals: No Fuel Like an Old Fuel," *The Research News* **XXIV**(12):3 (1974).

7 H. Eldon Sutton, *An Introduction to Human Genetics,* Holt, Rinehart and Winston, New York, 1965, p. 14.

8 "Control Techniques . . . ," pp. 4-21, 4-22.

9 Ray Barry, "Caution: Life May Be Hazardous to Your Health," *Michigan Technic* **XCVI**(4):10 (1978).

10 Nolan Van Gaalen, "Ultrasonic Synthetic Aperture Focusing as a Method of Non-Destructive Evaluation," *Michigan Technic* **XCVI**(4):18 (1978).

11 "Coal Technology . . . ," p. 9.

12 "Can We Understand the Universe?" *The Research News* **XXXII**(6/7):8 (1981).

13 Albert G. Wilson, "Astronomy and Eschatology," in *Frontiers in Science,* p. 207.

14 Guy Zimmerman, "The Pythagoreans: More Than Just the Theorem," *Michigan Technic* **XCVII**(3):9 (1978).

15 Van Gaalen, op. cit., p. 20.

16 Waldemar Kaempffert, *Explorations in Science,* Viking, New York, 1953, p. 190.

17 Mendenhall, op. cit., p. 51.

18 Steve Hannah, "The United States: Maintaining a Firm Grip on the World Computer Market," *Michigan Technic* **XCVII**(2):19 (1978).

19 David Varner, "Nuclear Cardiac Imaging," *Michigan Technic* **XCVII**(2):20 (1978).

20 James J. Duderstadt, "Some Thoughts on CAD, CAM and Computer Aided Engineering," *Michigan Technic* **C**(2):7 (1981).

21 "Control Techniques . . . ," p. 4-4.

22 Blanchard Hiatt, "Big Computing, Tiny Computers," *The Research News* **XXX**(4):5 (1979).

23 Van Vlack, op. cit., p. 29.

24 Hannah, op. cit., p. 18.

25 "Control Techniques . . . ," p. 3-2.

26 Gerald Nadler, *Work Systems Design: The IDEALS Concept,* Richard D. Irwin, Homewood, Ill., 1967, p. 23.

27 Alice S. Cohen (ed.), *Investigations in General Chemistry,* Department

of Chemistry, University of Michigan, Ann Arbor, Mich., 1978; revised by Nancy Konigsberg, 1979, p. 117.

28 George Gamow, *The Creation of the Universe,* Viking, New York, 1952, pp. 6–20.

17

COUNTABILITY AND THE INDEFINITE ARTICLE

In order to understand the use of articles, one must first understand a basic property of common nouns, namely, the property of "reference." We all know that nouns refer to objects or concepts *at different levels of generality:* the same noun that refers to an entire category of objects in one sentence may refer only to a particular, unique object in another. For example, consider the word *window* in a sentence like "Every room in this building has *a window.*" The writer of this sentence is probably thinking not of any particular window, but of windows in *general:* each room has an object belonging to the general category of window. On the other hand, in a sentence like *"The window* in Room 303 is broken," the writer is singling out a *particular* window, one that has a unique identity.

In this book, we shall use the term *nonunique reference* to describe examples of general reference (the first window example above) and the term *unique reference* to describe examples of particular reference (the second window example above). This chapter deals exclusively with cases of nonunique reference; it points out that no article is used unless the noun is a singular, "countable" noun (in which case *a* or *an* is used). Chapter 18 describes many cases of unique reference, showing that the definite article (*the*) is used in each case.

17.1

UNCOUNTABLE NOUNS

Many English nouns represent objects or concepts that do not have a distinct shape or form: *water, rice, gravity, magnetism, information, engineering,* etc. Such nouns are called uncountable nouns because they are customarily not counted; for example, one does not normally say *"5 waters"* or *"8 rices"* or *"15 magnetisms."* In fact one does not normally use the plural form at all for such nouns.[1]

Some common types of uncountable nouns are:

1   Those representing an amorphous physical mass (a mass without definite shape): *sugar, sand, salt, rice, flour,* any liquid or gas, etc.

2   Those representing an abstract concept: *gravity, information,*[2] *curiosity, satisfaction, magnetism,* etc.

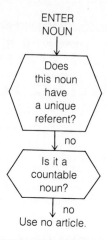

ENTER
NOUN

Does
this noun
have
a unique
referent?

↓ no

Is it a
countable
noun?

↓ no

Use no article.

FIGURE 17-1 Partial flow chart for choosing the correct article
(step 1).

3   Those representing a continuous process: *photosynthesis, pollution, osmosis, combustion, etc.*

4   Those representing a field of study: *mathematics, chemistry, physics, engineering, etc.*

RULE 1
Whenever an uncountable noun is used with a nonunique (general) referent in mind, no article is used.

EXAMPLES
Magnetism is the force that causes iron filings to be attracted to a magnet.

My niece says she wants to study *engineering.*

We can represent the above rule with the flow chart shown as Figure 17-1.[3]

17.2

COUNTABLE NOUNS

In contrast to uncountable nouns, many nouns in English represent concepts or objects that do have a distinct form. These are the so-called countable nouns. Some examples of countable nouns are *book, automobile, molecule, computer,* and *microscope.* Unlike uncountable nouns, words like these readily occur:

1   In the plural form *(books, automobiles, molecules)*

2   With numbers (*four computers, nine microscopes*)

3   With quantifiers such as *several, many,* and *few* (*several books, many microscopes, few molecules*)

Included among the countable nouns of English are words like *idea, concept, theory,* and *hypothesis*. Although nouns like these obviously do not have a physical form the way *book, automobile, molecule,* etc., do, they are nevertheless thought of as having a "form," even though that form is abstract. Accordingly, these more abstract nouns can also occur:

1   In the plural form (*ideas, concepts, theories*)

2   With numbers (*three hypotheses, two concepts, four theories*)

3   With quantifiers like *several, few,* and *many* (*several ideas, few theories, many hypotheses*)

EXERCISE 17-1   Mark each of these nouns as being countable or uncountable:

rocket	program
engine	pollution
inflation	satellite
lever	molecule of carbon dioxide
carbon dioxide	ecology
wheel	engineer
friction	radar
gravity	combustion
atom	motor
computer	engineering
aluminum	book
weather	meter
machine	sodium nitrite
information	photosynthesis

EXERCISE 17-2 Mark each of the italicized nouns as being countable or uncountable:

A   The acceleration of *gravity, g,* is the *acceleration* imparted to a *body* by its own *weight*.[4]

B   If *friction* could be eliminated, no force at all would be necessary to keep an *object* in motion, once it had been started.[5]

C   The *coefficient* of friction just before *motion* begins is larger than the coefficient of friction when there is actual sliding of one *surface* over the other.[6]

Of course, countable nouns are often used with a nonunique (general) referent in mind, as our window example showed. In such cases, if the noun is singular, one should use the indefinite article: *a book, an automobile, a theory,* etc. (The indefinite article *a* is used before words beginning with a consonant sound: *a meter, a handle, a unit,* etc.; the indefinite article *an* is used before words beginning with a vowel sound: *an atom, an hour, an uncontrolled reaction,* etc.)

RULE 2
Whenever a countable noun is used with a nonunique referent in mind:

1   The indefinite article is used if the noun is singular.

2   No article is used if the noun is plural.[7]

EXAMPLE
A betatron is *a device* used to accelerate *electrons*. The electrons travel around *a circular path* in *a vacuum tube* known as a "*doughnut*." The force of acceleration is supplied by *a magnet*.[8]

This example is taken from an introductory physics textbook. The author is trying to tell his readers what a betatron is, in very general terms; he is not talking about any particular betatron or vacuum tube or magnet, etc. To indicate this level of generality, he uses the indefinite article for the singular countable nouns and no article for the plural countable noun. (When *electrons* appears a second time, the definite article is used because now we are talking about particular electrons, namely, those referred to in the previous sentence.)

Notice that the first sentence of the above example constitutes a formal definition, where we mean by *formal definition* a statement conforming to the pattern:

Term = Class + Distinguishing Features

In this case, the term being defined is *betatron;* it belongs to the class of *devices;* it is distinguished from other members of this class by the fact that it is used to accelerate electrons. All definitions, of course, are general statements: they refer not to particular things but to *classes* of things. This is why we seldom find definite articles used in definitions. (One exception: the term being defined may have the generic definite article, e.g., "*The* betatron is a device used to accelerate electrons." See Chapter 18 for discussion of this use of the definite article.)

OTHER EXAMPLES OF NONUNIQUE REFERENCE

An ideal machine is one that has no friction.

Bill hopes to buy *a computer* someday.

An effective way of learning mathematics is by working on *problems.*

Valence electrons are those located in the outermost shell of *an atom.*

Adding Rule 2 to our previous flow chart, we have:

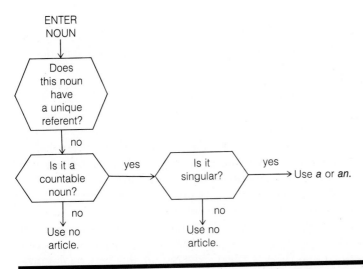

FIGURE 17-2   Partial flow chart for choosing the correct article (step 2).

To see how this flow chart works, let us take the sentence *Bill hopes to buy* _____ *computer someday.* Do we use *a, an, the,* or no article at all before the noun *computer?*

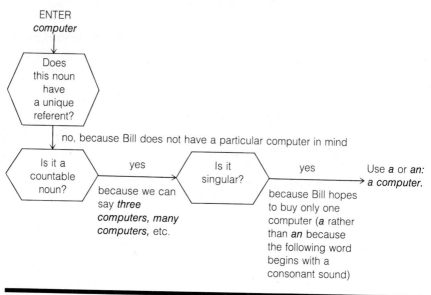

FIGURE 17-3 Use of partial flow chart for choosing the correct article.

By answering the questions in the flow chart and following the arrows, we thus arrive at the correct answer, *Bill hopes to buy a computer someday.*

EXERCISE 17-3 In the following passages, insert indefinite articles (*a* or *an*) where necessary. (Use Figure 17-2 as an aid.)

A **STORING ELECTRIC POWER**

In _____ space the sun is always shining, so there is no problem with storing _____ electricity. On earth, however, the sun shines only half the time in _____ good weather and not at all in bad. Thus, _____ electricity must be stored for _____ sunless periods. Currently this is done with _____ lead-acid storage batteries, similar to those used in _____ automobiles. _____ day's electricity for _____ average single-family house can be stored in _____ batteries occupying the space of _____closet; _____ row of such "closets" in the basement stores _____ power for _____ sunless periods.[9]

B **POWER REQUIREMENTS IN AN AVERAGE HOME**

_____ average single-family residence (_____ four-person, 1500-square-foot, non-air-conditioned house) uses about _____ 700 kilowatt-hours per month, the equivalent of _____ 1-kilowatt generator running continuously. Because this average house needs _____ 1 kilowatt of _____ average power, it requires _____ 5 kilowatts of _____ peak

power. This, in turn, would require about ＿＿ 500 square feet of
＿＿ solar cells at ＿＿ present efficiency levels under
＿＿ optimum conditions. The Energy Research and Development
Administration (ERDA) says 1500 square feet would now be required
in ＿＿ northern city like Boston.[10]

C   VECTORS AND SCALARS

Vectors are ＿＿ quantities that have both ＿＿ magnitude and
＿＿ direction. Examples of ＿＿ physical quantities that are vectors
are ＿＿ force, ＿＿ velocity, and ＿＿ acceleration. Thus, when
one states that ＿＿ car is moving north at 100 kilometers per hour,
with respect to ＿＿ coordinate system attached to the earth, one is
specifying a vector quantity, velocity, with ＿＿ magnitude (100 kilometers per hour) and ＿＿ direction (north).

Scalars are ＿＿ quantities that have ＿＿ magnitude only. Examples of ＿＿ physical quantities that are scalars are ＿＿ mass,
＿＿ distance, ＿＿ speed, and ＿＿ density. Thus, when one
states only the fact that ＿＿ car is moving at 100 kilometers per
hour one has specified ＿＿ scalar, ＿＿ speed, since only ＿＿
magnitude (100 kilometers per hour) is given (that is, no direction is
specified).[11]

17.3

COUNTERS

Although uncountable nouns themselves cannot be counted, they can usually
be modified by certain types of phrases—called "counters"—that *can* be
counted. One example of a counter is *cup of* ＿＿ . It can be used to
modify an uncountable noun like *water*, thereby creating a countable noun
phrase, *cup of water*. Thus, although one cannot say "a water" or "two
waters" or "three waters," etc., one *can* say

a cup of water

two cups of water

three cups of water

etc.

Counters are typically composed of a countable noun and the preposition *of,* so that the complete noun phrase has this structure:

Countable + *of* + Uncountable
  Noun   Noun
‿‿‿‿‿‿‿‿‿‿
   "Counter"

The countable noun should represent a unit of measure that is *appropriate* for the object or concept represented by the uncountable noun. For example, liquids are customarily measured either by putting them into a container having a known capacity, or by first putting them into a container and then weighing them. In the first case, the countable unit of measure might be, say, "cups" or "liters" or "quarts"; in the second case, it might be "ounces" or "grams" or "pounds."

Even uncountable nouns referring to more abstract concepts can usually be modified by counters. For instance, electric current can be measured in terms of amperes (*25 amperes of electric current*); information can be measured in terms of "pieces" (*a piece of information*); processes can be measured in units of time (*3 months of experimentation*). Once again, we must emphasize that the countable noun in these constructions should be an *appropriate* unit of measure for the uncountable noun involved.[12]

EXERCISE 17-4   A list of uncountable nouns follows. Insert appropriate counters or measure phrases so that you change these uncountable nouns to countable units. The first two examples are done for you.

Uncountable		Countable
oxygen	3 liters of oxygen 3 moles of oxygen 3 molecules of oxygen	
electric power	3 watts of _____	electric power
current	_____	current
force	_____	force
work	_____	work
pressure	_____	pressure
temperature	_____	temperature
liquid	_____	liquid
pollution	_____	pollution
chemistry	_____	chemistry
electrolysis	_____	electrolysis
radar	_____	radar
combustion	_____	combustion
friction	_____	friction
oxygen	_____	oxygen
electric potential	_____	electric potential

17.4

"TWO-WAY" NOUNS

Many nouns in technical English can refer to either countable or uncountable concepts, depending on the writer's focus. Such "two-way" nouns all have one thing in common: when they are used in the uncountable sense they refer to a *general* concept, but when they are used as countable nouns they refer to a *specification* of that general concept. Consider the word *metal*, for example: it is commonly used in an uncountable sense, to mean the general class of matter that is opaque, ductile, and fusible and typically consists of only one or two chemical elements, e.g., "Chisels are usually made of *metal*." However, it is also sometimes used as a countable noun, to mean a *type of metal*. For example, the sentence "Brass is an alloy composed of two *metals*, zinc and copper" means the same thing as "Brass is an alloy composed of two *types of metal*, zinc and copper." Some other two-way nouns that usually have the meaning *type of ___x___* in the countable sense (where *x* is the name of a class) are *fuel, acid, soil, material,* and *gasoline.*

Note, for instance, the following:

1   The vast majority of American automobiles run on *gasoline*. (Here *gasoline* is general, uncountable.)

Our local service station usually carries several *gasolines:* leaded regular, leaded premium, unleaded regular, and unleaded premium. (Here *gasolines* is specific, countable. Note that you could have said, Our local service station usually carries several *types of gasoline* (or *grades of gasoline*): leaded regular, etc.)

2   Railcars transporting *acid* must be specifically designed. (Here *acid* is general, uncountable.)

Electrolytic action in an automobile battery produces an *acid* that can lead to corrosion of the terminals. (Here *acid* is specific, countable. You could have said, Electrolytic action in an automobile battery produces a *type of acid* that can lead to corrosion of the terminals.)

In many other cases, however, the specification of a two-way noun does not mean *type of ___x___* but instead means *amount of ___x___*. Note this example:

The two circuits have different *resistances.*

Here *resistances*, a word that is commonly used as an uncountable noun referring to *general* opposition against electric current, refers specifically to the *amount* of opposition. The above sentence means, in effect, "The two circuits have different *amounts of resistance.*" Some other two-way nouns that usually mean *amount of ___x___* in the countable sense are *mass, velocity, force, pressure, power,* and *acceleration.*

Sometimes the specification of a two-way noun means, roughly, a *complete process of __x__*. For example, the word *distillation* normally is used to refer to the general process of distillation and is thus used as an uncountable noun. Sometimes, however, the writer wants to focus on one or more specific applications of the process:

At the refinery, the crude oil undergoes *two complete processes of distillation.*

This sentence can be shortened as follows:

At the refinery, the crude oil undergoes *two distillations.*

There are still other possible interpretations that can be attached to the specification of a two-way noun; it is advisable to exercise caution in your own writing and not to perform a specification that you are not sure of.

EXERCISE 17-5 Mark each of the italicized nouns as countable or uncountable. Examples 1 and 2 in C have already been done.

A 1 *Water* expands when it is heated in a *container.*
    2 Chemistry *students* study the *concept* that *matter* is made up of *particles.*
    3 *Molecules* occupy *space.*
    4 *Velocity* and speed are both the *ratio* of a length to a time.[13]
    5 The unit of *velocity* (or speed) in the English system is *feet* per second.[13]

B 1 Approximately 99 percent of the *matter* making up the *sun* and other *stars* is a *mixture* of the *elements* hydrogen and *helium.*[14]
    2 The earth is about 150,000,000 *kilometers* (93 million *miles*) from the *sun*—one astronomical unit in the *language* of *astronomers.*[15]
    3 If the *earth* were at some other *distance* from the sun, we would receive more or less of the sun's *energy* than we do.
    4 A positive *velocity* indicates *motion* toward the right, and a negative *velocity* indicates motion toward the left.[16]
    5 In the *year* 1624, Sir Isaac Newton (the famous *physicist*) was born and Galileo died.

C 1 Mechanics is the branch of physics and engineering which deals with the interrelations of *force,* matter, and motion.[17] (uncountable)
    2 We can exert a *force* on a body by muscular effort; compressed air exerts a *force* on the walls of its container; a locomotive exerts a *force* on the train which it is drawing.[17] (countable in all instances)
    3 The newton is equal to a *force* of 0.224 pounds, a *force* which

imparts to a mass of one kilogram an acceleration of one meter per second, per second.[18]

4 *Mass* is a quantitative measure of inertia.[19]

5 *Mass,* as used in mechanics, refers to that property of matter which in everyday language is described by the word inertia.[20]

6 The newton is a force which imparts to a *mass* of one kilogram an acceleration of one meter per second, per second.[21]

7 We may derive the relation between *pressure* and *elevation* in a fluid.[22]

8 *Pressure* is expressed in lbs/ft^2, newtons/m^2, or dynes/cm^2.[23]

9 A *pressure* of 1.013×10^6 dynes/cm^2, or 14.7 lbs/in^2, is called one atmosphere.[24]

17.5

INFORMAL USAGES

There are certain informal usages that you are apt to hear spoken in conversations or lectures but will rarely see in writing. It is important to avoid such usages in your writing, since they are stylistically inappropriate for anything but the most informal level.

For example, in the informal situation of the chemistry laboratory or small discussion group, student chemists sometimes talk about normally *uncountable* nouns, such as the names of elements and molecules, as if they were *countable* nouns. A student might read aloud this equation, for instance,

$$Ag_2O + H_2O_2 = 2\,Ag + O_2 + H_2O$$

as "*one silver oxide* plus *one hydrogen peroxide* yields *two silver(s)* plus *an oxygen* plus *a water.*" Similarly,

$$Cu + 2\,H_2SO_4 = CuSO_4 + SO_2 + 2\,H_2O$$

could be read aloud as "*one copper* plus *two sulfuric acid(s)* yields *one copper sulfate* plus *one sulfur dioxide* plus *two water(s).*" Although this type of rendition is frequently heard in informal settings, it would be highly unsuitable for formal purposes. A more formal reading of the same formula would be "*one <u>atom</u> of copper* plus *two <u>molecules</u> of sulfuric acid, etc.*"

A second example of informal usage frequently heard is the use of the indefinite article instead of *per* in measured-rate expressions like the following:

A small car should get at least 35 miles *a* gallon.

Northern Michigan averages 5.2 hours of sunlight *a* day.

More formal versions of these two sentences would be "A small car should get at least 35 miles *per* gallon" and "Northern Michigan averages 5.2 hours of sunlight *per* day."

REFERENCES

1   See the discussion of two-way nouns and informal usages (Sections 17.4 and 17.5) for some exceptions.

2   The word *information* deserves special attention from students whose native language is French, Italian, German, or Spanish since the corresponding noun in these languages is countable.

3   The flow chart should be used as follows: Taking the example, "Magnetism is the force that causes iron filings to be attracted to a magnet," let us consider the noun (properly speaking, the noun phrase) *magnetism*. We first ask of it, "Does it have a unique referent?" The answer to this question is no (since magnetism in a general, not particular, sense is meant here), and so we follow the arrow from the "no" cell to the next question, "Is it an uncountable noun?" The answer here is yes, and so we follow the "yes" cell to the final instruction, "Use no article." The other flow charts in this chapter are used in similar fashion.

4   F. W. Sears, *Mechanics, Heat, and Sound,* Addison-Wesley, Reading, Mass., 1950, p. 283.

5   Adapted from Sears, p. 20.

6   Sears, p. 24.

7   Certain verbs, especially verbs of volition, can allow a specific interpretation for a following noun with an indefinite article. For example, the sentence "John wants to marry a Norwegian" can be interpreted as meaning that John has a specific person in mind that he wants to marry, who is of Norwegian nationality (as opposed to an alternative reading according to which one of John's requirements for *any* future spouse is that she be of Norwegian nationality). This specific reading, focusing only on the nationality of John's intended spouse, should not be confused with cases of unique reference (as embodied, for example, in "John wants to marry *the* Norwegian"), where complete identity (of John's intended) is implied. For further discussion of this distinction between specificity and uniqueness, see J. Lyons, *Semantics,* Cambridge University Press, Cambridge, 1977, pp. 187–192.

8   C. H. Bachman, *Physics for the Non-Scientist,* privately published, 1968, p. 101.

9   Adapted from D. Morris, "Solar Cells Find Their Niche in Everyday Life on Earth," *Smithsonian,* October, 1977, p. 40.

10   Adapted from Morris, p. 41.

11   Adapted from Sears, p. 4.

12 Obviously, a writer must know what units of measure are customarily used if he or she wants to modify an uncountable noun by a counter. Providing such units of measure should pose no problem, however, since they are part of the basic vocabulary in any field. A writer who knows enough about a subject to write about it certainly knows the field's basic units of measure.

13 Sears, p. 51.

14 Adapted from P. F. Brandwein, *Matter: An Earth Science,* Harcourt Brace Jovanovich, New York, 1975, p. 200.

15 Adapted from Brandwein, p. 202.

16 Sears, p. 53.

17 Sears, p. 1.

18 Sears, p. 77.

19 Sears, p. 73.

20 Sears, p. 72.

21 Adapted from Sears, p. 77.

22 Adapted from Sears, p. 301.

23 Sears, p. 300.

24 Sears, p. 305.

18

THE DEFINITE ARTICLE

Recall the example given at the beginning of the preceding chapter, involving the word *window*. We pointed out that—in contrast to a sentence such as "Every room in this building has a window," where the writer is thinking of windows in a general sense—a sentence such as "The window in Room 303 is broken" refers to a particular window, one that has a unique identity. Notice that this distinction between generality and particularity is marked by the use of the indefinite article *(a window)* in the first sentence and the definite article *(the window)* in the second.

In general, the definite article *the* is used to show that what the noun (or noun phrase) refers to *is unique;* that is, it has a *unique referent.* Unlike the indefinite article, the definite article may occur with any type of noun—singular or plural, countable or uncountable.

There are many different circumstances under which a noun may be said to have a unique referent:

1   The noun may have a *special adjective* as a modifier (Section 18.1).

2   The noun may be a *special noun* referring to some unique time or place in our common existence (Section 18.2).

3   The noun may be *generic,* that is, it may refer to an entire species or type of something (Section 18.3).

4   The noun may have the same referent as some *previously mentioned* noun in the present context (Section 18.4).

5   The noun may have a *following modifier* that restricts it to a unique referent (Section 18.5).

6   The noun may have a unique referent by virtue of *shared knowledge* between writer and reader (Section 18.6).

7   The noun may have a unique referent by *implication* (Section 18.7).

18.1

SPECIAL ADJECTIVES
Certain adjectives modify a noun so that it has a unique referent in almost any context. In other words, such a noun is assumed to refer to a particular person, place, or thing.

One such type of highly restrictive adjective is the *superlative* adjective. Superlative adjectives are words like *tallest, fastest, heaviest, least important,* and *most useful.* In any given context, there is normally:

Only one tallest building (tree, person, etc.)

Only one fastest car (plane, runner, etc.)

Only one heaviest machine (person, book, etc.)

and so on. Thus, each of these nouns has a unique referent and must take the definite article: *the tallest building, the fastest car, the heaviest machine,* etc.

Similarly, ordinal adjectives—*first, last, second, fifth, nth,* etc.—each represent a unique position in an ordering: given any ordered set, there can be only one first position, only one last position, only one second position, and so on. Consequently, whenever a noun or noun phrase is modified by such an adjective, it almost always requires the definite article: *the first attempt, the last investigation, the second stage, the 19th century.*[1]

Some other adjectives by their very nature also restrict nouns to a unique referent and therefore normally occur with the definite article. Examples of such adjectives are *only, sole, exact, current,* and *present.* Three of these adjectives can be found in the following report on the surface sampler for the Viking mission to Mars.

> The Viking lander's surface sampler is *the only means* for acquiring small "bites" of Martian soil and then delivering them to the three analytical instruments located deep inside the lander. Without directly acquired samples of the Mars surface, *the sole use* for these instruments would be to analyze wind-blown dust that might accumulate in them over a long time—a very unattractive alternative. Thus, on the second day, when the surface sampler jammed during its initial operation, a team of experts at the Viking mission control center immediately sprang into action to remedy the situation.
>
> *The exact nature* of the problem became evident during tests of a model of the lander, known as the Science Test Lander (STL), at the Jet Propulsion Laboratory in Pasadena, California.[2]

18.2

SPECIAL NOUNS

Certain nouns are commonly used with the definite article to refer to periods of time or to certain physical features of our world: *the past, the present, the future, the 1940s, the early 1980s, the sky, the earth, the sun, the moon, the ground.* Such nouns take the definite article because each refers to some unique aspect of our common existence.

There are also many proper nouns that take the definite article. (Proper nouns are names, usually of persons or places: *Einstein, Mao Tse-tung, New York, Tokyo, Iran, Vietnam,* etc.; as can be seen from this sample list, they often occur without any article.) In cases where the "head noun" (usually the rightmost one) of the name is derived from a common noun, the definite article is often used: *the United States, the Twin Cities, the Soviet Union, the Merritt Parkway, the Ventura Expressway, the Mississippi River, the Indian Ocean, the Rocky Mountains.* Words such as *states, cities, union, parkway, expressway, river, ocean,* and *mountains* exist as common English nouns as well as in names such as these listed here.

One should be careful, however, for there are many exceptions to this pattern. Consider, for example, the following proper nouns: *Washington State, Salt Lake City, Lydecker Street, Fifth Avenue, Cripple Creek, Walden Pond, Bunker Hill, Lookout Mountain.* Each of these head nouns is derived from a common English noun *(state, city, street, avenue, creek, pond, hill,* and *mountain),* and yet *no* definite article is used. How can we account for such a difference?

In general, the distinction between these two classes of nouns seems to be based on relative *size:*

A group of states or cities	is usually larger than a single state or city.
A parkway or expressway	is usually larger than a street or avenue.
A river or ocean	is usually larger than a creek or pond.
A chain of mountains	is usually larger than a single mountain or hill.

Since large geographical objects like rivers, oceans, and mountain ranges tend to be uniquely identifiable for many people, the definite article is usually attached to their names. On the other hand, smaller objects like creeks, ponds, and hills are not so uniquely identifiable. Thus, one says *the* Atlantic Ocean (with the definite article) but *Walden Pond* (without one).

EXERCISE 18-1 In the following passages, find all the examples of the special adjectives and nouns just discussed in Sections 18.1 and 18.2. Be prepared to classify the various *definite article plus noun* and *indefinite article plus noun* units and to explain your classification.

A **ALUMINUM AND CHARLES MARTIN HALL**
   In the 1880's, if anyone had invented a way of making aluminum cheaply, that man could have benefitted the world—and made a fortune. At that time, aluminum was very expensive to make. Although

aluminum is the most abundant metal in the earth's crust, it was then very difficult to separate from the earth's crust. The first man to discover a way of separating aluminum from the earth's crust was a man named Charles Martin Hall. The way he did this used electricity to separate out the aluminum.[3]

B   **THE PLANETS: ORIGIN AND ORBIT**
The earth is a planet moving in orbit around the sun. It is not the sun's only planet. There are other planets, in other orbits, moving around the sun. The earth is merely the third planet from the sun. All the planets, including the earth, move in the same direction around the sun. All the orbits lie in nearly the same plane—except the outermost orbit (Pluto's), which is tilted.

How did this remarkable collection of moving objects get this way? How did the objects come to have orbits in nearly the same plane? How do the objects maintain the present order? Why does the farthest planet, Pluto, have a tilted orbit? Is there something special about an outermost orbit which makes it different from other orbits? Did the solar system come together bit by bit from different directions in space, at different times? Or did it have a single origin? These are some of the questions which astronomers try to answer.[4]

C   **AN INTRODUCTION TO 3-D RADAR**
Air-traffic radar of the 1950's could readily determine the direction of incoming and outgoing aircraft, but not their height above the ground. The planes might be a mile apart in altitude or on the verge of collid-ing—but air-traffic controllers on the ground had no way of obtaining this information from current radar data alone, or without some pre-vious radar history on both planes before they entered the same range-azimuth "capsule." Recognizing this obvious deficiency, the Airways Modernization Board (the forerunner of the FAA) decided to investigate the radar "height finder" as a likely solution to the problem.

The new air-traffic radar would have to resolve targets to within 1000 feet at 50 miles and be able to determine whether two aircraft, approaching each other from opposite ends of the sky, might pose a threat to each other. This meant that the radar had to measure the altitude of each aircraft before it entered the range of another and to allow for any distortions in the target's return signals due to blending in the atmosphere.

Looked at another way, the problem was to hold steady a lever 50 miles long—the length over which the radar beam had to be effec-tive to do its job. Designers had to contend with wind, temperature, and humidity variations, and even the light, imperceptible seismic mo-tions of the ground itself.

In addition to these structural problems, the major problems facing the designers were "ground clutter"—the bane of all radar—and, unexpectedly, system noise. However, an unexpected benefit of the system was the height finder's unique ability to track targets in the rain.[5]

EXERCISE 18-2   Insert *the, a,* or *an* where appropriate in the following passages and be prepared to justify your choice.

A   **ALUMINUM AND BAUXITE**
Aluminum is _____ element, _____ substance made up of one kind of _____ atom. Aluminum happens to be _____ element that joins with other elements easily and firmly. In other words, _____ aluminum easily and firmly combines with _____ other elements to form _____ compounds. _____ compounds are substances, as you might know, in which different kinds of _____ atoms are combined. And once it gets into a compound, _____ aluminum atom is difficult to get out.

This can easily be seen in bauxite. Bauxite is _____ claylike material containing aluminum. In _____ bauxite, aluminum is very tightly bound to _____ oxygen.[6]

B   **THE FORMATION OF OUR SOLAR SYSTEM**
The nebular hypothesis and _____ collision hypothesis each describe one way _____ solar system might have been formed. However, these hypotheses proved to be unsatisfactory for _____ number of _____ reasons. In _____ first place, they did not account for the fact that the orbits of the planets are nearly circular and in nearly _____ same plane. If the masses that became _____ planets were violently pulled out of _____ sun, their orbits would have been _____ long, narrow ellipses and each orbit would probably be in _____ different plane.

Another objection is that _____ astronomers have not found _____ star that could have met with _____ sun at about the time that _____ earth was formed. Here is _____ instance of how observations can upset _____ hypothesis. Scientists demand that hypotheses fit the facts, even _____ most attractive hypothesis.[7]

C   **THE BIG-DISH RADIO TELESCOPE**
"One of _____ most challenging engineering problems of _____ century" . . . "Dwarfing anything constructed in _____ past for the study of the universe" . . . " _____ largest movable land-based structure ever constructed in _____ world."

These were only three of the descriptors enthusiastically applied to the U.S. Navy's attempt in _____ late 1950's to build _____ mammoth radio telescope with _____ 600-foot-diameter antenna that would be fully steerable. The structure, which would have been twice as large as any fully steerable telescope built since, would have towered 66 stories above _____ ground on _____ 7-acre foundation near Sugar Grove, West Virginia. Unfortunately for eager astronomers, however, in 1961–62 original cost estimates of $52 million were re-evaluated and re-estimated to be between $200 and $300 million for _____ future. As a result, the so-called Big Dish was abandoned in 1962, _____ victim, proclaimed one engineering magazine, of galloping obsolescence. Now all that remains of that dream to see more of _____ sky and farther into space than ever before is the concrete foundation for the telescope tracks and the pintle bearing.[8]

18.3

GENERICS

The definite article is sometimes used generically to indicate that a countable noun or noun phrase refers to an entire *type* of something. In such usage, the noun or noun phrase is always in the singular. Here is an example:

> For simplicity and efficiency, *the Hawker Siddeley Harrier* is one of the best present-day VTOL aircraft. This plane uses the concept of "vectored thrust," where four rotating exhaust nozzles are used to deflect the exhaust from vertically down to directly behind.[9]

The author of this passage is referring not to a single plane but to a single *type* of plane. Since there is only one such type, the reference is unique. Some other examples are *the Ford Pinto, the IBM 3600, the Polaroid One-Step,* and *the Honda CVCC.*

It is possible, of course, to use a noun or noun phrase generically without using the definite article. One can say *Hawker Siddeley Harriers* or *a Hawker Siddeley Harrier* to refer to the entire type, just as one can say *the Hawker Siddeley Harrier.* In all three cases, the meaning is essentially the same, though it derives from different sources. With the indefinite forms (plural or singular), individual planes are being referred to as *representative* of the type; with the definite form, the type as a whole is being directly referred to.

18.4

PREVIOUS MENTION

Often a writer will refer more than once to the same object or concept. In such cases, the definite article is used with each mention beyond the first. (In some cases a pronoun or demonstrative can be substituted; see Chapter 11.) Consider the following example:

> Soil physicists have characterized the drying of *a* soil in three stages. They are:

The wet stage, where the evaporation is solely determined by the meteorological conditions;

An intermediate, or drying, stage, where *the* soil occurs in the wet stage early in the day but then dries off because there is not a sufficient amount of water in *the* soil to meet the evaporation rate; and

The dry stage, where evaporation is solely determined by the molecular transfer properties of water within *the* soil.

There is a striking change in the evaporation rate as *the* soil dries during the transition from the wet stage to the drying stage.[10]

In this example, the first mention of the noun *soil* does not have a unique referent. The author is randomly referring to any *soil* and so marks the noun with the indefinite article: *a soil* (line 1). The author then uses the randomly selected soil as a model for the soil-drying process. In doing so, he refers back to the model soil several times. Each of these later mentions of the word *soil* thus has a specific, unique referent—namely, the same soil sample randomly selected at the beginning. To make this reference clear to the reader, the author uses the definite article with each of these later mentions: . . . *the soil* . . . *the soil* . . . *the soil.*

Not all cases of "previous mention" *the* involve repetitions of the exact same noun. In many cases, *the* is used when the repetition involves a synonymous noun or variant form. In the following paragraph, for example, the three noun phrases in italics vary somewhat in form but nonetheless have the same referent.

The simplest approach to passive space heating is through direct gain of solar radiation by means of *a south-facing expanse of glass.* This approach works best when *the south window area* is double-glazed and when the building has considerable thermal mass in the form of concrete floors and masonry walls insulated on the outside. What results is, in effect, a live-in solar collector thermal storage unit. If *the south-facing window area* is vertical, seasonal temperature control is basically automatic. . . .[11]

Here, as in the previous example, the first mention of the noun phrase *south-facing expanse of glass* does not have a specific, unique referent. Hence, the indefinite article is used: *a south-facing expanse of glass.* Then, the author uses this *expanse of glass* as a model and refers to it twice. The first time it is called the *south window area;* the second time it is called the *south-facing window area.* In form, these are both slight variants of the original *south-facing expanse of glass,* but they both refer uniquely to that particular *expanse of glass.* Thus, the definite article must be used: . . . *the south window area* . . . *the south-facing window area.*

18.5

MODIFIERS FOLLOWING THE NOUN

Often a writer will designate a referent as unique by attaching one or more modifying phrases to a noun. For example, consider the noun *growth*. By itself, this noun does not have a unique referent, but if the writer adds appropriate modifiers to it, its meaning can be narrowed to a unique referent:

> *The growth of the American garment industry in the mid-19th century* was made possible by Isaac Singer's improvement of the sewing machine.

Now the word *growth* refers to a specific growth and thus requires the definite article: *the growth* . . .

This use of the definite article seems to occur most often when a noun is followed immediately by an *of* phrase:

Theory	*the* theory *of* relativity
Construction	*the* construction *of* Aswan Dam
Principles	*the* principles *of* thermodynamics
Cost	*the* cost *of* producing nuclear energy
Invention	*the* invention *of* the electric light-bulb
etc.	

It also occurs with other types of following modifiers, including relative clauses:

> At *the time the Mariner-9 spacecraft began its Martian orbit, in November 1971,* an intense, planet-wide dust storm was in progress. *The infrared spectroscopy experiment which was carried on the spacecraft* obtained information on the thermal structure of the atmosphere, both during and after the storm.[12]

In the first sentence, the noun *time* refers not to just any time but rather to a single, specific time, namely, the time when Mariner-9 began its Martian orbit (November 1971). Because of this unique reference, the definite article is used: *the time*. . . . Similarly, in sentence 2, the noun *experiment* is modified so that it refers not just to any experiment but to a specific one, namely, the one that was carried out on the spacecraft and used infrared spectroscopy.

EXERCISE 18-3   Find all the examples of *the* used for generics, previous mention, or a following modifier (the *the* discussed in Sections 18.3 to 18.5). Be prepared to classify the various *definite article plus noun* and *indefinite article plus noun* units and to explain your classification.

A   STEEL IN THE TIN CAN

The tin can is really made of steel, not tin. The can is made of steel because steel is strong. However, steel corrodes, and so the steel in the can must be protected from corrosion. The metal tin does not corrode. Thus, a thin coating of tin is put on the steel to make the "tin" can. The tin saves the steel from corroding.[13]

B   **THE COMPOSITION OF THE HEAVENLY BODIES**
The composition of the heavenly bodies is not uniform. About 99 percent of the matter making up the sun and other stars is a mixture of the light, gaseous elements hydrogen and helium. The other 1 percent of the matter consists of the heavier elements—oxygen, iron, aluminum, nitrogen, and others.

    The distribution of elements in our solar system is quite different. The distribution falls into two distinct groupings: the distribution in the inner planets (Mercury, Venus, Earth, and Mars) and that in the outer planets (Jupiter, Saturn, Uranus, and Neptune). For the inner planets, only about 1 percent of the matter consists of the lighter elements; the other 99 percent is made up of the heavier elements. In contrast, the outer planets contain a great deal of the lighter element hydrogen and relatively small amounts of the heavier elements. (The chemical makeup of the planet Pluto is not yet known.)[7]

C   **AN INTRODUCTION TO THE PHYSICS AND PHYSIOLOGY OF ACCELERATION**
In the days of the frail, canvas-covered aircraft which could not take stresses easily tolerated by the human body, acceleration was not much of a problem. Today, aircraft of much stronger construction travel at sonic and supersonic speeds and thus can impose tremendous forces for appreciable periods of time on the now relatively frail human occupant. Since aviation medicine has as yet little understanding of these important forces, a study of the fundamental principles involved in the physiology of acceleration is needed. Such a study should proceed through the following stages:

1   The history of acceleration and its relation to aviation medicine should be described.
2   The physiological effects and the clinical response to such forces should be understood by the flight surgeon.
3   The conventional terminology for discussing these forces and their effects must be established.[14]

EXERCISE 18-4   Insert *the, a,* or *an* where appropriate in the following passage and be prepared to justify your choice.

A   **AWARENESS OF TECHNOLOGICAL REVOLUTIONS**
It is easy to be aware of _____ revolution brought about by _____ internal combustion engine. _____ effects of _____ revolution are part

of our world. Most of us own _____ car and know something of _____ piston, _____ cylinder, and _____ horsepower. It is also easy to be aware of _____ revolution brought about by _____ electronic tube. Most of us own _____ radios and _____ television sets and know something about waves and _____ electrical interference. Unfortunately, it is harder to be aware of _____ chemical revolution because its products are hidden. Yet this revolution is important, because it is _____ basis for the other revolutions. _____ automobile could not run without fuel, which is a chemical. _____ car or _____ electronic tube could not be built without _____ metal, fabric, adhesive, glass, paint, and plastics, which are _____ chemicals or _____ results of chemical processes. If we are to understand our world, we must be aware of _____ chemical revolution.

B **NEWTON AND THE UNIVERSAL LAW OF GRAVITATION**

Isaac Newton was born the same year that Galileo died, 1642. Newton was only 24 years old when he solved _____ problem of what holds the solar system together. The story, as you probably know, is that _____ falling apple set Newton thinking in _____ right direction. It was not _____ question of why _____ apple fell that interested him. Everyone knew that gravity pulled things toward the earth. Gravity made _____ apple fall, of course.

It was at this point that _____ observant, questioning man produced _____ great scientist. Newton went on from _____ observation that gravity pulled things to the earth to ask other questions. Was the moon falling, too? Did _____ force of gravitation reach from the earth to the moon? Did it reach from the sun to the planets?

After much work, Newton concluded that _____ force of gravitation did reach from the earth to the moon and held them together. In fact, gravitation held together the whole solar system and universe. _____ now-famous conclusion is known as Newton's Universal Law of Gravitation.[15]

C **THE DYNAMICS OF ROTATION APPLIED TO THE CENTRIFUGE**

_____ classic treatment of rotational physics usually found in _____ engineering textbook and in _____ advanced text on gyrodynamics is ordinarily quite rigorous in its mathematical treatment of _____ subject. This paper is intended to present _____ subject in such manner, first, as to relieve _____ physicians and medical officers of _____ labor required to gain _____ rigorous insight into _____ subject and, second, to strip away _____ nonessentials associated with _____ classic developments, which are not pertinent to _____ work at hand.

To this end, _____ treatment used in this paper will be _____ largely intuitive, extensively graphical one, and _____ use of mathematics beyond algebra will be studiously avoided.[16]

18.6

SHARED KNOWLEDGE

Often the writer and intended readers share certain knowledge of the world because they belong to the same culture. (*Culture,* in this case, can be defined broadly or narrowly, depending on whether the writer is addressing a broad, general audience or a narrow, specialized one.) Thus, the writer will sometimes use a noun or noun phrase that has only one referent in the real world, believing that the readers already know about this uniqueness of reference, and he or she will accordingly use the definite article to mark the noun or noun phrase. Consider the following passage, for example, which begins a report presented to a group of aerospace scientists and engineers at a NASA symposium:

> This paper presents the preliminary results from *the Goddard-University of New Hampshire cosmic-ray experiment* during *the recent Pioneer-11 encounter with Jupiter.* Before continuing, however, I would like to say a few words about *the other Goddard experiment* on Pioneer-10, *the flux-gate magnetometer experiment.* This experiment performed flawlessly throughout the encounter, and it observed a maximum magnetic field strength of 1.2 gauss. . . .[17]

Now it happens to be a fact that each of the italicized noun phrases in this passage has a unique real-world referent; there was only one Goddard-University of New Hampshire cosmic-ray experiment on Pioneer-11, only one Pioneer-11 encounter with Jupiter, etc. Of course, not everybody knows these facts, but in the author's immediate culture—that is, in the circle of fellow scientists and engineers, the ones being addressed in this case—these facts seem to be *common knowledge:* everyone in the group knows that each of the noun phrases in question has a unique referent. Consequently, even at first mention, the author marks these noun phrases with the definite article.

If this writer had been addressing a broader, more general audience of readers, the indefinite article might have been chosen instead:

> This paper presents the preliminary results from *a Goddard-University of New Hampshire cosmic-ray experiment* during *a recent Pioneer-11 encounter with Jupiter.* Before continuing, however, I would like to say a few words about *another Goddard experiment* on Pioneer-10, *a flux-gate magnetometer experiment.* This experiment performed flawlessly throughout the encounter, and it observed a maximum magnetic field strength of 1.2 gauss. . . .

In deciding whether to use the definite or indefinite article, therefore, you sometimes have to consider *two* things: (1) whether or not the noun or noun phrase has a unique referent and (2) whether or not your reader *shares* this knowledge with you. If you do not know who your readers are likely to be or if there is a great variety of backgrounds among your readers—as is often the case with technical reports, for example—you cannot depend on the sharing of specialized knowledge. In such cases you may want to use the indefinite article.

18.7

IMPLIED UNIQUENESS

A less frequent use of the definite article occurs when the writer wants to imply that a noun has a unique referent even though the reader may not know of this uniqueness. For example, consider the following passage from a Volkswagen repair manual:

> Removing and Installing Regulator
> The regulator is mounted in the engine compartment, next to the battery. Always disconnect *the battery ground strap* before removing the regulator. . . .[18]

The manual had previously discussed the regulator, the engine compartment, and the battery, but it had not yet said anything about a "battery ground strap." Yet even a reader who knows nothing about cars would infer from the writer's use of the definite article that there is *only one* battery ground strap involved.

Here is another example of implied uniqueness (from the same repair manual):

> Valve Train
> *The camshaft* is gear-driven off the crankshaft and runs in three split-shell bearings. A Woodruff key positions *the crankshaft gear* while *the camshaft gear* is riveted in place. Solid lifters, pushrods, and adjustable rocker arms make up *the valve-operating linkage.*[19]

This passage is part of the manual's introductory description of the Volkswagen engine. There has been no previous mention in this manual of any "camshaft," "crankshaft gear," "camshaft gear," or "valve-operating linkage." Yet even a completely unspecialized reader would be able to infer from the writer's use of the definite article that there is only one of each of these items in a Volkswagen engine. In contrast, notice that the writer refers to *nonunique* items by using either the indefinite article (*a Woodruff key*) or the plural with no article (*three split-shell bearings, solid lifters, pushrods, adjustable rocker arms*).[20]

18.8

A FLOW CHART FOR ARTICLES

Reviewing the main points of this chapter and the preceding one, we find:

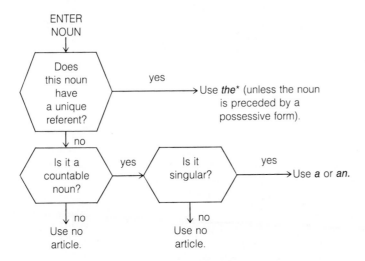

ENTER
NOUN

Does this noun have a unique referent? — yes → Use *the** (unless the noun is preceded by a possessive form).

no

Is it a countable noun? — yes → Is it singular? — yes → Use *a* or *an.*

no
Use no article.

no
Use no article.

*Or a demonstrative adjective (usually *this* or *these*) under certain special conditions (Chapter 20).

FIGURE 18-1   Flow chart for choosing the correct article.

1   The indefinite article (*a* or *an*) is used with a singular countable noun having a nonunique referent (Figure 17-2).

2   The definite article (*the*) is used with *any* noun (excluding most proper names) that has a unique referent.

3   A noun can be judged to have a unique referent for any one or more of the following reasons:
    a   The noun may have a *special adjective* as a modifier (Section 18.1).
    b   The noun may be a *special noun* (Section 18.2).
    c   The noun may be *generic* (Section 18.3).
    d   The noun may refer to some *previously mentioned* referent (Section 18.4).
    e   The noun may have a *following modifier* (Section 18.5).
    f   The noun may refer to *shared knowledge* (Section 18.6).
    g   The noun may have a unique referent by *implication* (Section 18.7).

If you are undecided as to which article to use, try using Figure 18-1. To see how this flow chart works, let's consider some of the noun phrases in the following introductory paragraph from a report by the U.S. Energy Research and Development Administration:

The continuing depletion of domestic fossil fuels may be a signal of one of the most significant long-term issues facing the United States as it enters its third

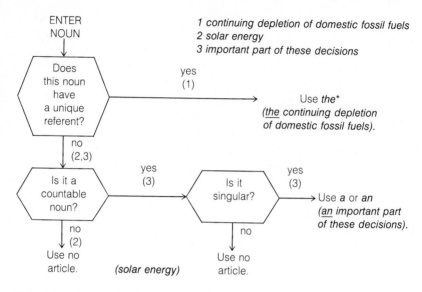

ENTER
NOUN

1 continuing depletion of domestic fossil fuels
2 solar energy
3 important part of these decisions

Does this noun have a unique referent?

yes
(1)

Use *the**
(*the* continuing depletion
of domestic fossil fuels).

no
(2,3)

Is it a countable noun?

yes
(3)

Is it singular?

yes
(3)

Use *a* or *an*
(*an* important part
of these decisions).

no
(2)

no

Use no article.

(solar energy)

Use no article.

*Or a demonstrative adjective (usually *this* or *these*) under certain special conditions (Chapter 20).

FIGURE 18-2   Use of flow chart for choosing the correct article.

century. The impacts of decisions made today concerning our remaining natural resources will persist for generations, beyond the lifetime of today's children. *Solar energy* will be *an important part of these decisions*.[21]

In the first sentence, the head noun of the italicized noun phrase is *depletion;* the other words of the phrase are its modifiers. If we were the authors of this paragraph and were not sure about which article to use, we could begin by asking ourselves the first question on the flow diagram in Figure 18-2 (in the upper left-hand cell): "Does this noun have a unique referent?" The answer is yes (1) because *depletion* has *a following modifier,* which identifies a particular kind of depletion, and (2) because this kind of depletion has been brought to public attention in recent years and is thus *shared knowledge.* Thus, we follow the "yes" arrow and arrive at the correct instruction: "Use *the.*"

Next, consider the noun phrase *solar energy* in the last sentence. We begin, "Does it have a unique referent?" The answer is no because *solar energy* is being used here to refer to solar energy in general, not to any particular form of it. Second, "Is it an uncountable noun?" The word *energy* is being used here in its usual uncountable sense, so we answer yes to this question and arrive at "Use no article." Again, this result agrees with the choice made by the ERDA authors.

Finally, let's analyze the noun phrase *an important part of these decisions.* The head noun in this case is *part,* and so we begin by asking the first question, "Does this noun (*part*) have a unique referent?" The answer is no, even though this noun has a following modifier (*of these decisions*); the writer wants to imply that there may be *other* important parts of these decisions, aside from solar energy. We therefore follow the "no" arrow and arrive at the next question, "Is it a countable noun?" The answer is yes since we customarily think of decision making as involving the consideration of many different factors, or "parts." Therefore we follow the "yes" arrow to the next question, "Is it singular?" The answer is obviously yes again, which yields the final instruction: "Use *a* or *an.*" (Since the first word following this article—*important*—begins with a vowel sound, the proper choice here is *an.*)

A word of caution in using the flow chart: be flexible and be prepared for occasional exceptions. No natural language can be completely precise or rule-governed, and many choices may depend on cultural knowledge or on subtle word meanings that happen to be unknown to you.

EXERCISE 18-5 Now look at the following text and determine, using the flow diagram, which article to use for each blank space where appropriate. Be prepared to explain and justify your choices.

A   **THE STEEL-MAKING PROCESS**
After iron has been made in _____ blast furnace, the iron contains _____ good deal of carbon. This makes _____ iron break easily. So _____ steel, which is stronger than iron, is made from _____ iron. This occurs in _____ steel-making process. At _____ beginning of _____ process, most of _____ carbon is burnt out of _____ iron. Then _____ definite amount of _____ carbon is added.

   To burn the carbon out of iron, different furnaces are used. The most important furnace is _____ furnace which converts iron to steel. This is _____ Bessemer converter. _____ Bessemer converter is shaped rather like _____ egg; it can be tilted for loading and unloading. It is loaded with molten (that is, melted) iron. Then _____ air is blown through _____ molten iron. The oxygen in _____ air combines with _____ carbon and some other materials in the iron. This makes the iron even hotter. It also makes _____ fiery blast of hot gases that spout from the converter.

   When _____ carbon has been burnt out of _____ iron, a measured amount of carbon is added. This is _____ final step of _____ steel-making process. The iron has become steel.[22]

B   **THE EFFECT OF VELOCITY ON THE PATH OF A MOVING BODY**
Imagine a cannon on top of _____ highest mountain on earth. It is firing _____ cannonballs horizontally. _____ first cannonball fired fol-

lows its path. As ____ cannonball moves, ____ gravity pulls it down, and it soon hits ____ ground. Now ____ velocity with which each succeeding cannonball is fired is increased. Thus, the cannonball goes farther each time. Cannonball 2 goes farther than cannonball 1 although each is being pulled by ____ gravity toward the earth all the time. ____ last cannonball is fired with such tremendous velocity that it goes completely around the earth. It returns to ____ mountaintop and continues around the earth again and again. ____ cannonball's inertia causes it to continue in motion indefinitely in orbit around ____ earth. In such a situation, we could consider ____ cannonball to be an artificial satellite, just like ____ weather satellites launched by ____ U.S. Weather Service.[23]

C **STANDARD ESSA WIND OBSERVATIONS**

By far ____ most complete wind data available in ____ United States reside in ____ records of ____ U.S. Weather Bureau (ESSA) stations, which number more than 1000. Many of ____ stations have been in operation for one or more decades, and ESSA has provided ____ excellent repository and processing center at Asheville, North Carolina, so that data can be retrieved and reprocessed readily. ____ typical "surface" wind observation is usually obtained from ____ sluggish cup anemometer and vane assembly, mounted in ____ well-exposed location about 30 feet above ____ ground or ____ building structure. ____ vast majority of ____ sites are at airports, and urban observations are comparatively rare. ____ standard observation procedure is to note ____ indicated wind speed and direction for ____ brief period once each hour and record them to 10° (formerly 16 compass points) and ____ nearest knot or mile per hour. Normally ____ contacting device is associated with ____ anemometer, from which one can obtain ____ time required for one mile of wind flow to pass ____ instrument. Observations called PIBALS or RABALS are taken at many of ____ stations, using either visual or electronic tracking of ____ rising balloon several times per day to determine ____ variation of wind with height. ____ data are processed to reflect ____ wind at 1000-foot intervals above sea level.[24]

18.9

POST-TESTS ON ARTICLES

A **MANNED FLIGHT: A SUCCESS STORY**

On December 17, 1903, Orville and Wilbur Wright successfully achieved ____ sustained flight in ____ power-driven aircraft. The first flight that day lasted only 12 seconds over ____ distance of 37 meters (120 feet). This is about ____ length of ____ Space Shuttle Orbiter. The fourth flight of ____ day (and ____ longest flight) traveled 260 meters (852 feet) in _____ 59 seconds. The initial notification of _____ event to _____ world was _____ telegram to _____ Wrights' father.

Sixty-six years later, _____ man first stepped on _____ lunar surface. An estimated 500 million people throughout _____ world saw the event on _____ television or listened to it on _____ radio as it happened. This was surely _____ historic event.

Historic events are spectacular. The space program, however, has always been much more than a television spectacular. Today, _____ space transportation is working in _____ many ways for us all, and we have come to expect this.

_____ whole new era of transportation will come into being in _____ 1980's with _____ advent of the Space Shuttle. As _____ transportation system to Earth orbit, it will offer _____ workhorse capabilities of such earthbound carriers as trucks, ships, and _____ airlines. It will be as vital to _____ nation's future in _____ space as the more conventional carrier of today is to _____ country's economic life and well-being.[25]

B   Insert articles (or *this* or *these,* if you wish) where appropriate. Be prepared to explain and justify your choices.

The Edsel: A Modern Anti-Success Story

Not since Ford's introduction of _____ Model A, 30 years earlier, had so much fanfare accompanied _____ arrival of _____ new car. When _____ same company formally unveiled _____ Edsel in September 1957, it had already invested _____ quarter of _____ billion dollars in development, production, and distribution, which, according to one account of _____ day, made _____ Edsel _____ costliest consumer product in _____ history. Ford counted on selling at least 2,000,000 Edsels _____ first year.

However, a little over two years and two months later, Ford had sold only _____ 100,000 Edsels and _____ automaker permanently discontinued production. _____ total loss to _____ company reached $350 million, according to some estimates. _____ loss was equivalent to that which would be incurred were _____ Ford simply to give away 110,000 models of its comparably priced car, _____ Mercury.

Conventional wisdom has held that _____ rapid decline of the Edsel was due to _____ company's overreliance on _____ results of _____ public-opinion polls and motivational research. According to _____ view, _____ results were slavishly adhered to in _____ way the Edsel was promoted, in _____ way it was named, and also in the way _____ car was designed. It is further argued that such efforts are doomed to failure, for when _____ car-buying public perceives itself pursued in _____ overly calculated manner, it will invariably turn away in favor of _____ more spontaneously attentive competitor.

When conceived in _____ late 40's, _____ the idea that eventually led to the Edsel was one of putting on _____ market _____ new and completely different medium-priced car. It would be designed to keep _____ upwardly mobile owners of _____ Fords, intent on trading in their symbols of low-income earnings, in _____ Ford family.

_____ Edsel design was certainly different. A novel radiator grill, in _____ shape of _____ horse collar, was set vertically in _____ center of _____ conventionally low, wide front end. _____ unique rear end was marked by widespread horizontal wings. Another striking aspect, within _____ driving compartment, was a cluster of _____ automatic-transmission push buttons on the hub of _____ steering wheel. _____ push buttons controlled what was then _____ most powerful engine (345 horsepower) for _____ automobile at _____ time of its introduction.

The Edsel's failure has been attributed to _____ rather long time lag between conception and market introduction. Forced by _____ Korean War in 1950 to postpone the car's development, Ford came out with the Edsel precisely at _____ time hindsight reveals the car-buying public was moving decisively toward _____ smaller, less powerful compacts. Moreover, within two years, _____ stock market would nose-dive, marking _____ beginning of _____ recession, and _____ automobile industry would end its season with _____ second largest number of unsold cars in history.[26]

C   Insert articles (or *this* or *these,* if you wish) where appropriate. Be prepared to explain and justify your choices.

The Role of Standard Reference Materials in Environmental Engineering

_____ provisions of _____ Clean Air Act Amendments of 1970 and _____ Federal Water Pollution Control Act of 1972 include _____ requirements to limit _____ emissions of _____ pollutants from various points of _____ discharge, such as _____ automobile tailpipe and _____ wastewater effluent. It is fairly obvious that such measures are required if we are to maintain our air and water in _____ sufficiently clean state for protection of _____ public health and welfare. What is not obvious is _____ exact extent to which it is necessary to limit _____ discharges in order to reach _____ sufficient purity. As long as _____ costs rise exponentially with degree of _____ purification, we may expect only enough public pressure to control discharges for adequate protection and no more. _____ economic fact underlines the critical necessity for ＿＿＿ development of ＿＿＿ valid environmental measurement system.

There are two fundamental elements to such ＿＿＿ system. First, we must establish as accurately as possible ＿＿＿ dose-response relationships so that ＿＿＿ most proper ambient air and water quality standards can be set. Secondly, we must establish ＿＿＿ relationship between ＿＿＿ pollutant concentrations at ＿＿＿ point of discharge and at ＿＿＿ point of human contact. Basic to ＿＿＿ measurement system are requirements that:

1)   ＿＿＿ health effect can be measured with requisite accuracy.
2)   ＿＿＿ accurate model is available.
3)   ＿＿＿ pollutant measurements at ambient and source concentrations are internally consistent.

＿＿＿ part of ＿＿＿ measurement system with which we at ＿＿＿ National Bureau of Standards have been primarily concerned is ＿＿＿ third of these. In the discussion that follows, it is presumed that ＿＿＿ other two requirements are met.[27]

REFERENCES

1   It should be pointed out that ordinal adjectives do occasionally occur with the indefinite article: "There are several problems that need to be resolved before we can make progress on this project. One problem is that. . . . A second problem has to do with the auxiliary generator. . . . " In such cases, we would argue, the adjective is being used not in a truly ordinal sense but rather in an enumerative one. In the example cited, for instance, the author seems to be presenting the problem not as a rank-ordering but rather as simply an unordered set. These comments are entirely speculative, however, and further study of the matter is called for, particularly with regard to the theoretical factors involved.

2   C. R. Spitzer, "Unlimbering Viking's Scoops," *IEEE Spectrum* **13**(10):92 (1976).

3   Adapted from P. F. Brandwein, *Concepts in Science*, 3d ed., Harcourt Brace Jovanovich, New York, 1972, p. 48.

4   Adapted from P. F. Brandwein, *Matter: An Earth Science*, Harcourt Brace Jovanovich, New York, 1975, p. 198.

5   Adapted from E. A. Torrero, "The Big Pan of 3-D Radar," *IEEE Spectrum* **13**(10):51 (1976).

6   Adapted from Brandwein, *Concepts*, p. 49.

7   Adapted from Brandwein, *Matter,* p. 200.

8   M. F. Wolff, "The Navy's Big Dish," *IEEE Spectrum* **13**(10):89 (1976).

9   T. A. Talay, *Introduction to the Aerodynamics of Flight,* NASA-SP-367, NASA, Washington, D.C., 1975, p. 149.

10   T. J. Schmigge, "Measurement of Soil Moisture Utilizing the Diurnal Range of Surface Temperature," in *Significant Accomplishments in Science and Technology: Goddard Space Center,* NASA, Washington, D.C., 1975, p. 2.

11   *Pacific Regional Solar Heating Handbook,* Los Alamos Scientific Laboratory, Solar Energy Group, University of California, Los Alamos, New Mexico, 1976, p. 82.

12   B. J. Conrath, "Dissipation of the Martian Dust Storm of 1971," in *Significant Accomplishments in Science and Technology: Goddard Space Center,* NASA, Washington, D.C., 1975, p. 135.

13   Adapted from Brandwein, *Concepts,* p. 82.

14   NATO: Advisory Group for Aerospace Research and Development, Biodynamics Committee, "An Introduction to the Physics and Physiology of Acceleration," in *Principles of Biodynamics: As Applied to Manned Aerospace Flight, Section A, Prolonged Acceleration: Linear and Radial,* Paris, 1967, p. 1.

15   Adapted from Brandwein, *Matter,* p. 208.

16   NATO: Advisory Group for Aerospace Research and Development, Biodynamics Committee, "The Dynamics of Rotation Applied to Centrifuges," in *Principles of Biodynamics: As Applied to Manned Aerospace Flight, Section A, Prolonged Acceleration: Linear and Radial,* Paris, 1967, p. 10.

17   B. J. Teegarden, "Late Results from the Pioneer-11 Flyby," in *Significant Accomplishments in Science and Technology: Goddard Space Center,* NASA, Washington, D.C., 1975, p. 157.

18   *Volkswagen of America Official Service Manual,* Robert Bentley, Cambridge, Mass., 1974, p. 4-17.

19   *Volkswagen of America Official Service Manual,* p. 5-4.

20   In cases of implied uniqueness, the writer's implication that there is only one referent will be accepted by the reader only to the extent that it conforms to that reader's experience in that particular culture. For example, "I took a stroll in the park yesterday and sat down for a while next to the person" sounds distinctly odd unless you have already

identified which person you mean. On the other hand, "I took a stroll in the park yesterday and sat down for a while next to the bandstand (drinking fountain, softball field, etc.)" sounds all right in U.S. culture because there many parks contain only one bandstand (drinking fountain, softball field, etc.).

21 *Solar Energy in America's Future: A Preliminary Assessment,* 2d ed., Energy Research and Development Administration (Division of Solar Energy), Washington, D.C., 1977, p. 2.

22 Adapted from Brandwein, *Concepts,* p. 77.

23 Adapted from Brandwein, *Matter,* p. 209.

24 I. A. Singer and M. E. Smith, "The Adequacy of Existing Meteorological Data for Evaluating Structural Problems," in *Wind Loads on Buildings and Structures.* U.S. Dept. of Commerce, National Bureau of Standards, Washington, D.C., 1970, p. 23.

25 Adapted from "A New Era in Space," in *Space Shuttle,* NASA, Washington, D.C., 1976, p. v.

26 D. Christianson, "The Edsel: A Modern Antisuccess Story," *IEEE Spectrum* **14**(11):94 (1977).

27 *Health, Environmental Effects, and Control Technology of Energy Use,* Report No. 600/7-76-002, U.S. Environmental Protection Agency, Washington, D.C., 1976, p. 58.

19

RELATIVE CLAUSES

A *relative clause* is a clause (or phrase) that is attached to a noun and serves to narrow the class of possible referents for that noun. Relative clauses typically begin with a *relative pronoun* (*which, that, who,* etc.), though the relative pronoun is often omitted. Here are some examples:

An exposure meter is a device *that converts light energy into electrical energy.*[1]

The ignition system of an automobile contains several components *which depend on the vehicle's 6- or 12-volt battery.*[2]

Any celestial body *that emits light waves* may be expected to emit radio waves because both wave forms are part of the same electromagnetic spectrum.[3]

The abstracts for these papers are then printed out, together with an identifying number *through which the article may be located.*[4]

A voltmeter is an instrument *used for measuring the differences of potential between different points of an electric circuit.*[5]

Sometimes two (or more) relative clauses overlap, that is, one relative clause may modify a noun which is itself part of another relative clause. Here is an example:

Simple gears have teeth *that are parallel to the axle or shaft <u>on which they turn</u>.*[6]

Relative clauses perform basically the same function as adjectives: they add meaning to a noun so that the noun refers to a more specific class of things. For example, the sentence "An exposure meter is a device" is so general as to be virtually useless in most contexts; it certainly does not work as an adequate definition of the term *exposure meter.* However, if we modify the noun *device* so that we are talking not about all devices but only about certain kinds of devices, then we are more likely to create a useful definition: "An exposure meter is a device *that converts light energy into electrical energy.*"[1]

Relative clauses are used in all languages of the world. In formal scientific and technical English they commonly occur in the ratio of one relative clause per three sentences of written text, sometimes even more frequently than that. Clearly, relative clauses are a very useful tool for adding precision to one's writing.

Nonnative users of English, however, often have trouble constructing grammatically correct relative clauses and thus often try to avoid using them. This is understandable since languages differ in the way they form relative clauses. If *you* have trouble making relative clauses in English, it is probably because your own native language does it differently.

The purpose of this chapter is to give you the basic rules for constructing grammatically correct and stylistically effective relative clauses in English.

19.1

GRAMMATICAL RULES

Positioning the Relative Clause

RULE 1
Put the relative clause directly behind the noun it modifies.

CORRECT
The energy *which is consumed in overcoming friction* is not lost but is converted into heat.[7]

INCORRECT
The energy is not lost *which is consumed in overcoming friction* but is converted into heat.

EXERCISE 19-1 Check each of the following sentences for proper positioning of the relative clause. Make appropriate changes where necessary.

A   Any celestial body may be expected to emit radio waves that emits light waves because both wave forms are part of the same electromagnetic spectrum.[8]

B   The concepts were known previously which underlie these methods but could not be economically utilized.[9]

C   A turbine is a device that converts the kinetic energy of a moving fluid to the rotational energy of a spinning shaft.[10]

D   Today a jet engine develops 40,000 pounds of thrust that weighs 5000 pounds.[11]

E   With a half-life of 700 million years, the quantity of uranium-235 which was initially present on earth has dwindled greatly.[12]

F   A handicapped person may have no way of interacting with his community who feels he cannot trust a transit system to get him out of the house and back again.[13]

Exceptions to the rule are allowed in cases where the rule would produce either an extremely unbalanced sentence (Chapter 14) or a confusing one. Consider, for example, this sentence:

EXCEPTION
A popular large computer system can do (computations) in one minute *that it would take approximately 50 years to do manually.*[9]

Here the relative clause modifies the noun *computations* but is separated from it by the phrase *in one minute.* Following the rule in this case would have produced a potentially confusing result: "A popular large computer system can do computations that it would take approximately 50 years to do manually in one minute." In this version, it sounds like computations are being done manually in one minute!

Here is another example where violating the rule is called for:

EXCEPTION
(Computer programs) have been written *which analyze activities in order to calculate which are critical to the timely completion of a project and which can take longer than expected without delaying the completion date.*[14]

In this case the relative clause modifies the noun *computer programs* but is separated from it by the verb phrase *have been written.* If the writer had followed the rule, the result would have been an extremely unbalanced sentence with the verb stuck at the very end: "Computer programs which analyze activities in order to calculate which are critical to the timely completion of a project and which can take longer than expected without delaying the completion date have been written." As is discussed in Chapter 14, such a lack of balance would cause readers more frustration than would the violation of the relative clause rule.

Do keep in mind, however, that exceptions like those just described are indeed exceptions! They occur only rarely in formal written English. Therefore, it is a good idea not to break the rule unless very unusual circumstances prevail.

Heading the Relative Clause with a Relative Pronoun

RULE 2
Make sure there is a relative pronoun at the head of the relative clause (optionally preceded by a preposition).

CORRECT
A hygrometer is a device <u>which</u> *is used to measure the relative humidity in the atmosphere.*[15]

INCORRECT

A hygrometer is a device *is used to measure the relative humidity in the atmosphere.*

CORRECT

The temperature *at which evaporation occurs* is called the saturation temperature.[16]

INCORRECT

The temperature *evaporation occurs* at is called the saturation temperature.

EXERCISE 19-2   Check each of the following sentences and see if you can spot grammatical errors involving relative clauses without relative pronouns. Make whatever corrections are necessary. (NOTE: Some of the sentences require two changes.)

A   Cryogenics is a term that refers to technology, processes, and research at very low temperatures.[17]

B   The flow of current through a wire conductor is equal to the number of electrons pass through its cross-sectional area in a particular unit of time.[18]

C   Thermodynamics is the science deals with heat and work and those properties of substances bear a relation to heat and work.[19]

D   A property can be defined as any quantity depends on the state of the system and is independent of the path by which the system arrived at the given state.[20]

E   A decade ago new electronic capabilities became available to increase greatly the kinds of information that could be gathered by remote means.[21]

F   For nonequilibrium processes, we are limited to a description of the system before the process occurs and after the process is completed and equilibrium is restored. We are not able to specify each state through which the system passes, nor the rate the process occurs at.[22]

G   This suggests that information-processing systems should be designed as man-machine systems in which each is assigned the part it is best suited for.[23]

H   All engines operate on the Carnot cycle between two given constant-temperature reservoirs have the same efficiency.[24]

I   A computer subroutine is essentially a program within a program: the main program transfers control to the subroutine, which accepts the necessary data, performs its function, and returns control back to the point in the main program from it came.[25]

Of course there are many cases where the relative pronoun can be omitted without making the relative clause ungrammatical. However, this can be done only with certain kinds of relative clauses and usually only if other changes are made at the same time (see Rules 5 to 7 below).

Avoiding a Duplicate Pronoun

RULE 3
Make sure you don't have a second pronoun duplicating the role of the relative pronoun.

CORRECT
The energy *which is consumed in overcoming friction* is not lost but is converted into heat.

INCORRECT
The energy *which it is consumed in overcoming friction* is not lost but is converted into heat.

Many languages of the world (e.g., Arabic, Persian) differ from English in this regard: they do allow a "duplicate" pronoun to appear in the relative clause. This is because, in those languages, there is not really a "relative pronoun" as such. Instead, there is only a "relative marker." Let us explain.

A relative pronoun actually performs two functions. One, it serves as a *relative marker,* signaling the presence of a relative clause. Two, it serves as a *pronoun,* playing a grammatical role in the relative clause and sharing the same referent as the main-clause noun. This double function can be seen most easily in sentences like the following:

The punched card concept *which Hollerith devised in the 1880s* turned out to be a major step in the development of the computer.

Since the verb *devise* requires a direct object, we sense that there is a "hole" in this relative clause, after the verb. In effect, the relative pronoun (*which*) represents the "missing" direct object at the same time as it marks the relative clause. It is as if the sentence went through these "stages of construction":

1   The punched card concept in the [Hollerith devised it in the 1880s] turned out to be a major step in the development of the computer.

2   The punched card concept in the which Hollerith devised — in the 1880s turned out to be a major step in the development of the computer.

Since relative pronouns in English serve not only as relative markers but also as pronouns, there is no need to fill the "hole" with another pronoun; in fact, it is grammatically wrong to do so.

INCORRECT

The punched card concept which Hollerith devised *it* in the 1880s turned out to be a major step in the development of the computer.

EXERCISE 19-3   Inspect each of the following sentences and cross out any pronouns that duplicate the role of the relative pronoun.

A   A pump is a device that it induces fluid flow through pipes, generally by means of a moving piston.[26]

B   Pulley systems provide a mechanical advantage which it is equal to the number of supporting ropes involved.[27]

C   Many different fuels and oxidizers have been tested, and much effort has gone into the development of fuels and oxidizers which they will give a higher thrust per unit mass rate of flow of reactants.[28]

D   Such a system would not provide the cooling power that people are now accustomed to it in air-conditioned cars.[29]

E   A difference between research and power reactors is in the kind of uranium that they use.[30]

F   The lowest weight of an element which it enters into chemical combination is called its atomic weight.[31]

G   Volta was the first experimenter in history who he was able to generate a continuous electric current.[32]

H   Dosimeters are small badges that people who work around nuclear radiation sources wear.[33]

19.2

STYLISTIC RULES

Once you have firm control of Rules 1 to 3 above, you can consider using additional rules for stylistic purposes. The rest of this chapter is devoted to four such rules, all of which serve to make sentences shorter, easier to pronounce, and often more emphatic. Unlike Rules 1 to 3, which must be followed for grammatical correctness, these next rules are strictly *optional*: if you do not feel comfortable using them, do not use them!

[NOTE: Rules 4 to 7 apply only to true relative clauses, not to appositive (nonrestrictive) relative clauses, which are set off by commas.]

Substituting *that* for *which*

RULE 4
You may substitute *that* for *which* if it is not preceded by a preposition.

CORRECT

The energy $\begin{Bmatrix} which \\ that \end{Bmatrix}$ is consumed in overcoming friction is not lost but is converted into heat.

INCORRECT
Multispectral sensing is a process *in that the earth's surface is surveyed from an airplane equipped with infrared and ultraviolet sensors.* (The correct form is *in which. . . .*)

Traditionally, the use of *that* as a relative pronoun was thought to be too informal for scientific and technical writing. Times have changed, however, and it is now fully accepted. Indeed, using *that* instead of *which* generally makes for a smoother, easier-to-pronounce sentence. It is also favored by many writers who feel it makes their writing less "stuffy" (i.e., less stiffly formal).

EXERCISE 19-4 Convert *which* to *that* wherever possible in the following sentences.

A   Technology assessment is an opportunity for universities to carry on studies which go beyond the bounds of individual fields of study.[34]

B   The transit industry can benefit from other studies which focus on operational questions and scrutinize their cost-effectiveness.[35]

C   A thermodynamic system is defined as a quantity of matter of fixed mass and identity upon which attention is focused for study.[36]

D   An economizer is simply a heat exchanger in which heat is transferred from the products of combustion (just before they leave the steam generator) to the condensate.[37]

E   Even though automatic computing is much faster and cheaper than it has ever been, there are still problems which cannot be computed owing to their size and complexity.[38]

F   Energy which is useful is energy which can be transformed into another form.[39]

G   Combustion is a useful kind of chemical activity by which potential energy in chemicals turns to heat energy.[39]

H   A steam generator is a device in which the hot water which emerges
under pressure from the reactor core transfers its thermal energy to a
second water loop, turning the water in this secondary loop into the
steam which will drive the turbine and electric generator.[40]

Omitting the Relative Pronoun and Auxiliary Verb

RULE 5

If the relative clause begins with a relative pronoun and some form of the
auxiliary verb *to be* followed by a verb phrase, you may omit the relative
pronoun and auxiliary.

CORRECT

The energy $\begin{Bmatrix} that\ is \\ \varnothing \end{Bmatrix}$ *consumed in overcoming friction* is not lost but is converted into heat.

In other words, you can omit any combination of relative pronoun and form
of the verb *to be,*

which is	*who is*
which are	*that was*
that is	*etc.*
that are	

provided it is not immediately followed by a noun phrase or adjective phrase.

INCORRECT

The technician *sick last year* has returned to work. (The correct form is . . . *who
was sick last year.* . . .)

EXERCISE 19-5   In the following sentences, reduce the relative clauses by
omitting the relative pronoun and auxiliary verb:

A   In 1911 Ernest Rutherford showed that the atom was largely empty
space and hence had to contain something that was smaller than the
atom.[41]

B   "Water treatment" is a term that is traditionally applied to municipal
treatment of water.[42]

C   In many power plants the air which is used for combustion is preheated
in the air preheater by transferring heat from the stack gases as they
are leaving the furnace.[16]

D   Simulation can thus offer a basis for decisions about a new system without the disruptions and the costs which are associated with trying a system which may fail.[14]

E   Cheapness of oil has been widely identified as a factor in the high standard of living that is enjoyed in many nations today.[43]

The fact that you *may* omit the relative pronoun and auxiliary does *not* mean that you always *should.* Good writers sometimes choose not to make such a reduction even though the opportunity presents itself. Why? It is mainly a question of whether or not you want to emphasize the word immediately following the auxiliary. If you do want to emphasize it, then you will probably want to retain the relative pronoun and auxiliary; this has the effect of giving greater prominence to the word following the auxiliary. If you do not want to emphasize it, you will probably want to omit the relative pronoun and auxiliary.

To see this, consider sentence E of Exercise 19-5 as it actually appeared in a well-written research article:

Cheapness of oil has been widely identified as a factor in the high standard of living enjoyed in many nations today.

The most important words and phrases of this sentence—those most likely to be stressed if one were to read the sentence aloud—are those given below in italics:

Cheapness of oil has been *widely identified* as a *factor* in the *high standard of living* enjoyed in *many nations today.*

Notice that there is a balanced spacing of these important words and phrases, so that they occur at rhythmic intervals. It is as if the sentence consisted of alternating peaks and valleys, with emphasis given only to the peaks. Now, this being the case, notice that the word *enjoyed* is located in between two peaks, i.e., in one of the valleys. This is appropriate, it seems to us, because the word *enjoyed* does not carry much importance in this sentence; in fact, it could even be omitted without really changing the meaning of the sentence: "Cheapness of oil has been widely identified as a factor in the high standard of living in many nations today."

In effect, the writer of this sentence deliberately deemphasized the word *enjoyed* by placing it in a valley. How was this effect achieved? By applying Rule 5, of course! If Rule 5 had not been applied, this would have been the result:

Cheapness of oil has been widely identified as a factor in the high standard of living that is enjoyed in many nations today.

The presence of the relative pronoun and auxiliary verb (*that is*) in this version of the sentence would have moved *enjoyed* into a position where, as a result of the rhythm principle described above, it would more likely be treated as a peak than as a valley:

> *Cheapness of oil* has been *widely identified* as a *factor* in the *high standard of living* that is *enjoyed* in *many nations today.*

In other words, the writer would have incorrectly put emphasis on a word that does not deserve it.

EXERCISE 19-6   In the following sentences, examine each relative clause headed by a relative pronoun and some form of the auxiliary verb *to be*. Is the next word an important one? If so, leave the relative clause as it is. If not, cross out the relative pronoun and auxiliary.

A   A steam-electric generating plant is a giant engine that is attached to an electrical generator.[44]

B   One type of refrigeration process involves passing the air through a throttle valve that is so designed and so located that there is a substantial drop in the pressure and temperature of the air.[45]

C   The high-pressure, high-temperature products of combustion expand as they flow through the nozzle, and as a result they leave the nozzle with high velocity. The momentum change that is associated with this increase in velocity gives rise to the forward thrust of the vehicle.[28]

D   The computer has made possible systems studies that were heretofore impractical.[14]

E   Decision simulations that are complex enough to be effective require a computer for making the computations that are involved.[46]

F   Much work has also been done on solid-propellant rockets. They have been very successfully used for jet-assisted takeoffs of airplanes, military missiles, and space vehicles. They are much simpler in both the basic equipment that is required for operation and the logistic problems that are involved in their use.[28]

G   As each electrical pulse reaches a synapse, chemicals which are called neurotransmitters come into play.[47]

H   After mining and milling, uranium ore is concentrated into an oxide form which is known as "yellow cake."[48]

I   We speak of substances as having different "phases." A phase is defined as a quantity of matter that is homogeneous throughout.[20]

J   The cloud chamber is a scientific device that is used to record the paths of subatomic particles, often in conjunction with collision experiments.[49]

Using the *-ing* Form

RULE 6
If the relative clause begins with a relative pronoun and a main verb (i.e., nonauxiliary verb), you may omit the relative pronoun and change the verb to its *-ing* form.

CORRECT

Persons $\left\{ \begin{array}{l} who\ do \\ doing \end{array} \right\}$ *research* find it very time-consuming and sometimes im-

possible to make an adequate search of the literature $\left\{ \begin{array}{l} which\ bears\ on \\ bearing\ on \end{array} \right\}$ *the*

subject they are investigating.[4]

This rule cannot be applied if the verb following the relative pronoun is an auxiliary verb, i.e., any form of the verb *to be,* any modal verb (*can, may, will, would,* etc.), or any form of the verb *to have* when it is used as an auxiliary. Here is an example:

A hologram is a photographic image $\left\{ \begin{array}{l} which\ is \\ *being \end{array} \right\}$ *produced without using any*

lens. (*The *-ing* form is incorrect.)

Sometimes the verb *to have* is used as a main verb, not an auxiliary, in which case the rule may be applied:

CORRECT

A pure substance is one $\left\{ \begin{array}{l} that\ has \\ having \end{array} \right\}$ *a homogeneous and invariable chemical*

composition.[50]

INCORRECT
It is not possible to discuss experiments in particle physics without first considering the theoretical context *having spawned the experiments.* (The correct form is . . . *that has spawned.* . . .)

EXERCISE 19-7   In the following sentences, reduce whatever relative clauses you can by omitting the relative pronoun and changing the verb to its *-ing* form. Make sure the verb is a main verb!

A   The model allows those who use it to compress real time, so that years of operation can be studied in a few minutes.[14]

B   Some of the liquid refrigerant that emerges from the condenser can also be used for air-conditioning.[51]

C   Today, marine engineers are responding to the ocean pollution problems that have now been widely acknowledged.[52]

D   Tankers that ply traditional shipping lanes in ocean waters can run aground while far out of sight of land.[53]

E   There is concern today over the potential dangers which attend the disposal of radioactive by-products produced in nuclear plants.[54]

F   The solution to a complex linear programming problem involves calculations which would be impractical without a computer.[9]

G   In a technical information-retrieval system which uses key words, the contents of each technical paper are identified by key words.[4]

H   All the things that can be done with rational numbers may be succinctly expressed in the form of the equation $ax + b = c$.[55]

I   The real and complex numbers may be subdivided into those that are algebraic and those that are not.[56]

J   The mission of the Pioneer Venus atmospheric probes was to determine at each level the amount of radiation which enters the atmosphere and rises from it.[57]

As in the case of Rule 5, good writers do not always apply Rule 6 whenever the opportunity presents itself. In fact, they avoid applying the rule more often than not. It depends mainly on how much emphasis or focus the writer wants to give the relative clause: a full relative clause tends to command more attention than an -*ing* type of relative clause.

To see the difference, examine this excerpt from a report on coal technology, paying particular attention to the relative clause in italics:

Coal differs from other familiar energy sources in its chemical makeup. Fuels like natural gas or gasoline consist of relatively small and uniform molecules *that contain carbon and hydrogen*. Natural gas is mainly methane (CH_4), and gasoline contains iso-octane (C_8H_{18}) and similar molecules. Heating-oil molecules are similar to gasoline molecules, but they have a longer chain of carbon atoms; lubricating oil and asphalt molecules are longer still. When these substances burn, their carbon and hydrogen atoms combine with oxygen to form carbon dioxide, water, and heat.

Coal, on the other hand, is not described by chemical formulas like the above. Coal contains varying amounts of carbon, volatile organic compounds, incombustible minerals, sulfur, nitrogen, and moisture. . . .[58]

The writer chose to use a full relative clause in this case because he wanted to emphasize the importance of chemical composition in differentiating between coal and other fuels; specifically, he wanted to focus attention on the fact that carbon and hydrogen characterize fuels such as natural gas and gasoline but do not characterize coal. If the writer had chosen to use the *-ing* form of relative clause instead of the full form, he would not have succeeded in putting as much emphasis on the idea that carbon and hydrogen are *defining features* of these noncoal fuels.

EXERCISE 19-8   In each of the following paragraphs there are one or more relative clauses in italics. Look at each and decide whether or not it should be reduced to the *-ing* form.

A   The first class of numbers to be discovered was the natural numbers: 1, 2, 3, 4,.... They are familiar as the numbers used in counting. There is evidence that this class began very modestly. Some tribes still have only a vocabulary of three number words: "one," "two," and "lots." Mathematics, like any other language, had to enlarge its vocabulary to enable mathematicians to give full expression to their ideas. Thus, in the more advanced civilizations, the vocabulary of three terms was long extended to a vocabulary *that consisted of an infinite sequence of natural numbers.* But even this infinite vocabulary was eventually found to be insufficient for mathematical purposes, indeed even for everyday affairs, and the number system had to be extended repeatedly to include other classes of numbers.[55]

B   Particle physics is the study of the tiny fundamental constituents of matter. Another name for this science is high-energy physics, because to study the smallest parts of matter one must employ the highest levels of energy. High-energy devices are the synchrotrons and other particle accelerators *that impart energy and light-like speeds to beams of particles in a vacuum.* Such beams, which require vast amounts of electric power to create and control, are the probes or tools by which subatomic events are made to occur. What happens in such events, properly observed, tells physicists what the particles are like.[41]

C   In technical information retrieval using key words, the contents of each technical paper are identified by key words. For example, an article about the effect of radiation on metals in the Van Allen belt might be identified by "metals," "radiation," "Van Allen belt," and "space." In addition to the key-word identifiers, a short abstract is prepared and stored on secondary storage medium. A person *who wishes information on a topic* chooses a set of identifiers *which best describes the topic.* The computer is programmed to compare this set of identifiers with the list of all identifiers and to select papers *which have sets of identifiers which contain one or more identifiers common to the topic identifiers.* The abstracts for these papers are then printed out, together with an identifying number through which the article may be located.[4]

Omitting the Relative Pronoun

RULE 7
If the relative clause begins with a relative pronoun and another pronoun, you may omit the relative pronoun.

CORRECT
The free market system ~~that~~ *we are all familiar with* is increasingly under scrutiny.

EXERCISE 19-9   In the following sentences, reduce whatever relative clauses you can by omitting the relative pronoun where appropriate.

A   At the conclusion of a cycle all the properties have the same value that they had at the beginning.[59]

B   A number of energy sources have serious potential and merit the research attention which they are getting, but most are not yet ready for practical use.[54]

C   The products of combustion from power plants are discharged to the atmosphere, and this constitutes one of the facets of the air pollution problem that we now face.[16]

D   Neuroscientists have in the last decade discovered that synapses take on a variety of characters, which are largely determined by the chemical neurotransmitters which they employ.[60]

E   Harbor safety requirements must be stiffer than ever in view of the widespread effects that a tanker accident may bring about.[53]

F   Mathematicians created larger and larger classes of numbers so as to have more room to move around in and to do all the things that they wanted and needed to do.[55]

EXERCISE 19-10 (REVIEW EXERCISE)   In the following example there are a number of relative clauses in italics. Examine each and make stylistic changes (*that* for *which*, omission of relative pronoun and/or auxiliary verb, or change of verb to -*ing* form) where appropriate.

A   An investigation into the behavior of a system may be undertaken from either a microscopic or a macroscopic point of view. Let us briefly consider the problem *which we have* if we attempt to describe a system from a microscopic point of view. Consider a system *which consists of a one cubic inch volume of a monatomic gas at atmospheric pressure and temperature.* This volume contains approximately 10^{20} atoms. To describe the position of each atom, three velocity components must be specified.

Thus, to completely describe the behavior of this system from a microscopic point of view, it would be necessary to deal with at least 6×10^{20} equations. Even with a large digital computer, this is a quite hopeless computational task. However, there are two approaches to this problem *which reduce the number of equations and variables to a few which can be handled relatively easily in performing computations.* One of these approaches is the statistical approach, *in which,* on the basis of statistical considerations and probability theory, *we deal with "average" values for all particles which are under consideration.* This is usually done in connection with a model of the atom *which is under consideration.* This is the approach *which is used in the disciplines which are known as kinetic theory and statistical mechanics.*[61]

REFERENCES

1   *How Things Work,* Bantam/Britannica, New York, 1979, p. 56.

2   *How Things Work,* p. 186.

3   *How Things Work,* p. 147.

4   G. B. Davis, *Introduction to Electronic Computers,* 2d ed., McGraw-Hill, New York, 1971, p. 42.

5   Adapted from *Webster's Seventh New Collegiate Dictionary,* G. C. Merriam, Springfield, MA, 1969, p. 997.

6   *How Things Work,* p. 19.

7   *How Things Work,* p. 13.

8   Adapted from *How Things Work,* p. 147.

9   Adapted from Davis, p. 39.

10   *How Things Work,* p. 24.

11   Adapted from B. Hiatt, "Heat into Work: The Second Law of Thermodynamics," *The Research News* **28**(11/12):13 (1977).

12   B. Hiatt, "Nuclear Power, Nuclear Safety," *The Research News* **30**(6):9 (1979).

13   Adapted from B. Hiatt, "Transportation and Energy," *The Research News* **30**(1/2):11 (1979).

14   Davis, p. 41.

15   *How Things Work*, p. 137.

16   R. E. Sonntag and G. J. Van Wylen, *Introduction to Thermodynamics: Classical and Statistical*, Wiley, New York, 1971, p. 4.

17   Sonntag and Van Wylen, p. 12.

18   Adapted from *How Things Work*, p. 89.

19   Adapted from Sonntag and Van Wylen, p. 17.

20   Adapted from Sonntag and Van Wylen, p. 20.

21   B. Hiatt, "Technology Assessment: Creating the Future," *The Research News* **28**(9/10):4 (1977).

22   Sonntag and Van Wylen, p. 22.

23   Davis, p. 35.

24   Adapted from Sonntag and Van Wylen, p. 197.

25   Adapted from Davis, p. 200.

26   Adapted from *How Things Work*, p. 21.

27   Adapted from *How Things Work*, p. 17.

28   Adapted from Sonntag and Van Wylen, p. 16.

29   Adapted from Hiatt, "Heat into Work," p. 12.

30   Hiatt, "Nuclear Power, Nuclear Safety," p. 14.

31   Adapted from N. B. Reynolds and E. L. Manning (eds.), *Excursions in Science*, McGraw-Hill, New York, 1939, p. 278.

32   Adapted from W. Kaempffert, *Explorations in Science*, Viking, New York, 1953, p. 229.

33   Adapted from Hiatt, "Nuclear Power, Nuclear Safety," p. 7.

34   Hiatt, "Technology Assessment," p. 10.

35   Hiatt, "Transportation and Energy," p. 14.

36   Sonntag and Van Wylen, p. 17.

37   Sonntag and Van Wylen, p. 2.

38   B. Hiatt, "Big Computing, Tiny Computers," *The Research News* **30**(4):5 (1979).

39 Hiatt, "Heat into Work," p. 3.

40 Hiatt, "Nuclear Power, Nuclear Safety," p. 5.

41 B. Hiatt, "What's the Matter: Particle Physics at the University of Michigan," *The Research News* **26**(3/4):3 (1975).

42 J. Wei, "Towards Cleaner Water: Studies in Toxic Substances and Other Contaminants," *The Research News* **29**(8/9):6 (1978).

43 Adapted from Hiatt, "Transportation and Energy," p. 5.

44 Hiatt, "Heat into Work," p. 7.

45 Sonntag and Van Wylen, p. 13.

46 Davis, p. 43.

47 B. Hiatt, "Signals and Synapses: The Brain's Neurotransmitters," *The Research News* **30**(10–12):4 (1979).

48 Hiatt, "Nuclear Power, Nuclear Safety," p. 8.

49 *How Things Work*, p. 152.

50 Sonntag and Van Wylen, p. 43.

51 Hiatt, "Heat into Work," p. 10.

52 B. Hiatt, "Oil on Troubled Waters," *The Research News* **26**(2):12 (1975).

53 Hiatt, "Oil on Troubled Waters," p. 5.

54 B. Hiatt, "Coal Technology for Energy Goals," *The Research News* **24**(12):3 (1974).

55 J. Wei, "Pure Mathematics: Problems and Prospects in Number Theory," *The Research News* **30**(3):10 (1979).

56 Wei, "Pure Mathematics," p. 13.

57 B. Hiatt, "Planetary Exploration," *The Research News,* **29**(10–12):10 (1978).

58 Hiatt, "Coal Technology for Energy Goals," p. 6.

59 Sonntag and Van Wylen, p. 23.

60 Hiatt, "Signals and Synapses," p. 5.

61 Sonntag and Van Wylen, p. 19.

20

COHESION

When you write or speak about something, you transfer information about it to someone else. One important ingredient in this transfer of information is the creation of a text that is cohesive, that is, a text in which the individual sentences belong together, are clearly related, and flow from one to another.

This is very important in communication in English. Although some cultures consider it rude or inappropriate for a writer or speaker to create cohesion in the sense defined above, in English this creation of cohesion (clarifying relationships and establishing a flow between sentences) is an important function of the writer or speaker. If it isn't done, the communicator may be seen as inadequate or even unskilled, the communication will probably be unclear to its audience, and the goals of the communicator (to transfer information, to request equipment or support, etc.) will probably not be met. Thus, it is quite important that communicators understand what creates or destroys cohesion and that they build cohesion into their text.

One technique for establishing cohesion (that is, for making clear the relationship between sentences) is discussed in Chapter 14; this technique involves ordering noun phrases so that given information comes before new information. Other techniques are covered in other chapters. Chapter 15 discusses the contribution to cohesion of proper emphasis (achieved by combining sentences, by using signal words and intensifiers, and by cutting out excess words), Chapter 16 discusses the contribution of connectives, and Chapters 17 and 18 discuss the contributions of articles and, in passing, demonstrative pronouns and adjectives.

Still another technique for creating cohesion is discussed in this chapter; it involves creating repeated references to a thing or concept. This repetition, especially of noun phrases, establishes a strong chain of words binding the sentences to each other. Such chains can be formed

By repeating a noun or noun phrase exactly from one sentence to another (full-form repetition)

By repeating a shortened form of a noun phrase (short-form repetition)

By using noun compounds

By using pronouns, articles, and demonstratives

By using a different word which means the same thing (synonymous repetition)

By using a noun phrase very closely associated with an earlier noun phrase (repetition of associated terms)

Let us consider each of these. (For a review of the concept of the noun phrase, see the first sections of Chapter 14.)

20.1

FULL-FORM NOUN PHRASE REPETITION

Often a writer needs to refer to something repeatedly in a series of sentences. In doing so, he or she is usually adding details or qualifications to an important idea. At these times, the writer must decide how to phrase the repetition so that the relationship between the sentences is clear without sounding too repetitive. One option is to repeat the entire word or phrase (the full form) which originally expressed the idea. This is an appropriate option when the writer needs the full form *to clarify meaning* or *to establish special emphasis*. For instance, consider the following example, a version of which was discussed in Chapter 15:

NEGATIVE EXAMPLE

The flue dampers shall be tied with the inlet dampers of the fans located at the exit side of the electrostatic precipitators. *These dampers* will open automatically in the event of any failure in the electrostatic precipitator system. The contractor shall provide proper instrumentation and control for the actuation of *these dampers*.

In this situation, it is not clear what *these dampers* refers to. It could refer to the flue dampers, to the inlet dampers, or to both sets of dampers. To avoid this confusion, the author should have repeated the full noun phrase of the original reference as follows:

REWRITTEN VERSION

The flue dampers shall be tied with the inlet dampers of the fans located at the exit side of the electrostatic precipitators. *The flue dampers* will open automatically in the event of any failure in the electrostatic precipitator system. The contractor shall provide proper instrumentation and control for the actuation of *the flue dampers*.

In addition to repeating the full form of a noun phrase to avoid confusion, a writer can also repeat the full form to establish special emphasis. For instance, the writer of the following example wants to emphasize that it is a proposed system which meets the criteria set by the design team, not the current system:

The proposed system meets all of the criteria specified by the design team:

1. *The proposed system* is compact enough to fit into a 4' × 5' area adjacent to the conveyor line.

2. *The proposed system* is economically competitive with the manual system.

3.   *The proposed system* is a pneumatic system and thus will be able to use the air supply line located approximately 10 feet above the conveyor line.

This sort of repetition for emphasis should not be used too frequently; if it is, too many full noun phrases would be repeated, the text would be annoyingly cluttered, and emphasis would be lost.

20.2

SHORT-FORM REPETITION

If a writer does not need full-form repetition for clarity or emphasis, he or she should try to repeat the noun phrase in a shorter form. Such short-form repetition will be the most appropriate form much of the time because it eliminates awkward full-form repetition but still provides a clear reference back to the original noun phrase. For instance, in the following example, the shortened form *these standards* repeats and refers back to *the same safety standards that gasoline cars must meet*. Once the specific standards have been defined in sentence 1, the reference to those standards in sentence 2 can be more compact:

> Electric cars must be able to meet *the same safety standards that gasoline cars must meet* as set up by the Department of Transportation. *These standards* are derived from an established crash test, in which the car is propelled against a solid wall at 30 mph.

Notice that there is only one set of standards under discussion in sentence 1 and thus only one possible referent for *these standards* of sentence 2. Under such conditions, using short-form repetition creates no confusion and allows the writer to avoid such distracting repetition as the following:

NEGATIVE EXAMPLE

Electric cars must be able to meet *the same safety standards that gasoline cars must meet* as set up by the Department of Transportation. *The same safety standards that gasoline cars must meet* are derived from an established crash test, in which the car is propelled against a solid wall at 30 mph.

Many times a short-form repetition will use a quite short form to repeat a very long one. This is illustrated in the example above, where two words, *these standards,* repeat the nine words of the original noun phrase, *the same safety standards that gasoline cars must meet.* However, just as frequently a short-form repetition will use a relatively short form to repeat a form only slightly longer:

> Repairing the fly ash returns is the optimal solution because of economic and downtime considerations. *The three return units* can easily be repaired by three

of our employees with a downtime of no more than two days and a cost of less than $1375, including labor. To have *the units* replaced, on the other hand, would take at least two months' downtime and, because of our limited staff, would have to be done by outside contractors at a cost of $70,000 or more.

In this example, a four-word unit (*the three return units*) has been shortened to a two-word unit (*the units*). This is a small gain in number of words saved, but a larger gain in removing distracting repetition.

One short-form repetition which appears frequently in some kinds of documents is the acronym/abbreviation. An acronym is a word formed from the first letter of each word in a name (as *COMP* for Committee on Management Practices) or a word formed from the first letters of each word in a name (as *radar* from *radio detecting and ranging*). The use of acronyms or abbreviations provides effective compression in many instances. Consider, for instance, the following example:

> Light Emitting Diode (LED) displays have been in use in many electrical devices for approximately fifteen years. Now another type of display, the Liquid Crystal Display (LCD), has been discovered and is starting to become competitive with the LED. In comparing the advantages and disadvantages of the LCD over the LED, it is clear that the LCD is more appropriate for use in our equipment which currently uses the LED.

20.3

SHORT-FORM REPETITION USING RELATIVE CLAUSES

Another important type of short-form repetition involves the short-form relative clause. Sometimes a noun phrase will appear as a noun-phrase-plus-relative-clause unit. Such a unit may be repeated as the noun phrase followed by either a short-form relative clause or a modifying phrase, as illustrated below:

> *The increased radiation dosages that would be experienced at sea level* would be so small that a significant effect on population levels would be extremely unlikely.
>
> The effect of reducing the intensity of the geomagnetic field to zero on *the radiation dosages experienced at sea level* may be regarded as three-fold. . . . Each of these effects is considered after a discussion of the role of energetic particles falling on the top of the earth's atmosphere in determining *the total radiation dose experienced by organisms at sea level.* . . .
>
> . . . *the radiation dose at sea level* is thus almost entirely due to the secondary particles created in the atmosphere.[1]

The original noun phrase being repeated in this passage includes a full relative clause: *radiation dosages that would be experienced at sea level.*

When the noun phrase unit is repeated at the beginning of the second paragraph, the relative clause is repeated as a short-form relative clause: *radiation dosages experienced at sea level.* This short-form repetition appears again at the end of the paragraph as the slightly varied repetition *radiation dose experienced by organisms at sea level.* (Notice that there is a variation between *dosages* and *dose* as well as the added *by organisms.*) Finally, in the third paragraph, the noun phrase appears again in an even shorter form, *the radiation dose at sea level.*

Thus, as the reader progresses through the article, he or she is introduced to this complex noun phrase in a very gentle way. The phrase first appears as the most expanded and easy-to-process form, the full relative clause. After the reader has seen and understood this version, he or she sees progressively more compressed forms of the phrase, each relying on the familiarity with the concept created in earlier stages. This is a good way to set up complex noun phrases formed with a relative clause.

20.4
REPETITION USING NOUN COMPOUNDS[2]

An even more compressed type of noun phrase which can be used to establish cohesion in a text is the *noun compound.* A noun compound is a noun phrase consisting of two or more nouns functioning together as a unit. In the following example, there are four compounds (italicized), each consisting of two words:

> The author provided an overview of the *research program* and described the *research objectives,* the potential *research opportunities,* and the impact of some past *research accomplishments.*

In all noun compounds, the rightmost noun is the head noun and all nouns preceding it serve as qualifiers in some way. Thus, the full meaning of a noun compound can often be expressed by simply "unwinding" it from right to left and inserting the appropriate preposition(s). For example, a *research program* is a *program of research* and a *water purification system* is a *system for the purification of water.*

However, the exact relationship between qualifiers or between one or more qualifiers and the head noun is not always clear—at least not to a nonspecialist. Take, for example, the noun compound *wall stresses:* does it mean the stresses *on* a wall (from outside) or the stresses *produced by* a wall or the stresses *inside* a wall? Unless you're a construction engineer, you might not know for sure. (It's the last.) Or how about *mission suitability:* does that mean suitability (of something) *for* a mission or suitability *of* a mission (for something)? Unless you're an aerospace specialist, you probably are not sure which. (It's the former.)

Constructing Noun Compounds

As suggested above, the correct interpretation of noun compounds depends heavily on the reader's prior knowledge of the subject being discussed. For this reason, noun compounds work best in writing intended for specialists; indeed, it is not uncommon to find that 20 percent or more of the words in a specialized text are noun compounds. For nonspecialist readers, on the other hand, noun compounds should be used with caution. If you have any doubts about the reader's ability to correctly interpret a noun compound, do not use the compound form until you have first used a full-form noun phrase to express that same meaning. For example, the author of the article discussed above uses the compound form *sea-level radiation* only after first giving a full relative clause version and then gradually compressing it down:

> The increased radiation dosages that would be experienced at sea level . . . the radiation dosages experienced at sea level . . . the total radiation dose experienced by organisms at sea level . . . the radiation dose at sea level . . . sea-level radiation.

Even when writing for specialists, however, be sure to observe the following precautions with noun compounds. First, use only standard, well-established compounds, ones that your readers will immediately recognize. And use them only and precisely in the standard form, taking note even of such seemingly minor details as the presence or absence of plural forms in the qualifier nouns. Generally, only the head noun of a noun compound can take the plural -s marking; qualifier nouns are usually unmarked for plural even though they may have a plural meaning. For example, an *electron beam* is a beam of electrons (plural), yet the word *electron* is not marked for plural when used as a qualifier noun in the compound form. Similarly, *pipe installation* is the installation of pipes, a *passenger ship* is a ship for passengers, *semiconductor devices* are devices using semiconductors, and so on; yet none of the qualifier words in these forms has the plural -s marking. There are many exceptions, though, where the plural form is used: *weapons system, materials science, parts shortage, emissions sampling.* Presumably the plural form is used in these cases to make it clear to the reader that the qualifier noun represents a diversified set of things: there is a diverse array of weapons in a weapons system, diverse materials are studied in materials science, etc. Nonetheless, these are all standard forms and should be used accordingly.

A second precaution to follow in constructing noun compounds is to avoid making them too long. Three or four nouns in a row should be about maximum. If you find yourself constructing a noun compound longer than that, try to break it up into smaller groups by inserting prepositions. For example, instead of writing *the lift arm front bearing cup retainer,* you could write *the retainer for the front bearing cup of the lift arm.* This is a longer form but it is easier to understand, even for a specialist.

Interpreting Noun Compounds

As a reader, you may sometimes be confronted with long noun compounds and may have trouble decoding them. In such cases, the best strategy is to "divide and conquer." This is basically a two-step procedure:

> **First step:** Using your knowledge of the field, try to identify familiar smaller compounds within the longer one.

For example, suppose you were reading an article on aeronautics and came across the term *nozzle gas ejection ship attitude control system.* You might immediately recognize *ship attitude* and *control system* as familiar compounds used in aeronautics. You might also isolate *gas ejection* as a separate compound, either by having seen it before or by having seen such analogous forms as *fuel injection* or *rocket propulsion.* (If there are any hyphens in the term, they are usually there to indicate that the two words connected by the hyphen belong together as a unit; for example, some writers would probably hyphenate *gas-ejection*.) Thus, at this point, you would have four units to deal with instead of the original seven: [*nozzle*] [*gas ejection*] [*ship attitude*] [*control system*].

> **Second step:** Try to find plausible meaning relationships between these subunits and combine subunits accordingly.

The meaning relationship between units of a noun compound in technical and scientific English is almost always one of the following: purpose, composition (material), principle of operation, mode of operation, shape, size, location, name of creator, or restricted reference. These relationships are all illustrated below, using the head noun *brakes.*

TYPE OF MEANING RELATIONSHIP	EXAMPLE
purpose	*emergency brakes:* brake designed for emergency use
composition	*carbon steel brakes:* brakes made of carbon steel
principle of operation	*air brakes:* brakes designed to operate according to the compressibility of air
mode of operation	*hand brakes:* brakes operated by hand
shape	*disk brakes:* brakes whose most important component is in the form of a disk

size	*six-inch brakes:* brakes with a diameter of six inches
location	*front brakes:* brakes located in the front
name of creator	*Ghirling brakes:* brakes designed and manufactured by the Ghirling Company
restricted reference	*car brakes:* the brakes of a car

If you keep in mind these nine types of semantic relationships as you look at the subunits of a long noun compound, you will probably be able to guess the intended meaning of the compound as a whole. For example, let's take our earlier example: [*nozzle*] [*gas ejection*] [*ship attitude*] [*control system*]. Working from right to left, let's look at the first two units: *control system* and *ship attitude*. A plausible guess would be that they share a purpose relationship: the purpose of the control system is to control the ship attitude. Are there any alternative relationships to consider? None that we can see, in this case. So let's bracket those two subunits together and move on to the other two: *gas ejection* and *nozzle*. Are these related in meaning? If so, how? Well, if you know what a *nozzle* is (a device used to spray liquid out the end of a hose) and you have some idea of what *gas ejection* might be, you can guess that these two terms share a mode-of-operation relationship: gas is ejected by means of a nozzle.

At this point, then, we have combined subunits so as to have only two: [*nozzle gas ejection*] [*ship attitude control system*]. Can these two groups now be combined according to one of the types of relationships listed above? Yes, it appears that they share a principle-of-operation relationship: the ship attitude control system operates according to principles of fluid dynamics embodied in the ejecting of gas. Thus we arrive at a plausible interpretation of our seven-term noun compound. (Whether or not it's the *correct* interpretation can be determined with certainty only by consulting an authority.)

EXERCISE 20-1 For practice, convert the following full-form noun phrases into noun compounds. (The first one has already been done for you.)

FULL-FORM NOUN PHRASE	**NOUN COMPOUND**
A A screw jack operated by a machine is	a machine screw jack
B Junctions made of polymer semiconductors are	
C A program of research in biomechanics is	

D The control of costs in building
projects is

          _____

E A device for jamming the fuses on
warheads is

          _____

F Tasks involving the handling of
materials at low risk are

          _____

G An award given for discovering a
formula for succeeding at
managing the feeding of people in
large volumes is

          _____

EXERCISE 20-2 Reduce the length of the following writing samples by converting long noun phrases to noun compounds where appropriate.

A   At the Administrator's request in December 1979, a program to test exhaust emissions was initiated to gather data on approximately 20 passenger cars which get high mileage and which are equipped with a diesel engine.

B   The 360X diesel engine was selected for the purpose of determining the deterioration of its emissions over the life of the vehicle while the vehicle is in use.

C   In an attempt to predict the overheating of generators more accurately, I investigated alternative methods of monitoring based on the degradation of electrical insulation at elevated temperatures.

D   The Swales Equipment Company is seeking to reduce the high rate of recalls for maintenance of the 3-B pump.

E   Sterilization of media used for fermentation can be done by using the traditional method of treating batches or by using a continuous process.

EXERCISE 20-3 If you can, try to unwind these noun compounds and present them as full-form noun phrases instead. This will require some educated guesses! (The first one has already been done for you.)

NOUN COMPOUND	**FULL-FORM NOUN PHRASE**
A A coal liquefaction process is	a process by which coal is liquefied
B Roadside breath-testing surveys are	
C An auto entertainment center is	

D Toluene insolubles removal is   _____

E A land disposal system is   _____

F High latitude radio transmission is   _____

G Diffusion convection mass transfer
resistance is   _____

H A chrome nickel steel forged
connection rod assembly is   _____

Using Noun Compounds to Promote Cohesion

Noun compounds have several functions in scientific and technical English. They can be used as types of short-form reference, as names for concepts being introduced into a discussion, and as a means for more easily shifting noun phrases around in a sentence to follow the principles outlined in Chapter 14 (concisely putting old information in the subject position).

Noun Compounds Used as Short-Form References

The noun compound is related to the short-form relative clause, which was discussed earlier. It is the final stage of compression of a full-form reference using a relative clause. Examples of this use appear in earlier parts of this section on noun compounds.

Noun Compounds Used as Names

Sometimes a noun compound is used for the first mention of a concept. In such a situation, the compound functions as a name and is not a shortened form of some other naming unit. For instance, in the following passage notice that all of the italicized noun phrases are names for newly introduced concepts:

> While thousands of weary commuters scurry for homeward-bound transportation, a mountain of *letter mail* generated by the day's *business activities* is just beginning its own trip through culling, facing, canceling, enriching, and *sorting equipment* at Manhattan's main *post office.* . . .
> And the freshest innovations—computer-controlled optical *character recognition* for sorting letter mail and the recently completed *bulk-mail system* for handling packages exclusively—are prime targets for critics grumbling about missent letters, *parcel damage*, and ever-increasing postal rates. Yet, like modern society's many other afflictions, the cure probably lies in pursuing even more imaginative technology. Jacob Rabinow, inventor of the *12-operator letter-sorting machine* now widely deployed throughout the *U.S. Postal Service* said as much in testimony before the Postal *Rate Commission.*
> Chartered as a replacement for the original *U.S. Post Office Department* in 1971, the U.S. Postal Service (USPS) was launched with the expectation

that *mail delivery* eventually could become a self-sufficient operation rather than continue forever as an outdated, heavily subsidized *Government bureaucracy*. . . . During the early 1970s, several "state of the art" computer-controlled optical *reading/sorting systems* were contracted for, tested, and then installed at *USPS letter-mail facilities* in New York City, Boston, and Cincinnati.[3]

This passage has several interesting uses of noun compounds as names. Sometimes the noun compound appears only once, as a name, as with *bulk-mail system* in paragraph 2. Sometimes a noun compound appears once as a name and then again in another, larger noun compound. This occurs with *letter mail* and *USPS letter-mail facilities*. Finally, sometimes a multiword noun compound or name is shortened to an acronym, as with *USPS* from *U.S. Postal Service*. This new name (*USPS*) can then be used by itself or used to form other names: *USPS letter-mail facilities*. Acronyms can be used quite frequently if you are writing for an audience who knows what they mean.

Noun Compounds Used to Ease Shifting of Noun Phrases

Chapter 14 outlines a method for establishing coherence in a text. This method involves ordering noun phrases so that given information comes before new information, topical information is placed in the subject position, and "lighter" units are placed before "heavier" ones.

It is sometimes easier to achieve this ordering if noun compounds are used, since they can be more easily shifted into different positions than can longer phrases and clauses. Noun compounds are especially useful for presenting the given information at the beginning of the sentence since there is a premium on keeping things short and light there. The flexibility offered by the noun phrase is illustrated below:

> The smooth flow of requests through the hospital's Correspondence Unit is marred by a number of operational problems. First, the Correspondence Unit has an average worker productivity of 59.1%, well below the suggested 65% level. Second, this low worker productivity has caused the *backlog of active and "waiting" requests* to rise above 2000 requests. Simply having this large a *backlog size* causes excessive processing delays. Third, these processing delays are often further increased by our method of assigning processing priority on the basis of patient registration number.

Notice how much easier it is to use the noun compound *backlog size* than it is to use the longer reference:

> Simply having this large a backlog of active and "waiting" requests causes excessive processing delays.

This sentence is too heavy before the verb and requires further cutting to make it acceptable.

EXERCISE 20-4 Identify all of the noun compounds in the following passage and be prepared to comment on their effectiveness and appropriateness. In its original location, this passage followed the one quoted on pages 416–417.

> The New York City 33rd Street/Eighth Avenue Post Office was chosen as the scene for a parallel development effort to produce a new generation of "read-sort" equipment. Both International Business Machines (IBM) and Recognition Equipment Inc. (REI) were selected in 1970 from proposals generated during an earlier design-study contract to build advanced optical character recognition (AOCR) equipment and put it into daily service at the Manhattan location. Each company attempted to take two existing 12-operator/305-pocket letter-sort machines (LSMs) and consolidate the input function to a single transport. REI used five small computers while IBM used one computer to control each transport pair. Total LSM speed was to be maintained at 12 letters per second— not an excessive rate for the optics and electronics. *Spectrum* visited the New York AOCR one evening while it was operating and saw that mail literally runs through the transports at machine-gun cadence.
>
> From talks with REI and postal employees, *Spectrum* learned that the REI AOCR effort in New York was finally terminated when major problems interfacing REI's transport with the LSMs persisted. Eventually the IBM system was brought on-line for everyday use. However, the IBM "success" apparently has proved to be extremely expensive. Jacob Rabinow estimates the cost at around $10 million. And it's important to note that instead of buying more IBM AOCR equipment to handle the LSMs originally assigned to REI, the USPS has adopted other options.[3]

20.5
REPETITION USING PRONOUNS, ARTICLES, AND DEMONSTRATIVES

Pronouns, articles, and demonstratives have already been treated at length in other chapters, and so they will just be reviewed here. For more complete discussions, see Chapters 17 and 18 for articles and demonstratives and Chapter 15 for pronouns and demonstratives.

The main point to be made here is that pronouns, articles, and demonstratives all help establish cohesion in a text. They allow a writer to repeat or refer to a previous noun phrase in a concise way. For instance, consider the following passage:

> A survey of the field was made to determine *possible choices for the new analysis technique.* I was assigned to review *the choices,* select *the best one,* submit *it* to a feasibility test, and, if *it* seemed feasible, to identify initial development and implementation concerns of *the selected technique.* This report presents my solution to our problem and provides initial information on development and implementation of that solution.

Two different noun phrase chains are italicized in this passage: the chain dealing with *possible choices* (repeated as *the choices, the best one, it,* and *it*) and the chain dealing with *the new analysis technique* (repeated as *the selected technique*). The last sentence introduces a third noun phrase chain, *my solution to our problem,* and its repetition, *that solution.* Without using the pronouns and demonstratives, this passage would be impossibly repetitive:

NEGATIVE EXAMPLE
A survey of the field was made to determine possible choices for the new analysis technique. I was assigned to review the possible choices for our new analysis technique, select the best choice for our new analysis technique, submit the best choice for our new analysis technique to a feasibility test, and, if the best choice for our new analysis technique seemed feasible, to identify initial development and implementation concerns of the selected choice for our new analysis technique. This report presents my solution to our problem and provides initial information on development and implementation of the solution to our problem.

This example should remind you that pronouns, articles, and demonstratives build cohesion while allowing a desirable cutting of unnecessary repetition.

It vs. This
Do not make the common mistake of freely substituting *it* for *this* or vice versa. If the repeated unit is the whole idea of a sentence or clause, a demonstrative pronoun, preferably *this,* must be used:

> Computers can solve problems at the rate of 100,000,000 steps per second. *This* is one reason why they have become so popular.

Here, the repeated idea is the entire first sentence. Thus, a demonstrative (*this*) must be used; the pronoun *it* cannot be used:

NEGATIVE EXAMPLE
Computers can solve problems at the rate of 100,000,000 steps per second. *It* is one reason why they have become so popular.

On the other hand, *it* can be used to repeat a single noun or noun phrase:

> I was assigned to evaluate *the proposed analysis technique.* I was to compare *it* to other potential techniques, and, if *it* seemed to be the best technique, I was to submit *it* to a feasibility test.

You should note that such a single noun or noun phrase can also be repeated by a demonstrative pronoun, such as *this:*

A simple measure of the rate of increase of the output of the scientific community can be obtained by looking at the Royal Society Catalogue of Scientific
Literature. *This* eventually covered all the scientific literature in all subject fields
in the 19th century. *It* occupies about the same shelf space as last year's output
in chemistry alone.[4]

In this example, both italicized words (*this* and *it*) repeat the single noun
phrase *the Royal Society Catalogue of Scientific Literature.* When is one
used rather than the other? Usually, *this* is used to highlight a noun phrase
newly introduced as the theme, the thing being talked about. Notice that in
the example above the theme and subject of the first sentence is *a simple
measure* (and its modifiers). By the end of sentence 1, a new theme has
been introduced, *the Royal Society Catalogue of Scientific Literature,* which
then becomes the theme and subject of the next two sentences. Notice also
that *this* would not usually be used for a second repetition of the original
noun phrase (as in the example above) and usually would not be used at all
if the first repetition is an *it.*

An interesting combination of the two types of repetition appears in
the following example:

> Legal decisions awarding damages to consumers injured by faulty products
> have encouraged public criticism of both industry and government. *This* has
> somewhat changed the design situation within industry. *It* has introduced a
> higher concern for safety into the design process and given more power to
> consumers trying to correct a problem.

Here, *this* is used in the second sentence because the repeated unit is the
idea of the entire first sentence. *It* is used in the third sentence because it is
repeating the *this,* meaning *this fact.*

As Follows

Another type of demonstrative that is often misused by nonnative speakers
is the phrase *as follows.* This is a standard, all-purpose, fixed phrase to use
when presenting equations, data, tables, etc., in the body of a text. It should
be used in exactly the form *as follows,* regardless of whether the noun it
occurs with is singular or plural or whether the context it occurs in refers to
past time, future time, or any other time:

> The equation is *as follows.*
>
> The equations were *as follows.*

It should be a simple matter to remember this phrase, but for some reason,
many nonnative speakers fail to do so, creating instead a variety of inflected
forms for *follows:*

NEGATIVE EXAMPLES
The equations are *as follow.*

The equation is *as will follow.*

One phrase you may confuse *as follows* with is *the following:* "The previous equation may be related to *the following.*" Here, *the following* is a short-form reference of *the following XXX,* where *XXX* is the category of the thing to come: "The previous equation may be related to *the following equation.*" Note, however, that you could NOT say

NEGATIVE EXAMPLE
The equation is *as the following.*

If you want to say "the data are" or "the equation is," you must use *as follows:*

The equation is *as follows.*

EXERCISE 20-5 The following passages contain many full-form noun phrases. Try to replace as many of these as possible with shorter forms— either short-form noun phrases, pronouns, or demonstratives.

A   The main conveyor line in Building 9 is used to transport three of Construction Equipment's products. The main conveyor line in Building 9 has three stations, which are located at various positions along the main conveyor line in Building 9. At each station of the three stations located along the main conveyor line of Building 9, a particular worker is assigned to one of Construction Equipment's three products. Each time the worker's one out of three assigned products passes by, it is the worker's responsibility to lift his one out of the three assigned products off the main conveyor line and to place the product in the appropriate chute, which transfers the product to another conveyor line.

B   Digital Systems Division (DSD) is currently involved in the design and implementation of an advanced Digital Signal Processor Unit (DSPU). The advanced Digital Signal Processor Unit (DSPU) will be used in applications such as run-way roughness measurements or highly accurate spectroscopy experiments. The Digital Signal Processor Unit (DSPU) design was completed last October by the Digital Systems Division (DSD) design team and is now being implemented. A prototype is expected to be completed by April. I was assigned to investigate and then propose the type of logic component best suited for use in the advanced Digital Signal Processor Unit (DSPU). The purpose of this report is to present the results of my investigation and the proposals based on the results of my investigation.

C   The present worth of the oxygen process is \$9.5 million whereas the present worth of the nitrogen process is only \$2.3 million.

D   To treat the sanitary sewage generated in the sewer service area, the City operates a wastewater treatment plant which utilizes the trickling filter process for secondary treatment and rapid sand filters for tertiary treatment. The wastewater treatment plant which utilizes the trickling filter process for secondary treatment and rapid sand filters for tertiary treatment is capable of treating a sustained flow of 6.5 million gallons per day through the primary and secondary treatment units.

20.6

REPETITION USING SYNONYMOUS TERMS

Another frequent but potentially misleading type of repetition involves the use of a different word meaning the same thing (a synonymous term) as the original noun phrase. This sort of repetition appears in the example in Section 20.2 on *fly ash returns,* where the original noun phrase *the fly ash returns* of sentence 1 reappears as *the three return units* of sentence 2. In this case the key word *return* is repeated in the second reference and helps make clear that the two phrases are equivalent. However, such repetition of a key word is not necessary if the intended audience clearly sees that one term is in some sense equivalent to the other. Consider, for instance, the following example:

> Because requestors desire medical information as soon as possible, hospital management feels that requests should be processed within *two weeks.* Since the system performance is presently falling well short of *this standard,* a study of request processing procedures was undertaken.

In this example, the writer has implicitly equated *two weeks* and *this standard.* If the reader understands that two weeks is a standard, and in this context the reader probably does, then the writer has chosen a good repetitive form for *two weeks.*

Notice, however, that a writer cannot freely use synonymous terms if the reader does not know enough to see that the terms are synonymous. Consider, for instance, the following example:

> It is the combination of *an inorganic polymer backbone* with a short organic (methyl) group in the structure that accounts for many of the unique properties of PDMS. The *siloxane backbone* is extremely flexible and, coupled with the short stable methyl group, gives many useful properties.[5]

Unless you have a good background in chemistry, you will probably not see immediately that the two italicized noun phrases are synonyms, even though *backbone* appears in both phrases. The compound under discussion is polydimethylsiloxane (PDMS), which has an inorganic polymer backbone consisting of alternating silicon and oxygen atoms with methyl groups attached to the silicon. Thus the backbone is a *sili*(con) + *ox*(ygen) + *ane,* or a *siloxane, backbone.* If you were writing this passage to a group of silicon chemists, who would already know these things, then the use of the synonymous term is probably fine. However, if you were writing to nonchemists or even to chemists who did not specialize in silicon chemistry, you might want to provide some extra information before using the synonymous term *siloxane backbone.*

Sometimes a writer will choose a synonymous form that is about the same length as the original term and closely following it, as illustrated in the preceding example. At other times, however, a writer will use a synonymous form that is much shorter than the original and several sentences removed from it. For instance, note the use of *evacuation time* in the Summary of the following passage:

FOREWORD

In your letter of January 23, 1979, to Dr. D. Newman, you expressed concern about the numerous delays in the positron beam experiment. Dr. Newman attributes most delays to the *ten hours required to pump the vacuum system down from atmospheric pressure to an acceptable operating level* and has asked me to investigate means of reducing these delays. This report recommends the purchase and installation of a gate valve for the Vac-Ion vacuum pump.

SUMMARY

The purchase of a gate valve for the Vac-Ion pump was found to be the optimal solution to the delay problem in the positron beam experiment. Addition of the gate valve will reduce *evacuation time* to three or four hours, cost no more than $600, and present no installation problems. The use of a larger fore pump was seen as the only alternative approach, but this would require extensive modification of the vacuum system, cost approximately $1100, and give no improvement in the *evacuation time.*

In this case, *evacuation time* obviously repeats the original noun phrase and so will probably cause no trouble for most readers of the report even though it is quite far from the original phrase and uses very different words.

In closing this section, we note that you should keep in mind the following WARNINGS:

1   If you want to use synonymous repetitions, you must be very careful that the repeated form is one the reader will easily understand, espe-

cially if it is several sentences away from the original full form of the noun phrase.

2   If you need to repeat a noun phrase several times and feel an overwhelming urge for variety in your repetitions, try to use short-form repetition combined with at most one synonymous form, rather than using several synonymous forms. Readers easily get confused, and much shifting of terminology is bound to create confusion.

20.7

REPETITION USING ASSOCIATED NOUN PHRASES

Many of the forms of repetition discussed in this section will confuse a reader badly if they are not used *very* carefully. As you read the examples in this section, consider their effect on several different audiences and notice that some of the repetitions might not seem at all repetitive to nonspecialist readers. Thus, instead of being cohesive for such readers, these repetitions might in fact destroy cohesion.

Sometimes a writer repeats a noun phrase by "repeating" it at a more general or more particular level. For instance, consider the following passage, in which the first mention of a noun phrase is as a general term, *the thermal properties of glassy materials at low temperatures*. The next three repetitions of the idea *thermal properties* appear as more specific or particular noun phrases, *the thermal conductivity* and *the specific heat below 4 K*. (The third of these repetitions appears as *it*, which repeats the idea *the specific heat below 4 K*.) Finally, the idea of *thermal properties* is repeated a fourth time as the general term *the thermal behavior*.

> *The thermal properties of glassy materials at low temperatures* are still not completely understood. *The thermal conductivity* has a plateau which is usually in the range 5 to 10 K, and below this temperature it has a temperature dependence which varies approximately at T^2. *The specific heat below 4 K* is much larger than that which would be expected from the Debye theory, and *it* often has an additional term which is proportional to T.
>
> Some progress has been made towards understanding *the thermal behavior* by assuming that there is a cut-off in the phonon spectrum at high frequencies . . . and that there is an additional system of low-lying two-level states. . . .[6]

In addition to using a general-to-particular or particular-to-general repetition, a writer sometimes repeats a noun phrase by using the whole-part or part-whole relationship. If the part is very closely associated with the whole in the reader's mind, naming the part will probably recall the idea of the whole, and vice versa. Then, if the original noun phrase names the whole, naming the part could be considered a repetition of the whole.

This form of repetition can be effective only if the audience already clearly knows the relationship between the whole and the part. For instance, consider the following passage from a speech to a group of scientists and engineers working in manufacturing companies. The noun phrase *a manufacturing company* in sentence 1 is "repeated," or recalled, by the mention of its subunits, or parts, in the subsequent sentences:

It is difficult to coordinate and direct *a large manufacturing company* so successfully that the company makes a profit and keeps its employees happy. *Sales, manufacturing, and design engineering* are all at odds with the executive group. *The sales department* wants low selling prices to make the goods easier to market, *the design engineering group* tends to overdesign the product to insure certainty of operation, while the *manufacturing group* tends to make frequent purchases of new equipment and to stockpile materials in order to maintain uninterrupted plant operations. Now it happens that *the executive arm of the firm* is also interested in low selling prices, quality design, and continuous plant operation, but not to the extent of running the business into bankruptcy. Consequently, *the management's* simultaneous desire to set prices which are certain to return a profit, to limit the quality of the design, and to keep a tight rein on equipment expenditures and inventories makes every problem solution a compromise with respect to the interests involved.[7]

Obviously, this type of repetition produces cohesion only if the audience is aware of the relationship between the whole and its parts.

Another type of repetition sometimes used by writers involves the use of a noun phrase strongly related to or associated with the original noun phrase. The association will usually not be as strong as a part-whole or whole-part association, but it will have to be significant if it is to create cohesion for the reader. Some common powerful associations in our culture are cause and effect or effect and cause, an item and some quality or characteristic of the item, and an item and some other item which shares its class or type. Again, the association of the recalling noun phrase must be obvious to the reader if it is to establish cohesion. For instance, consider the following:

A similar satisfactory report of tests made on the shell roof of a military establishment in the southeastern United States substantiates the findings of earlier tests. *An intense fire of $1\frac{1}{2}$-hour duration on the partly cured concrete shell* subjected the shell not only to extreme heat but also to sudden load changes. *The only damage visible to the $3\frac{1}{2}$-in. thick shell,* supported on columns 50 ft and 60 ft apart, consisted of minor cracking and spalling of the concrete. To confirm this observation a full-scale loading test was ordered. About a month after the fire, and before any repairs were effected, *a uniform load of 40 lb per sq ft* was applied on one of the damaged panels. *The maximum deflection* was 0.1 in. and disappeared on removal of the load. *The theoretical deflection of edge members under the live load* was calculated to be 0.027 in., whereas *the* *actual deflection* was 0.025 in. *The structural strength of the building* was thus shown not to have been impaired by the fire.[8]

This passage is quite interesting from several points of view. In sentence 2, the idea is introduced of *an intense fire of* $1\frac{1}{2}$*-hour duration on the partly cured concrete shell.* This idea is then recalled by the subject noun phrase of sentence 3, *the only damage visible to the* $3\frac{1}{2}$ *in. shell.* This does not directly repeat *fire,* the main noun in the original noun phrase, though it does repeat *shell,* the target of all of the tests conducted. The main sense of cohesion, however, arises not from the repeated *shell* but from the strong association we have between *an intense fire of* $1\frac{1}{2}$*-hour duration* and the resulting *damage* usually following such a fire.

This same pattern of establishing cohesion appears again between sentences 5 and 7, where an original noun phrase is recalled by using others associated with it. Here the original noun phrase in sentence 5, *a uniform load of 40 lb per sq ft,* is recalled by several associated noun phrases in sentences 6 and 7: *the maximum deflection, the theoretical deflection,* and *the actual deflection.* Finally, the subject of the last sentence, *the structural strength of the building,* recalls the ideas of deflection, load, and damage; the association here is that *structural strength* is the underlying cause of the small *actual deflection.*

EXERCISE 20-6 Identify the various forms of repetition in the following passages. If you find forms which do not involve direct repetition of some key word, be prepared to explain why you have identified the form as a type of repetition.

A   The first experiment in our series with mice showed that total removal of the adrenal glands reduces aggressiveness. Moreover, when treated with corticosterone, mice that had their adrenals taken out became as aggressive as intact animals again. These findings suggest that the adrenals are necessary for animals to show full aggressiveness.

But removal of the adrenals raises the levels of adrenocorticotrophic hormone (ACTH), and P. Brain found that ACTH lowers the aggressiveness of intact mice. Thus, the reduction of aggressiveness after this operation might be due to the higher levels of ACTH which accompany it.[9]

B   We are standing by the 1.4-mile test track of Ontario's Urban Transportation Development Corporation. Around us are empty fields that the UTDC hopes will become the foremost center of urban mass transit in North America. The elevated train I've come to hear and see is the vehicle that could bring that plan to fruition. It's tomorrow's means of mass transportation today. Subways, of course, are ideal—fast, unobtrusive, impervious to bad weather. But at $80 million per double-track mile, precious few cities can afford to build them. Few can even afford to maintain the ones they have.[10]

C   The formation of ice layers underneath the pavement causes frost heave. Although the temperature deep in the ground remains constant throughout the year, the temperature in the ground near the surface fluctuates with air temperature. When the pavement surface freezes, the temperature in the ground under the pavement falls below freezing. The freezing and low temperatures induce capillary tension, which sucks up water from the water table below. This capillary tension greatly increases the amount of water in the first zone under the pavement. When this water freezes, the soil expands far more than it otherwise would and causes the pavement to heave.

REFERENCES

1   C. J. Waddington, "Paleomagnetic Field Reversals and Cosmic Radiation," *Science* **158**:913–915 (1967).

2   The approach taken in this section was inspired largely by the work of David McNeill and that of Lierka Bartolic, referenced in the Additional Reading to this chapter.

3   D. Mennie, "A Try at Automated 'Zip'," *IEEE Spectrum,* October 1976, p. 41.

4   D. J. Urqhart, "Times Literary Supplement," May 7, 1970.

5   P. G. Pape, "Silicones: Unique Chemicals for Petroleum Processing," paper delivered at the 56th Annual Fall Technical Conference and Exhibition of the Society of Petroleum Engineers of AIME, October 1981.

6.   Quoted by John Swales, *Aspects of Article Introductions,* Aston ESP Research Reports, No. 1, Language Studies Unit, The University of Aston, Birmingham, England, 1981.

7   Adapted from R. R. Raney, "Why Cultural Education for the Engineer?" speech to the Iowa-Illinois Section of the American Society of Agricultural Engineers at East Moline, Ill., April 1953; quoted by J. G. Young, "Employment Negotiations," *Placement Manual: Fall 1981,* College of Engineering, University of Michigan, Ann Arbor, 1981, p. 50.

8   American Society of Civil Engineers, *Design of Cylindrical Concrete Shell Roofs,* ASCE Manuals of Engineering Practice, No. 31, 1951, p. 11.

9   C. Turk, "Do You Write Impressively?" *Bulletin of the British Ecological Society* **9**(3):5–10 (1978).

10 C. Maurer, "Tomorrow's Train Is Running Today in Canada," *Popular Science,* December 1980, p. 75.

ADDITIONAL READING

L. Bartolic, "Nominal Compounds in Technical English," in M. Todd-Trimble, L. Trimble, and K. Drobnic (eds.), *English for Specific Purposes: Science and Technology,* English Language Institute, Oregon State University, Corvallis, 1978, pp. 257–277.

J. M. Carroll, "The Role of Context in Creating Names," *Discourse Processes* **3:** 1–24 (1980).

J. M. Carroll and M. K. Tanenhaus, "Prolegomena to a Functional Theory of Word Formation," in R. E. Grossman, L. J. San, and T. J. Vance (eds.), *Papers from the Parasession on Functionalism,* Chicago Linguistics Society, Chicago, 1975, pp. 47–62.

P. Downing, "On the Creation and Use of English Compound Nouns," *Language* **53**(4):810–842 (1977).

L. Gleitman and H. Gleitman, *Phrase and Paraphrase,* Norton, New York, 1970.

M. P. Jordan, "Some Associated Nominals in Technical Writing," *Journal of Technical Writing and Communication* **11**(3):251–264 (1981).

R. B. Lees, "The Grammar of English Nominalizations," *International Journal of Applied Linguistics,* vol. 26, no. 3, 1960.

J. Levi, *The Syntax and Semantics of Complex Nominals,* Academic Press, New York, 1978.

D. McNeill, "Speaking of Space," *Science* **152**:875–880 (1966).

M. C. Potter and B. A. Faulconer, "Understanding Noun Phrases," *Journal of Verbal Learning and Verbal Behavior* **18**:509–521 (1979).

K. Zimmer, "Some General Observations about Nominal Compounds," in *Stanford Working Papers on Language Universals,* Stanford University, Stanford, CA, 1971, pp. C1–C21.

K. Zimmer, "Appropriateness Conditions for Nominal Compounds," in *Stanford Working Papers on Language Universals,* Stanford University, Stanford, CA, 1972, pp. 3–20.

21

MODAL VERBS

It is important that the statements you make as a scientist or engineer be very precise. You do not want to exaggerate a claim, understate or overstate a conclusion, or otherwise make logically inconsistent assertions.

One important way of being precise is to modify main verbs with appropriate modal auxiliary verbs (or modal verbs, for short) whenever the occasion calls for it. In scientific and technical English, there are eight such modal verbs in common use: *may, can, must, should, could, would, will,* and *might* (several "semimodal" verbs can sometimes substitute for some of these, for example, *have to, be going to, be able to,* and *need to*). The difficulty in using these modal verbs correctly is that they all have two or more different meanings, depending on how they are being used. For example, the word *must* can mean that the statement in which it is used is a logically necessary statement, i.e., that it follows logically from previously mentioned facts: *Since $2x + y = 11$ and $5y = 15$, x must equal 4.* Or, it can mean that the statement in which it is used represents some kind of obligation or duty: *We must thank Dr. Wilbur for her help.* In all, these 8 modal verbs can be used to express at least 16 different meanings.

Our purpose in this chapter is to make it as easy as possible for you to learn and use all 16 of these modal meanings. To do this, we shall divide them into three distinct groups, each of which constitutes a scale of degrees. These groups we shall label "obligation," "probability," and "ability."

21.1

OBLIGATION

Modal verbs are commonly used to indicate degree of obligation or prohibition, as in the sentence given above, *We must thank Dr. Wilbur for her help.* Here the obligation is a strong one, equivalent to "It *is required* that we thank Dr. Wilbur for her help." The writer or speaker could reduce the degree of obligation by using a weaker modal, such as *should* or *may,* or even go to the opposite extreme by using a form such as *should not* or *must not,* indicating prohibition.

The scale as a whole is given in Figure 21-1. Each modal verb is positioned on this scale according to the approximate degree of obligation or prohibition it imposes on the meaning of a sentence. *Must* is at the top end of the scale because it means strongly obligatory, whereas *must not* is at the opposite end because it means obligatory in the opposite (prohibitive) sense. The modals *can, may,* and *could* are located near the middle of the scale because their meaning lies midway between these two extremes; when

FIGURE 21-1 Modals indicating degrees of obligation.

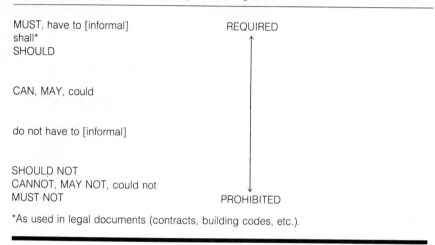

MUST, have to [informal]   REQUIRED
shall*
SHOULD

CAN, MAY, could

do not have to [informal]

SHOULD NOT
CANNOT, MAY NOT, could not
MUST NOT   PROHIBITED

*As used in legal documents (contracts, building codes, etc.).

you use one of these three modals, you are saying that the action of the main verb is neither required nor prohibited, but optional. Finer distinctions can be made, too. For example, the modal phrase *do not have to* also has an optional sense to it, but it differs from *can, may,* and *could* in that it means optional in a negative sense.

The most frequently used modals of obligation in scientific and technical English are *must, should, can,* and *may;* accordingly, we have written these with capital letters in Figure 21-1. The left-to-right ordering is also an indicator of frequency. For example, *can* is more often used than *may,* which in turn is more often used than *could.* It should be kept in mind, however, that frequency of use varies according to situation and context. In a situation requiring extensive use of the past tense, for instance, *could* (the past-tense form of *can*) is more likely to be used than *can.* In oral communication, *have to* might be heard more frequently than *must.*

In scientific and technical English the modals of obligation are most often used with statements describing procedural correctness. In the following extract, for example, the writer has used *may* to indicate that his statements describe legitimate options that one can take in explaining what an arithmetic progression is:

To explain arithmetic congruence, let us start with an arithmetic progression. An arithmetic progression is a sequence of integers such as

$$3, 5, 7, 9, 11, 13, \ldots$$

which we *may* also represent as

$$a_1, a_2, a_3, a_4, a_5, \ldots$$

in which each term a_n (n stands for an integer) is generated by adding a common difference (here 2) to the preceding term. Hence we *may* define an arithmetic progression by the formula

$$a_n = qn + r$$

where n stands for the rank of the term a_n and has values ranging from 1 to infinity, q for the common difference, and r for the remainder obtained by dividing the term by the common difference.[1]

This use of modals in descriptions of procedural correctness is particularly prevalent in textbooks. In fact, most of the modal verbs used in science, mathematics, and engineering textbooks are modals of obligation used in describing procedures.

One other feature of the modals of obligation is that they are usually used in passive sentences with the agent not mentioned. In the extract just cited, for example, most writers would have written this sentence:

Hence *we may define* an arithmetic progression by the formula

$$a_n = qn + r$$

like this:

Hence an arithmetic progression *may be defined* by the formula

$$a_n = qn + r$$

After all, there is no real need to mention that it is "we" who are doing the defining; it is standard algebraic procedure, and anyone familiar with algebra could do the same. So, by putting the sentence in passive form, we can omit the reference to "we."

EXERCISE 21-1 Below are three excerpts: one from a basic physics laboratory exercise book and two from a basic chemistry textbook. For each of the blank spaces, choose the modal of obligation that you think best fits the context.

A   **INDEX OF REFRACTION BY A MICROSCOPE**
Because of refraction, objects appear to be closer when viewed through a dense medium, such as glass or water, than when viewed through air. This phenomenon _____ be used to find the index of refraction n of the medium by means of the equation

$$n = \frac{t}{t - d}$$

where *t* is the thickness of a transparent medium and *d* is the apparent shortening of the perpendicular line of sight through the medium.

PROCEDURE

1. Index of refraction of glass

The microscope _____ be used to measure distances normal to the stage by attaching a vernier scale to the side of the rack as shown in Fig. 34-1. Using a very sharp pencil, *carefully* construct a centimeter scale about 6 cm long with 1-mm subdivisions on the edge of a piece of white paper and a vernier scale on another piece of white paper. The vernier _____ consist of 10 equal divisions in a space of 9 mm. Figure 1-4 provides a suitable pattern for making these scales. Mount the scales on the microscope as shown in Fig. 34-1 so that the zero index of the fixed vernier scale is at or below the zero index of the centimeter scale when the body tube is fully lowered.

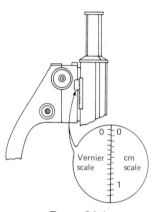

Figure 34-1

Determine whether the 16-mm objective provides the *working distance* required for the thickness of the glass plate. This _____ be done by making a pencil mark on a white index card, mounting the card on the stage, and focusing on it with the 16-mm objective in place. Without shifting the position of the card, place the glass plate over the mark and *cautiously* lower the body tube. If the mark _____ again be brought into sharp focus, the working distance of the lens is adequate and it _____ be used in this part of the experiment.

CAUTION: *The objective _____ not be forced down in contact with the glass plate.* If this second focus is not achieved, the 32-mm objective _____ be used; or if the 16-mm objective is divisible, the front lens section _____ be removed to provide the equivalent lens.[2]

B   **ERRORS AND SIGNIFICANT FIGURES**

All quantitative science is based on measurements, but measurements are usually afflicted by errors, which _____ be taken into account. Several strategies _____ be used to serve this purpose. . . .

An important aspect of the handling of data is that intermediate computations are required. This raises two questions. One concerns the relation of the uncertainty in the quantity of interest to the errors in the original measurements. In other words, how do the errors propagate through the equations that connect the quantity of interest with the data? The other question concerns the accuracy with which intermediate calculations _____ be carried out. To do justice to the data, the accuracy of the calculations _____ be sufficiently high that no new errors are introduced, but excessive accuracy is pointless and inefficient.[3]

C   **CHEMICAL EQUATIONS**

The reactions of chemical substances _____ be represented by chemical equations that show the formulas and relative numbers of the reactants and products involved. Equations _____ , by definition, be balanced: they _____ show the same number of atoms of a given kind on each side, and the net charge _____ be the same on each side. The condition, sometimes slavishly insisted upon, that all coefficients in an equation _____ be integral is not essential; coefficients _____ be fractional or even have a common factor. . . .

Many equations are easy to balance, once the important reactants and products are known. For example, when dissolved in water, silver nitrate, $AgNO_3$, and sodium sulfate, Na_2SO_4, react to form solid silver sulfate, Ag_2SO_4, which precipitates from the solution. An unbalanced equation for the reaction is

$$AgNO_3 + Na_2SO_4 \rightarrow Ag_2SO_4 + \cdots$$

which _____ be balanced to

$$2\, AgNO_3 + Na_2SO_4 \rightarrow Ag_2SO_4 + 2\, NaNO_3$$

The arrow is often replaced by an equal sign for typographical reasons, but this _____ be done only when the equation is balanced.[4]

21.2

PROBABILITY

The modal verbs are also used to indicate degree of probability, as in the following sentence:

The x-ray count data were corrected for background but were not corrected for effects of absorption, so reported concentrations *may* be in error by up to ± 20 percent of the measured amount.[5]

FIGURE 21-2 Modals indicating degrees of probability.

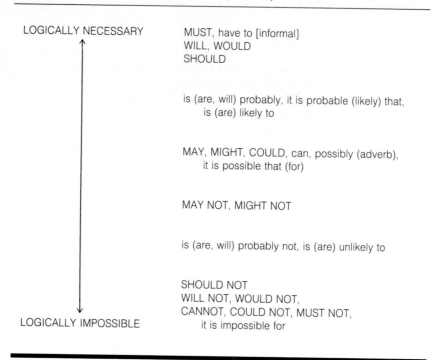

LOGICALLY NECESSARY

MUST, have to [informal]
WILL, WOULD
SHOULD

is (are, will) probably, it is probable (likely) that,
    is (are) likely to

MAY, MIGHT, COULD, can, possibly (adverb),
    it is possible that (for)

MAY NOT, MIGHT NOT

is (are, will) probably not, is (are) unlikely to

SHOULD NOT
WILL NOT, WOULD NOT,
CANNOT, COULD NOT, MUST NOT,
    it is impossible for

LOGICALLY IMPOSSIBLE

Here the writer is implying that an error of ± 20 percent is *possible,* i.e., only moderately probable. If he had thought there was a high probability of error, he would likely have used a stronger modal, such as *should, will,* or *must* (or some adverbial form like *are probably*). The complete scale of choices, ranging from logically necessary to logically impossible, is given in Figure 21-2.

In scientific and technical English, the modals of probability are most often used in conclusions, summaries, abstracts, etc.—wherever tentative generalizations are being made. They are also commonly used in reviews of literature, where the findings of previous investigations are being described. In all of these cases, statistics reveal that *may, will,* and *would* are the most frequently used modals, followed by *might, must, can,* and *should,* in that order.

Figure 21-2 also includes a number of modal phrases that can be used to imply degrees of probability—phrases such as *is probably, is likely to, it is possible that,* and *it is impossible for.* Since most of these expressions occupy important positions on the scale and do not have exact modal verb equivalents, they should be learned and used as well.

Making Hypothetical Statements

From time to time you may find it necessary to make hypothetical or contrary-to-fact statements. For example, if you are describing the possible consequences of some action or theorizing about something or drawing an analogy to explain a difficult concept, you may want to make it clear to your reader that you are merely hypothesizing. In such cases, *would* is by far the most commonly used modal verb, followed by *might* and *could*.

Hypothetical statements are always made with regard to certain conditions. Consider the following sentence, for example:

> If the earth's magnetic field were reversed, the earth's surface would be exposed to a greater cosmic ray intensity than normal.[6]

Here the hypothetical statement (*the earth's surface would be exposed to a greater cosmic ray intensity than normal*) is directly preceded by a statement of conditions (*If the earth's magnetic field were reversed. . .*). Note that the verb in the *if* clause is in the simple subjunctive form, *were reversed*, in keeping with the simple conditional form of the main clause verb, *would be exposed*. It is important to maintain correct pairings of verb forms in such constructions and to not be confused by the similarity of subjunctive and past-tense forms. In this case, for example, *were reversed* looks like a past tense, yet has nothing to do with past time. (If the writer had wanted to refer to the past, he would have used the past subjunctive form, *had been reversed*, in the *if* clause and the past conditional, *would have been exposed*, in the hypothetical statement.) To make conditional statements that are less hypothetical (i.e., more likely to come true), use the present tense for the verb in the *if* clause:

> If the earth's magnetic field *is reversed,* the earth's surface *will be exposed.* . . .

Alternatively, you can use *should* in place of *if* and a simple nonfinite form of the verb:

> *Should* the earth's magnetic field *be reversed,* the earth's surface *will be exposed.* . . .

Although hypothetical statements are always made with regard to certain conditions, these conditions are not always stated explicitly in an *if* clause. For example, the meaning of the sentence given above is essentially retained in these versions without an explicit *if* clause:

> *It is possible for the earth's magnetic field to someday be reversed. In such an event,* the earth's surface would be exposed to a greater cosmic ray intensity than normal.

A reversal of the earth's magnetic field would have the effect of exposing the earth's surface to a greater cosmic ray intensity than normal.

In the first instance, a full sentence is used to present the hypothetical condition and a connective phrase (*in such an event*) is then used to link this condition to the subsequent *would* statement. In the second instance, a noun phrase is used. The indefinite article preceding this noun phrase and the conditional *would* immediately following it make it clear to the reader that such a reversal of the earth's magnetic field is only a hypothetical possibility.

Once the conditions are established—whether by *if* clause, by full sentence, or by noun phrase—it is often possible to make a number of hypothetical statements without having to restate the full set of conditions each time. Continued use of the modal verb in these succeeding statements will make it clear to the reader that they are hypothetical statements, and the reader will then infer that the same conditions hold. Consider this passage, for example:

> If the earth's magnetic field were reversed, the earth's surface *would* be exposed to a greater cosmic ray intensity than normal. Some scientists argue that the consequent increase in radiation dosage at sea level *could* have a serious impact on animal life; they claim, in particular, that it *might* cause an increased mutation rate. Others argue, however, that this *would* be most unlikely, since the increased radiation dosages *would* be relatively small.

Here the basic condition is established in the first sentence, and it is then assumed to hold for the subsequent series of hypothetical statements. Because the writer has been careful to use the modals *would, could,* and *might* repeatedly and consistently, we as readers have no trouble understanding that the discussion as a whole is meant to be purely hypothetical.

EXERCISE 21-2   Fill in the blanks with an appropriate modal verb or phrase.

A   An employer plans to interview eight equally qualified people, including Robert, for possible employment. From these eight she will select two for job offers. Robert _____ be selected.

B   Maria drove from Chicago to Detroit, a distance of 240 miles, in just under 4 hours. At some point, she _____ have been driving more than 60 mph.

C   A new plastic underwent preliminary strength testing. The breaking load for each of the eight trials was found to be (in 1000 pounds per square inch) 8.0, 8.7, 7.2, 7.9, 7.8, 8.4, 7.8, and 8.1. If we subject this plastic to a load of 8.6 thousand lbs/in.^{2}, it _____ break.

D   A certain transistor-manufacturing process has been known to have a
     defect rate in the 4–7% range. This means that as many as 70 transis-
     tors out of 1000 made by this process _____ be defective.

E   If the number 4,294,967,297 is equal to the product of 641 times
     6,700,417, it _____ be a prime number.

EXERCISE 21-3   Write a one-paragraph description of what you would do
if you were awarded a $2 million unrestricted research grant.

21.3

ABILITY

The final category of modal usage is that referring to ability, or capability.
The dominant modal of this category is *can*, followed by its past tense/
hypothetical variant *could*. The phrase *be able to* is also used quite frequently, combining with modals such as *should, may,* and *might* to indicate
various degrees of ability.

   Below is an example showing modals of ability being used in a scientific
report about the possibility of constructing a gigantic "cosmic lift" in the
earth's upper atmosphere. Notice how the writer artfully shifts back and forth
between *can* and the more hypothetical *could.*

> The energy liberated by the lift *can* be used in different ways. First of all, it *can*
> be used for space flights. The energy developed over the upper section—above
> 36,000 km—*could* be fed to a power station supplying the lower section of the
> lift. This would result in an interesting situation. The expenditure of energy for
> ascending the "cable-way" to outer space *could* be reduced to a minimum.
> According to Artsutanov's calculations, a height of 144,000 km *can* be reached
> without wasting any energy at all. The amount of work obtained along the
> upper part of the path would equal that spent along the lower part. At heights
> greater than 144,000 km, operation of the cosmic lift would turn into pure
> profit. The lift would become a sort of power station.[7]

   The complete scale of choices for the modal verbs and phrases of ability
is given in Figure 21-3.

EXERCISE 21-4   Below are three passages from chemistry and physics
textbooks. For each of the blank spaces, choose the modal verb or phrase
that you think best fits. (NOTE: This exercise tests your ability to use modals
from all three of the categories discussed above.)

A   **SIGNIFICANT FIGURES AND RELATIVE PRECISION**
     In general, results of observations _____ be reported in such a way that
     the last digit given is the only one whose value _____ be in doubt. The

FIGURE 21-3 Modals indicating degrees of ability.

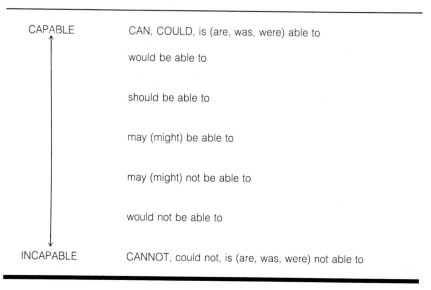

CAPABLE	CAN, COULD, is (are, was, were) able to
	would be able to
	should be able to
	may (might) be able to
	may (might) not be able to
	would not be able to
INCAPABLE	CANNOT, could not, is (are, was, were) not able to

digits that constitute the result, excluding leading zeros, are then termed significant figures. The number of significant figures is the same in 23.5 mg and 0.0235 g; this is reasonable, for the uncertainty in a result _____ not depend on the units in which the answer is expressed. Note that there is some uncertainty about the number of significant figures in a weight reported to be 2350 kg, because it is unclear whether the final zero is significant or not. The matter is clarified by reporting the weight either as 2.350×10^3 kg or as 2.35×10^3 kg, whichever applies.

The concept of significant figures does not apply to quantities known to be integers; for example, in a molecular formula, such as C_6H_{14}, the number of figures in each subscript does not imply that the composition is uncertain. We _____ assume that there are exactly 6 mol carbon atoms per mole of compound, even though the formula _____ to the unsophisticated seem to imply that this ratio is known only to low precision. Similar considerations apply to integral numbers of objects; the integers are regarded as infinitely precise.[8]

B **INHIBITION OF CORROSION**

Corrosion _____ be curbed in a number of ways. The most obvious method is to cover the metal surface with a coating impervious to air and moisture. Because corrosion _____ begin in even microscopic cracks, this coating _____ itself be completely intact. It _____ be a layer of paint, plastic, or some similar material, a coating of a strongly adherent oxide, or a layer of some metal that is itself resistant to corrosion.

An interesting phenomenon is *passivation,* in which a metal surface is made inactive by covering it with a very thin oxide layer. Iron _____ , for example, be passivated by treatment with concentrated nitric acid or dichromate. Once in this state it does not react with acid or reduce cupric ions. Passivation is easily destroyed, however, and if the surface is scratched anywhere, the protection breaks down—not only at the location of the scratch but in a steadily increasing area with a fast-moving boundary.[9]

C   THERMIONIC EMISSION

If an electron is to escape from the surface of a conductor, it _____ do the work necessary to overcome the surface forces, and the energy _____ come from the electron's kinetic energy resulting from its motion. When an electron's kinetic energy exceeds the work it _____ perform to overcome the surface forces, it _____ escape into the space beyond the surface. At high temperatures, where the average kinetic energy of the free electrons is large, an appreciable number _____ be able to escape from the surface of the material. . . .

Electron emission in practically all receiving-type vacuum tubes is derived from cathodes coated with a mixture of barium and strontium oxides over which is formed a surface layer of metallic barium and strontium. Such emitters _____ be heated to their operating temperature either indirectly by radiation from an incandescent tungsten filament or directly by the conduction of filament current. . . .

PROCEDURE: Set up a data table on separate paper using the heading shown at the end of the procedure. Include this data sheet with your report of this experiment.

One section of a 6H6 twin diode or a 6J5 _____ be used. The triode _____ be operated as a diode by connecting the grid directly to the plate. The socket connection diagrams (bottom views) for both tube types are given in Fig. 61-1. If another tube type is to be used, you

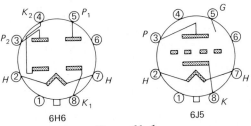

Figure 61–1

———— have modifying instructions from your instructor before proceeding. [10]

REFERENCES

1   J. Wei, "Pure Mathematics: Problems and Prospects in Number Theory," *The Research News* **30**(3):18 (1979).

2   J. E. Williams, F. E. Trinklein, and H. C. Metcalfe, *Exercises and Experiments in Physics,* Holt Rinehart and Winston, New York, 1976, p. 241.

3   J. Waser, K. N. Trueblood, and C. M. Knobler, *Chem One,* McGraw-Hill, New York, 1976, p. 33.

4   Waser, Trueblood, and Knobler, p. 41.

5   E. W. White and W. B. White, "Electron Microprobe and Optical Absorption Study of Colored Kyanites," *Science* **158**(3803):915 (1967).

6   Adapted from C. J. Waddington, "Paleomagnetic Field Reversals and Cosmic Radiation," *Science* **158**(3803):913 (1967).

7   J. D. Issacs, et al., "Sky-Hook: Old Idea," *Science* **158**(3803):947 (1967).

8   Waser, Trueblood, and Knobler, p. 35.

9   Waser, Trueblood, and Knobler, p. 524.

10   Williams, Trinklein, and Metcalfe, p. 315.

22

VERBS

Although nouns and noun phrases are the principal carriers of information in scientific and technical English, verbs are the principal means by which these units of information are tied together in a sentence. Thus it is important that you have full command of the English verb system and of the most frequently used verbs in scientific and technical English.

This chapter addresses the most common areas of difficulty for non-native speakers with regard to the use of verbs: (1) major tense distinctions, (2) the use of the progressive (-ing) aspect of the verb, (3) subject-verb number agreement, and (4) grammatical irregularities of particular verbs.

22.1
MAJOR TENSE DISTINCTIONS IN SCIENTIFIC AND TECHNICAL ENGLISH

Of the 12 traditional verb tenses in English, only 5 are used with any frequency in scientific and technical writing: simple present, simple past, present perfect, future (*will*), and present progressive. Each of these has particular meaning(s) associated with it and so should be learned thoroughly and used carefully. (*Will*, which is not a tense in the strict sense, is treated in Chapter 21.)

The Simple Present Tense

In formal scientific and technical English, the simple present is used primarily to express "timeless" generalizations, i.e., general statements which do not specify any particular time frame. For example, the statement that "Water boils at 212°F" is not restricted to a particular time frame and so is properly cast in the present tense (*boils*). Similarly, this recommendation in a company report—"I *recommend* that we continue to use the lime reactant-agent in our desulfurization process"—is given in the simple present tense because it represents a general statement that holds true not only at the time of writing but also indefinitely into the future.

Be sure to take full advantage of the simple present tense in your own writing. It is the most useful verb tense in scientific English, predominating in almost every type of writing situation except those explicitly set in the past (e.g., historical reviews, laboratory writeups) or in the future (e.g., proposals, speculations, recommendations). Even in these circumstances, the simple present tense can be used occasionally in an important contrastive way to indicate a generalization that is not restricted to the past or to the future. Such generalizations often represent evaluative judgments or interpretations on the writer's part and thus are crucially important to good writing. Unfortunately, many technical people often fail to make interpretive statements in situations where such statements are called for, thinking that their role as an objective scientist or engineer requires them to report only the facts. They consequently tend not to use the simple present tense as often as they should, preferring instead to use long sequences of past tense forms. This sort of thinking is misguided, however. Scientists and engineers and other professionals are trained and employed precisely to make educated judgments and interpretations, not simply to report facts. Therefore, look for appropriate opportunities to make generalizations, and when the opportunity arises, use the present tense.

The Simple Past Tense

In contrast to the present tense, the simple past tense specifies a particular event or condition which occurred or existed at some time in the past but which no longer occurs or exists. For example, if you carried out a procedure as part of an experiment and are now reporting on it, you would probably use the simple past tense in your description:

> Commercial cholesterol *was recrystallized* three times from 95% ethanol. Radioactive cholesterol monohydrate *was prepared* by mixing 5 g of the recrystallized cholesterol. . . .

The immediate results of carrying out an experimental procedure are also usually reported in the simple past tense:

> NMR studies quantitatively *confirmed* the monohydrate nature of the crystals. TLC studies *indicated* the absence of any impurities.

Similarly, if you are writing a report in response to some request, you might begin by referring, in the past tense, to that request:

> On 15 December 1982 you *asked* us to help you find ways to lower your energy costs.

You might then continue with a description of the actions you undertook, also in the past tense.

> Accordingly, we *visited* your plant and *inspected* the complete heating system.

In short, the simple past is the tense to use when you are simply reporting facts.

The Present Perfect Tense

Nonnative speakers often confuse this tense and the simple past tense, thinking that the two are more or less interchangeable. This is wrong, however: each has its own meaning and range of uses. Basically, the simple past tense is used for completed actions whereas the present perfect is used for actions which were begun in the past but which are still going on. In the Review of Literature section of a scientific report, for example, single isolated studies are usually referred to in the simple past tense (Adler *reported* that . . .) whereas multiple studies, suggesting an ongoing sequence of studies, are usually referred to in the present perfect (Adler, Pierce et al., and other researchers *have reported* that . . .).

Similarly, the present perfect is used to report on actions that were carried out in the past but are still producing effects in the present. Consider this example taken from the opening paragraph of a company memo:

> Our patent department lawyers *have asked* this research group to provide them with data for a patent infringement suit. A competitor *has* recently *marketed* a new tower packing which they claim has superior pressure drop characteristics compared to our product.

In the first sentence, the use of the present perfect tense indicates that even though the lawyers may have asked only once, their request is still in effect and the research group is still expected to act on it. In the second sentence, the use of the present perfect not only reports on a past action (the marketing of a new product by a competitor) but also implies that this product is currently being actively promoted by the competitor. The choice of the present perfect tense in both cases thus serves to emphasize the immediate ongoing nature of the threat posed by the competitor's past actions.

EXERCISE 22-1 In each of the following passages, fill in the blank spaces with the most appropriate form of the verb given in parentheses. Be prepared to justify each of your choices in accordance with the guidelines provided above. The first one has already been done for you.

A   (beginning of a memo in response to the memo discussed above)

On September 14, 1982, you **requested** (1) a
(request, requested, have requested)

comparison of our patented ⅜″ raschings rings with a competitor's packing to support a patent infringement suit. Specifically, you

requested (2) that data on the packing factor
(request, requested, have requested)

and pressure drop characteristics be compared. Accordingly, we

have carried out (3) your request, using a column of
(carry out, carried out, have carried out)

the competitor's packing in our research facility. The purpose of this report _____*is*_____ (4) to provide our lawyers with the requested data
(is, was, has been)
for the patent infringement suit.

The results _____*show*_____ (5) that the packing charac-
(show, showed, have shown)
teristics of the competitor's packing differ significantly from those of our packing. Our packing factor _____*is*_____ (6) 1000; the competitor's
(is, was, has been)
packing factor _____*was*_____ (7) determined to be 681.
(is, was, has been)

Justification:

(1) Simple past tense because the request was made in the past and has been carried out
(2) Simple past tense for the same reason as #1
(3) Present perfect because it relates past actions (measuring and comparing) to present purpose
(4) Simple present because the purpose of the report is not restricted to a particular time period but applies whenever anyone reads the report
(5) Simple present because this is a generalization that is not restricted to a particular time frame
(6) Simple present because this is a general fact about the packing factor of the company's product that applies at all times
(7) Simple past because it refers to the action of measuring the competitor's packing factor, which was done in the past

B   Before the Voyager I encounter with Saturn we _____
(have, had, have had)
far less information about its satellites than we _____ about Jupiter's
(do, did)
before 1979. This _____ due to Saturn's greater distance
(is, was, has been)
from Earth and the sun, the smaller diameter of its satellites, and their
closeness to Saturn. Only Titan _____ comparable in size
(is, was, has been)
to the Galilean satellites. The icy satellites _____ inter-
(are, were, have been)
mediate in size between the largest asteroids and the moon. Voyager I
_____ the first close look at bodies in this size
(provides, provided, has provided)
range.[1]

C   (the introductory paragraph of a company memo)
At our meeting on October 15, 1982, I _____ asked to
(am, was, have been)
investigate the noise-related problems that one of our production supervisors, D. P. Stonear, _____ having on the job lately. As
(is, was, has been)

you may recall, Mr. Stonear _____ that he
(complains, complained, has complained)
_____ having increasingly more difficulty understanding peo-
(is, was, has been)
ple on the telephone in his small office located in the middle of the
wheel stamping room. He _____ that his office
(believes, believed, has believed)
_____ too noisy and _____ continually apprehen-
(is, was, has been)   (is, was, has been)
sive about answering the phone. Most important, his nervousness over
this communication problem _____ to be under-
(seems, seemed, has seemed)
mining his effectiveness as a supervisor. The purpose of this report
_____ to determine the cause of Mr. Stonear's communica-
(is, was, has been)
tion problem and to suggest corrective measures.

D   (the introductory paragraph of a science journal article)
Recently great efforts _____ made to develop geochem-
(are, were, have been)
ical techniques to estimate environmental parameters such as temperature and salinity. While the oxygen isotope method for paleotemper-
atures _____ employed extensively, efforts to refine methods
(is, was, has been)
for paleosalinities _____ less successful. Most paleosalin-
(are, were, have been)
ity methods _____ based on Goldschmidt's classical ob-
(are, were, have been)
servations on the occurrence and distribution of trace elements in different sedimentary environments. For example, Degens et al. (1957)
_____ marine from freshwater shales
(differentiate, differentiated, have differentiated)
on the basis of spectrochemical analysis for boron, gallium, and rubid-
ium. Similarly, Potter (1963) _____that
(demonstrate, demonstrated, has demonstrated)
each of the elements boron, chromium, copper, nickel, and vana-
dium _____ more abundant in marine than in freshwater ar-
(is, was, has been)
gillaceous sediments.[2]

22.2

THE USE OF THE PROGRESSIVE (-*ING*) FORM OF THE VERB

Another common area of difficulty for nonnative speakers is the use of the
progressive (-*ing*) aspect of verbs. This form should be reserved for situations
where you are describing an event in the process of occurring—that is, an
incomplete, ongoing event. It is a particularly useful form in progress reports,
in letters, and in introductions to technical reports. Here is an example:

In this day and age, the computer *is finding* more applications than were ever
conceived possible. Computers *are* now *controlling* heating in buildings, traffic

signals in cities, and guidance control in commercial and military jets. The feasibility of using an onboard computer to control fuel flow in personal transportation has also been a point of interest because a substantial amount of fuel can be saved with the proper equipment controlling fuel allocation.

Global Design Corporation *is* currently *developing* a computer-controlled vehicle that uses both an internal combustion engine and lead-acid batteries as power sources.

This passage appears at the beginning of a technical report about electric hybrid vehicles. By using the progressive form of the present tense, the writer focuses attention on the here-and-now, on the state of the art in computer-controlled vehicles. This emphasis on actions that are not yet complete makes the description more dynamic and attention-getting than it would otherwise be.

In using the progressive form of the verb, however, you must keep in mind an important restriction: generally only "event verbs" can take the progressive form, not stative verbs. This makes sense since the progressive form, as mentioned above, is used to describe events. Notice, for example, that the three verbs in the progressive form in the above example are all event verbs: *find, control,* and *develop.* The only main verb—*be,* in sentence 3—is a stative verb and thus does not appear in the progressive form. It would sound very much like a nonnative speaker error to read, *"The feasibility of using an onboard computer . . is also being* a point of interest. . . ."[3]

Below is a list of all the stative verbs found among the 200 most common verbs in scientific and technical English. Commit them to memory so that you do not make the mistake of using any of them in the progressive (*-ing*) form.

Verbs That Do Not Occur in the Progressive Form

afford	correspond	represent
appear	differ	result in
be	exist	satisfy
believe	involve	seem
concern	known	suppose
consist of	mean	understand
constitute	need	yield
contain	possess	

Note, however, that although stative verbs cannot be used in the progressive aspect, they *can* be used in short-form (*-ing*) relative clauses, as described in Chapter 19. For example, the sentence *"The procedure is consisting of five main steps"* is ungrammatical because *consist of* is a stative verb and cannot be used in the progressive aspect; however, a sentence like

"This is a procedure consisting of five main steps" would be perfectly correct because the -ing form in this case is not the progressive but rather part of a short-form relative clause.

22.3
SUBJECT-VERB NUMBER AGREEMENT

A basic rule of English grammar is that the subject noun phrase and the main verb of a sentence or clause must agree in number—that is, they must both be singular or both be plural. No singular-plural mixes are allowed. Unfortunately, this rule is too often violated, especially by nonnative speakers whose native language does not have such a rule (e.g., Chinese, Korean, Japanese, Vietnamese). The fact that other languages do things differently, however, does not mean that the rules of English can be freely violated. It not only makes you look bad to have such grammatical errors in your writing, it can also create difficulties for the reader trying to comprehend your meaning. For example, if you wrote *"The other method for determining K use the Arrhenius Equation," it would not be clear to the reader whether you meant one method or more than one method because you've put the subject noun phrase (*method*) in singular form but the main verb (*use*) in plural form. In such a case, the reader will simply be forced to guess your meaning—and he or she may guess wrong.

If you're using a grammatically complex subject noun phrase, be sure to use the head noun in determining subject-verb number agreement. Otherwise, you may commit errors like these:

*The above procedures in the dark stage is based on tracer studies.

*A lot of research programs has already been carried out.

In the first sentence, *procedures* is the head noun in subject position, and so the main verb should be in the plural; the fact that there is an intervening noun in singular form (*stage*) has no bearing on the form of the verb because this intervening noun is not the head noun. In the second sentence, *research programs* is the head noun in subject position, and so the main verb here, too, should be in the plural. Do not be misled by the singular noun *lot*: it is simply part of the quantifier phrase *a lot of*, equivalent in meaning and in use to *many* (just as *a number of* is equivalent in meaning and in use to *quite a few*).

Another common mistake in subject-verb agreement involves the word *each*, e.g., *"Each of the components have a different function." In this type of sentence, *each* is the head noun of the subject phrase; since it is singular, the main verb should also be singular: "*Each* of the components *has* a different function." On the other hand, if the sentence were constructed like this—"The components each have a different function"—*have* would be the correct form of the verb because the head noun is now *components*, not *each*.

22.4

GRAMMATICAL IRREGULARITIES OF PARTICULAR VERBS

Many mistakes made by nonnative speakers in the use of verbs derive from the irregular features those verbs have. We saw an example of this in Section 22.2, where it was pointed out that a small set of English verbs—stative verbs—do not occur in the progressive. In this section, we shall discuss several other irregularities associated with particular verbs: (1) verbs that do not occur in the passive form, (2) verbs taking *-ing* or unmarked complements, and (3) verbs with irregular forms of inflection.

Verbs That Do Not Occur in the Passive Form

A passive sentence (as discussed in Chapter 14) is one in which the recipient of the action appears in subject noun phrase position, not in its normal direct object position. For example, instead of saying, "We then measured the reaction rate," we could put the direct object phrase, *reaction rate,* in subject position and say, "The reaction rate was then measured (by us)." The passive is a particularly common and useful form for describing experimental procedures, chemical processes, cause-and-effect relationships, etc.

The problem is that a fairly large number of English verbs do not take direct objects to begin with; these are the so-called intransitive verbs, such as *exist, fall, differ,* and *occur.* Using the passive form with such verbs is therefore completely ungrammatical, e.g., *"This contamination is occurred frequently." Quite a few other verbs exist, too, which logically could be used in the passive form but which virtually never are, at least not in scientific and technical English: *yield, suffer, possess,* etc. Below is a complete list of those verbs from among the 200 most frequently used verbs in scientific and technical English that do not occur in the passive form.

appear	fall	possess
arise	flow	proceed
be	function	remain
become	get	result in
begin	go	rise
come	happen	seem
consist of	have	suffer
correspond	lead	tend
depend on	let	travel
differ	lie	undergo
enable	live	work
exist	occur	yield

Verbs Taking *-ing* or Unmarked Complements

Most complement-taking verbs in scientific and technical English take *to* or *that* complements:

> Recently you *asked* us *to undertake* an investigation of. . . .
> Our findings *indicate that.* . . .

Thus, many nonnative speakers tend to rely exclusively on these two types of complements. This strategy works in most cases, but not in all, for there is a small set of verbs which take *-ing* or unmarked complements. If you fail to observe these exceptions, you'll make mistakes like *"We recommend to choose the first option" and *"This modification should make the system to function better." *Recommend* takes an *-ing* (or *that*) complement, and *make* takes an unmarked complement; therefore the correct versions of these two sentences would be "We recommend choosing the first option" and "This modification should make the system function better."

Below is a list of frequently used verbs in scientific and technical English which take *-ing* or unmarked complements. Commit them to memory so that you do not repeatedly make mistakes with them.

Verbs taking *-ing* or *to* complements:	*try, start*
Verbs taking *-ing* or *that* complements:	*suggest, recommend, emphasize*
Verbs taking *-ing* complements only:	*discuss, consider, resist, reject, consist of, insist on, depend on, result in*
Verbs taking unmarked complements only:	*make, let, help*

NOTE: This last set of verbs also requires a noun or noun phrase between the verb and its complement, as in "This modification should make *the system* function better." If you do not observe this rule, the result will be ungrammatical: *"This modification should make function better the system."

Verbs with Irregular Forms of Inflection

Many of the most common verbs in English came originally from German, which typically changes the vowel of the verb stem to indicate past tense or participial forms rather than simply adding a suffix to the end of the word. Thus, whereas most English verbs simply add *-ed* or *-d* to the stem to form the past tense and participial forms (e.g., *demonstrate, demonstrated, have demonstrated*), many of our most common verbs make other changes instead (e.g., *give, gave, given*).

These irregular verbs are indeed irregular, and for that reason they cause all sorts of problems for nonnative speakers. However, since they are among the most frequently used verbs in the language, you should make every effort to learn and use the correct forms. Below is a list of 50 of the most common ones.

Uninflected form	Simple past	Past participle
arise	arose	arisen
bear	bore	borne
become	became	become
begin	began	begun
bend	bent	bent
break	broke	broken
bring	brought	brought
build	built	built
catch	caught	caught
choose	chose	chosen
come	came	come
cut	cut	cut
do	did	done
draw	drew	drawn
fall	fell	fallen
feed	fed	fed
find	found	found
get	got	gotten or got
give	gave	given
go	went	gone
grow	grew	grown
have	had	had
hold	held	held
keep	kept	kept
know	knew	known
lead	led	led
leave	left	left
let	let	let
lie	lay	lain
lose	lost	lost
make	made	made
mean	meant	meant
put	put	put
read [pron. "reed"]	read [pron. "red"]	read [pron. "red"]
rise	rose	risen
run	ran	run
say	said	said

see	saw	seen
send	sent	sent
set	set	set
show	showed	shown
shut	shut	shut
spend	spent	spent
spread	spread	spread
stand	stood	stood
sweep	swept	swept
take	took	taken
tear	tore	torn
think	thought	thought
write	wrote	written

REFERENCES

1   B. A. Smith et al., "Encounter with Saturn: Voyager 1 Imaging Science Results," *Science* **212**(4491):170 (1981).

2   N. A. Nugent and R. C. Fuller, "Sedimentary Phosphate Method for Estimating Paleosalinity," *Science* **158**(3803):917 (1967).

3   The asterisk (*) indicates that the sentence following it is ungrammatical.

ADDITIONAL READING

Practical
   S. McKay, *Verbs for a Specific Purpose,* Prentice-Hall, Englewood Cliffs, N.J., 1981.
Theoretical
   B. Comrie, *Aspect,* Cambridge University Press, Cambridge, 1975.
   B. Gorayska, "The English Verb System," *Interlanguage Studies Bulletin* **3**(2):234–249 (1978).
   M. Joos, *The English Verb: Form and Meanings,* University of Wisconsin Press, Madison, 1964.
   A. Ota, *Tense and Aspect of Present-Day American English,* Kenkyusha, Tokyo, 1963.

23

PROOFREADING

The last step in preparing any piece of writing—be it a letter, a memo, a formal report, or whatever—is to proofread it for errors. Proofreading doesn't take much time, and it can really pay off in terms of conveying both accurate information and an image of credibility. Don't forget that readers will often judge you by the quality of your writing. If your writing is full of typographical errors, misspellings, and other kinds of easily noticeable mistakes, your readers may well judge you to be not only a careless writer but a careless person in general. This may lead them to question the quality of your technical work: the accuracy of your measurements, the preciseness of your calculations, the soundness of your judgment. In short, your professional image can be damaged if you allow your writing to circulate without first checking it over for errors.

Even more serious than errors of form are errors of substance: miscalculations, misrepresentations of data, misleading claims. Errors of this type may easily go unnoticed by your readers and thus may genuinely misinform them, sometimes to the point of having very serious consequences. Needless to say, such errors of substance must be eradicated from any piece of writing.

Given the fact that there are several different kinds of errors to look out for, it makes sense to divide the proofreading process into different phases so that you can "sweep" the manuscript, concentrating on one type of error at a time. We recommend the following sequence of steps: first, proofread for substantive errors; second, proofread for sequencing errors; finally, proofread for spelling errors.

SUBSTANTIVE ERRORS   Are your calculations correct? Are your data correct? Have you plotted your curves accurately? Have you got the right dates, times, places, job titles, model numbers? Remember, this kind of information may be of critical importance to your readers—so make sure it's right!

SEQUENCING ERRORS   There may be one or more numerical or alphabetical sequences in your manuscript: page numbering, references, illustrations, footnotes, report sections, lists, figures, steps in a procedure, etc. Make sure these sequences are all correct and complete.

MISSPELLINGS   For many writers, this is where proofreading is particularly fruitful. The English spelling system is notoriously irregular, causing writers no end of trouble. Despite this, many readers—including many readers in supervisory positions—attach great importance to spelling, seeing it as a sign of how careful and even of how well educated the writer is. So don't take any chances! Correct all spelling errors!

Most spelling errors occur with ordinary words that can be found in any pocket-size dictionary: words like *foreword, receive,* and *vacuum.* So, as you proofread for spelling, keep a small dictionary near you and use it for any words you're not sure of. If you're using a word processor with a text editor, it may have a built-in dictionary and spelling program; you can save considerable time by using that. Another way to save time, of course, is to ask someone who's known to be a good speller to proofread your manuscript for you.

Although any of these stratagems will work for you, try not to become overly dependent on them; after all, you may someday have to write a report in a hurry and may not have a text editor or friend available to help you out, or the time to look up every other word in a dictionary. Also, no text editor or ordinary dictionary is likely to be of much help when it comes to technical terms or proper names. So try to learn the correct spelling of words as you encounter them, especially those words that you use often in your writing. Practice writing them down, so that you can see them in your own handwriting as well as in print. Make a list of words that give you particular trouble, and then test yourself on them from time to time.

EXERCISE 23-1 Below is a list of some of the most frequently misspelled words in scientific and technical writing. Study the list; then have someone dictate the words to you while you write them down. Repeat the exercise until you've spelled all the words correctly.

accidentally	commission
accommodate	definitely
achieved	dependent
acoustic	desirable
address	develop
alignment	development
allot	different
allotted	environment
analyze	exaggerate
apparent	exceed
appropriate	excel
argument	exhaust
auxiliary	existence
basically	feasibility
beneficial	February
calendar	fluoride
ceiling	foreign

foreword   parallel
government   personnel
irrelevant   precedent
judgment   probably
knowledgeable   quantity
laboratory   receive
likelihood   recommend
lose   refer
maintenance   reservoir
manageable   retrieve
mathematics   schedule
misspell   seize
necessary   separate
occasionally   therefore
occurred   truly
omit   unnecessary
omitted   vacuum

Many writers have trouble with homophones—words that have the same pronunciation but are spelled differently and have different meanings. Probably the most frequently misused homophones in English are *its* and *it's*. Are you sure which is which? (Answer: *its* is a possessive form; *it's* = *it is*.) It's especially important to learn the difference between homophones because a text editor generally cannot. That is to say, if a writer uses *it's* where *its* is appropriate, most word processor spelling editors will fail to notice the error because they don't take usage into account; they look only at the form of the word and see if that form is found in the dictionary. So, if you are using a word processor to check your spelling, don't expect it to pick up spelling mistakes when homophones are involved.

Here are some other homophones that often cause writers trouble:

affect   (The new policy will not affect us.)

effect   (We decided to effect a change.)
          (The extra dosage had no effect.)

all ready   (They are all ready to begin.)

already   (She has already made her decision.)

complement   (The illustrations nicely complement the written text of the report.)

compliment   (Linda's supervisor complimented her for having devised such a workable system.)

foreword   (Many technical reports begin with a foreword.)

forward   (We look forward to your visit.)

lead   (That idea went over like a lead balloon.)

led   (past tense of the verb *to lead:* Our company has led the way in this field.)

plain   (Use plain English whenever possible.)

plane   (The two lines formed a plane angle.)

principal   (The principal feature of this approach is its simplicity.)

principle   (The basic principle in this design is the conversion of heat energy to work.)

stationary   (A land-based missile system would provide a stationary target for the enemy.)

stationery   (Please don't use company stationery except for official business.)

their   (We would like to try some of their products.)

there   (There are too many terminals already in place there.)

they're   (What they're trying to do is capture the market.)

weather   (We'll launch at 700 hours, provided the weather's OK.)

whether   (Whether we succeed or not probably depends on our marketing strategy.)

who's   (Who's there?)

whose   (Whose jacket is this?)

your   (Do you have your calculator handy?)

you're   (When you're ready to begin, give the signal.)

EXERCISE 23-2

A   Keep a list of any words you misspell and review them from time to time. After you've accumulated 20 of them, have someone test you as in Exercise 23-1.

B   Proofread the letter of application for employment shown as Figure 23-1 for misspellings and other errors.

FIGURE 23-1 Letter of application for employment, to be proofread for misspellings and other errors.

225 Nugent Hall
Brookfield Polytechnic Institute
Ames, NY 12181
(518) 270-4391

Febuary 25, 1982

Mr. Bill Dahlen
Sperry Univac Semiconductor Divisions
U2X26, P.O. Box 3525
St. Paul, MN 55165

Dear Mr. Dahlen:

In response to your advertisement in IEEE Spectrum (August 1982), I am writing to apply for the position of Bi-polar Developement Engineer. After considering your challenging requirements; I believe that my experience and educational background would enable to make significant contributions to your expanding division. My backround includes:

Employment by the Electron Physisc Laboratory, Brookfeld Polytechnic Institute. I participated in designing an innovative computer modeling process of a MOS submicron devise. I was responsible for designing the numerical analysis subroutines of the computer program.

Completion of a senior desing project in designing and fabricating bi-polar device chips using state-of-the-art technics. I was also responsable for testing and characterizing the device chips.

Completion of several graduate courses in semi-conductor devices, intergrated circiuts and digital logic circuit design.

I will recieve my B.S.E.E. from BPI in August 1982 and would appreciate being considered for permanant employment starting therafter. I will call on you next Wed. during your office hours to set up an interview at your convenience. If you desire additional information, I can be reached at (518) 270-4381 during the mornings.

Thank you for your kind consideration.

Yours truely,

Perry C. Culbert

Perry C. Culbert

ADDITIONAL READING

H. E. Kirn, *All Spelled Out: Basic Spelling Patterns for Learners of English*, ELS, Portland, Ore., 1981.

24

VOCABULARY BUILDING

It goes without saying that anyone who wishes to be an effective communicator of scientific or technical information must possess a good vocabulary. Without a good vocabulary, you simply cannot put into practice all the principles we have been discussing. We are talking not about a passive, recognition vocabulary but about an active, production vocabulary, one from which you can select words to suit different topics and different types of readers under varying circumstances. Such a vocabulary must be fairly large, it must include both technical and nontechnical words, and it must be constantly incorporating new words.

The best way to increase one's vocabulary, of course, is to read and hear English as much as possible. The more you read and listen, the more exposure you'll have to new words. Given plenty of leisure time, of course, this would be no problem. If you're like most people, however, you don't have plenty of leisure time, and so you need to optimize your use of time, that is, you need to build your vocabulary in the most efficient way possible. This chapter is designed to help you do that. It describes a strategy that allows you to expand your vocabulary at the same time as you pursue your technical studies, without memorizing word lists or constantly looking up words in a dictionary. Instead, you'll be using contextual clues to infer the meaning of new words.

24.1
USING CONTEXTUAL DEFINITIONS

Psychological research has demonstrated that people learn best when the items being learned are presented in a meaningful context. Apparently, the context allows our minds to form associations between the word being learned and other, related words so that we have a more vivid image of the word's meaning to store in our memories. This suggests that the best way to expand one's vocabulary is not by studying words in isolation—as is done with the traditional word-list method, for example—but by studying words as they occur in meaningful contexts, such as in textbooks, articles, and reports. No doubt you already do this, at least to some extent. Whenever you read something in English, you probably come across some words you don't know, and you probably consult your dictionary to find out what they mean. Consulting a dictionary in this manner can be an effective technique, but it is also an enormously time-consuming one. It not only takes up a lot of time but also slows down your reading, making it hard to maintain the continuity you require in order to gain a single, coherent picture from the reading passage.

What you need is a technique containing many time-saving shortcuts, specifically, a technique that allows you to deduce the meaning of a new word without always having to look up the definition in your dictionary. One such shortcut, which you probably already use, consists of taking advantage of definitions contained within the text itself. For example, suppose you are taking a course in one of the natural sciences and have begun to read the textbook for the course. It starts like this:

> *The natural sciences.* The development of modern science during the last three centuries has been at once so broad and so deep that a great deal of learning and thought is required for even the best of human minds to encompass a portion of it. This is in large part because progress in science is cumulative— science builds on and extends what has been observed and understood earlier. It grows by the interplay of experimental observations, imaginative reasoning, predictions based on this reasoning, and experiments to test the predictions. As the body of systematized observations and generalizations about the natural world has increased, it has been artificially subdivided into "different" scientific fields, and these have in turn been partitioned, all because human life is too short and the mind too limited to be able to learn all that has been observed and postulated.[1]

Suppose that when you began reading this passage you did not know the meaning of the word *cumulative.* Would you interrupt your reading and pick up your dictionary to find the definition for this word? Not if you're alert to the fact that the authors themselves provide a definition for you, right there in the text! Not a formal definition such as you might find in a dictionary, of course, but a definition nonetheless: Progress in science is "cumulative" because it "builds on and extends what has been observed and understood earlier. . . ." This is not an all-purpose definition of the word, but it is certainly vivid and memorable for most readers, especially since it is followed by further description and elaboration ("It grows by the interplay of experimental observations, imaginative reasoning. . . .").

Informal definitions abound in textbooks, and you should always take advantage of them rather than wasting time consulting your dictionary. Sometimes they are easy to locate because the authors have used typographical devices to help you: parentheses, footnotes, italics, etc. At other times they may be less obvious but are there nevertheless. In the following passage—a continuation of the preceding one, incidentally—four informal definitions are provided, two of which are easy to find and two of which are not quite so easy. See how quickly you can find all four:

> The different natural sciences were once considered to be physics, chemistry, biology, geology, and astronomy, but now the subdivisions and extensions of these fields have become far more numerous. Not only do they include areas

in which several of these disciplines overlap, such as geophysics (the physics of the earth) and biochemistry (the chemistry of living things), but, increasingly now, they encompass areas in which the methods of many of the classical natural sciences are integrated and brought to bear on a particular portion of our world. For example, oceanography includes the related studies of the physical, chemical, biological, and geological aspects of the ocean and planetary science involves a correlated study in all the foregoing disciplines of the members of our solar system.[1]

If you're like most readers, you probably spotted the informal definitions for *geophysics* and *biochemistry* immediately, thanks to the authors' use of typographical aids (in this case, parentheses). Although you probably took a little longer to find the informal definitions of *oceanography* and *planetary science,* you took less time than if you had looked these words up in a dictionary.

PRINCIPLE 1
Whenever you come across a new word, first try to find an informal definition of it somewhere in the immediate context.

24.2

USING CONTEXTUAL CLUES
Of course, you are likely to encounter in your reading many unfamiliar words that are not defined for you. In such cases, it is even more tempting to reach immediately for the dictionary. Before you do, however, we would like you to consider the following three facts:

1 Every time you interrupt your reading to consult a dictionary, you lose valuable time and concentration.

2 It is not always necessary to know the exact meaning of every word in order to understand the meaning of a passage as a whole.

3 The context surrounding an unfamiliar word often contains enough clues for you to be able to guess the meaning of the word—if not the precise meaning, then at least a general meaning.

Given these facts, we urge you to use the context when you come across a new word, before you do anything else. If the context doesn't contain an informal definition of the word, it probably contains enough clues for you to deduce the meaning.

What sorts of clues? Word-formation clues, syntactic clues, semantic clues, and rhetorical clues. The following is a step-by-step procedure that allows you to systematically inspect the context and find the clues you need.

Grammatical Function of the Word

Determine the grammatical function of the word. Is it being used as a noun? As a verb? An adverb? An adjective? To build your vocabulary, you must be able to properly identify the grammatical functions that words perform in sentences and phrases. Do not rely on a dictionary to do this for you! A dictionary will tell you what function a word *normally* plays, not what it plays in any particular context. In science and technology especially, words often perform a grammatical function different from that in ordinary English.

You can begin to diagnose a word's grammatical function by looking at its form, specifically at whatever ending, or suffix, it may have. For example, the *-tion* ending attached to a word normally indicates that the word is a noun (e.g., *optimization*) and an *-ize* ending normally indicates that a word is a verb (e.g., *optimize*). Also, of course, the vast majority of adverbs in English are marked by the *-ly* suffix (e.g., *optimally*). Below are several lists of suffixes commonly associated with grammatical functions; you should know them all by heart. A knowledge of these suffixes and others like them is a great help in determining the grammatical function of some word you've never seen before. You cannot rely entirely on such knowledge, however. Notice in particular, that there are several cases where the same suffix can be used for either of two grammatical functions: *-al* can indicate either a noun or an adjective, *-ate* can indicate either an adjective or a verb, and so on. In addition, there are many words of every grammatical category that do not have any function-marking suffix at all, such as *comet, platform, shuttle, quick, stop, dark,* and *flood,* to mention but a few.

SOME COMMON NOUN-FORMING SUFFIXES

Suffix	Examples
-age	leverage, mileage, tonnage
-al	dismissal, approval, refusal
-an (-ian)	technician, mathematician, historian
-ance (-ence)	conductance, acceptance, equivalence
-ation	filtration, refrigeration, determination
-ency (-ancy)	efficiency, transparency, buoyancy
-ent (-ant)	emollient, solvent, lubricant
-er (-or)	worker, manager, supervisor
-icide	pesticide, insecticide, fungicide
-ing	engineering, manufacturing, painting
-ist	physicist, biologist, typist
-itude	magnitude, amplitude, exactitude

Suffix	Examples
-ment	equipment, shipment, easement
-ness	hardness, brittleness, dampness
-s	mathematics, physics, economics
-sion (-tion)	precision, corrosion, construction
-sis	osmosis, electrolysis, pyrolysis
-th	length, width, strength
-ture	mixture, furniture, expenditure
-ty (-ity)	certainty, activity, plasticity
-us	corpus, locus, terminus
-y	accuracy, recovery, synchrony

SOME COMMON VERB-FORMING SUFFIXES

Suffix	Examples
-ate	activate, detonate, enumerate
-en	strengthen, weaken, worsen
-ify (-fy)	fortify, solidify, liquefy
-ize	sterilize, optimize, minimize

SOME COMMON ADJECTIVE-FORMING SUFFIXES

Suffix	Examples
-able (-ible)	variable, malleable, flexible
-al	original, hexagonal, optical
-ant (-ent)	resultant, insistent, dependent
-ary	secondary, sanitary, temporary
-ate	indeterminate, duplicate, intermediate
-ful	careful, powerful, useful
-ic	probabilistic, deterministic, magnetic
-ile	mobile, ductile, tensile
-ive	quantitative, negative, indicative
-less	careless, odorless, useless
-ous	porous, dangerous, hazardous
-y	oily, greasy, rubbery

A second and even more important way of diagnosing the grammatical function of a word in a particular sentence is to look at the other words in the sentence and see how they all fit together. For example, consider this sentence from an article on planetary exploration:

The probe will jettison the chute and continue its descent into oblivion.[2]

Even if you had never seen the words *probe, jettison, chute,* and *oblivion* before, you could accurately determine their grammatical function by observing how they fit with other words in the sentence. *Probe* must be a noun because it occupies the subject noun position ahead of the main verb and because it is preceded by an article (*the*). *Jettison* must be a verb because it follows the modal auxiliary *will;* modal auxiliaries are always followed by main verbs in full sentences. *Chute* must be a noun because it occupies the direct object noun position in the first part of the sentence, immediately after the main verb, and because it is preceded, as is *probe,* by a definite article. *Oblivion* must be a noun because it occupies the object-of-a-preposition position immediately following *into*. Notice that you could not have made these determinations by looking at these words in isolation: they do not carry any of the characteristic suffixes described earlier (in fact, most native speakers mistake *jettison* for a noun when they first encounter it).

EXERCISE 24-1 Without referring to the lists of suffixes given above, see if you can guess the likely grammatical function of each of the following words:

bionic	seepage
duplicity	darken
exonerate	perusal
tertiary	optimist
fortitude	miscible
chalky	victimize
rapidity	aqueous
insolvency	prehensile
deliberately	verify
disruptive	fibrosis

EXERCISE 24-2 Using whatever clues you can, try to determine the grammatical function of each of the italicized words in the following sentences:

A   There is no *impediment* to bringing spacecraft and comet into line and allowing the craft to drift in and *bump* down.[2]

B   The same gas molecules and atoms that absorb ultraviolet light may *emit* ultraviolet airglow, again at characteristic wavelengths that can be detected *spectroscopically*.[3]

C   Much *Appalachian* coal appears to be low in organic sulfur and hence apt for *beneficiation*. Nonetheless, the *inapplicability* of beneficiation to all types of coal is an important *drawback* to the process.[4]

D   During periods between meals, when the supply of nutrients in the *lumen* is minimal, the weaker *pumps* on the plasma side *revert* to their original direction and *pump* nutrients from the *plasma* into the cell, protecting the cell's nutritional level.[5]

Meaningful Parts of the Word

See if you can reorganize any meaningful parts of the word. Many words are made up of two or more parts: a basic part, or root, and one or more suffixes or prefixes. For example, the word *inaudible* is made up of three parts: a prefix *in-* (meaning "not"), a root *-aud-* (meaning "hear"), and a suffix *-ible* (meaning "able to be"). Putting these meanings together, we arrive at a meaning for the word as a whole: "not able to be heard." Thus, even if you had never seen the word *inaudible* before, you could make an educated guess about its meaning if you knew the meaning of its parts.

Most English words, unfortunately, are not as readily analyzable as *inaudible* is. The English language is the product of thousands of years of development, having borrowed numerous words from other languages, which in turn may have borrowed them from still other languages, and significant changes of form and/or meaning may have occurred along the way. For example, the word *manufacture* is derived from the Latin word *manu* ("by hand") and *facere* ("to make"), and so it means literally "to make by hand." Centuries ago, that's what the word meant. Today, however, in the wake of the Industrial Revolution, the meaning of the word has been extended to include the making of something not only by hand but also by machines, by chemical processes, or by any other agency. Thus, in using roots and prefixes to help you figure out the meaning of an unfamiliar word, do not assume that the meaning of the parts will always give you the meaning of the whole. Often they will only give you a partial meaning, and you must use other devices to get the whole meaning.

The following lists contain a number of roots and prefixes that commonly occur in scientific and technical English. You should memorize as many of them as you can.

SOME COMMON WORD STEMS

Stem	Original meaning	Examples
aer (air)	air	aerial, aerodynamic, airport
ann (enn)	year	annual, anniversary, centennial
aqua	water	aqueous, aqueduct, aquifer
aud	hear	audience, audible, auditorium
bio	life	biology, biography, antibiotics
chron	time	chronology, chronic, synchronized
clos (clud)	shut	close, disclose, enclosure
corp	body	corporation, incorporated, corpuscle
cred	believe	credit, credo, incredible
dict	say	dictate, predict, contradict
duc	lead	product, reduce, transducer
fac (fic)	make, do	manufacture, satisfaction, efficient
flu	flow	fluid, effluence, influence
gen	birth	generate, genus, hydrogen
geo	earth	geology, geography, geometry
hydro	water	hydraulic, dehydrate, hydrofoil
ject	throw	eject, reject, injection
labor	work	laboratory, collaborate, elaborate
log	reason, science	logic, technology, analog
manu	hand	manual, manufacture, manuscript
mit (miss)	send	emit, omit, missile
nov (neo)	new	novel, innovate, neoplasm
ped (pod)	foot	pedal, podium, expedite
photo	light	photograph, photoelectric, photon
plic (ply)	fold	complicate, implicate, multiply
port	carry	portable, export, report
pos	put	deposit, composition, propose
puls	drive, push	impulse, propulsion, pulse
rota	wheel	rotation, rotary, rotate

Stem	Original meaning	Examples
scend	climb	ascend, descend, transcend
scop	look	microscope, telescopic, scope
sequ (secu)	follow	sequence, consequence, consecutive
solv	release	solve, solvent, resolve
spect	look	inspect, aspect, perspective
stat (stit)	state	state, static, constitute
strain	bind	restrain, constraint, stringent
tact	touch	tactile, contact, intact
tele	distant	television, telephone, telecommunication
tempor	time	temporal, contemporary, extemporaneous
therm	heat	thermodynamics, thermometer, thermal
tract	draw	traction, attract, contract
val	strength	value, valence, evaluation
vert	turn	invert, divert, conversion

SOME COMMON PREFIXES

Prefix	Meaning	Examples
a-	not	atypical, asymmetric, abnormal
ad- (at-, ap-, etc.)	to, toward	adhere, ad infinitum, attraction
anti-	opposite	antibiotic, anticatalyst, antimatter
back-	behind	background, backlog, backfire
bi-	two	biweekly, bilateral, bivalent
centi-	1/100	centimeter, centigram, centiliter
co- (col-, con-, etc.)	together with	cooperate, collaborate, consolidate
counter- (contra-)	opposite	counterbalance, counterweight, contradict
de-	reverse, undo	deflate, decrease, defrost
di-	two	dimethyl, divide, dilemma
dia-	across, through	diameter, diagonal, diagnose

Prefix	Meaning	Examples
dis-	not	discontinuous, disproportionate, dishonest
dis-	reverse, undo	discharge, dislodge, dismantle
down-	down	downturn, downslope, downtime
e- (ex-)	out of	evaporate, eject, expel
en-	in, within	enclose, encapsulate, engage
extra-	beyond	extrapolate, extraordinary, extracurricular
in- (im-, ir-)	in	insert, immerse, irrigate
in- (im-, ir-, il-)	not	independent, inelastic, immobile
inter-	between	interpolate, intercontinental, interfuse
iso-	similar, same	isotope, isomorphic, isosceles
mal-	bad, badly	malfunction, malformed, malpractice
mis-	bad, badly	miscalculate, misfire, mispronounce
mono-	one	monomer, monorail, monotone
multi-	many	multiple, multiplex, multitude
non-	not	noncollapsible, nonmalleable, nonflammable
on-	on	ongoing, onlooker, on-line
out-	surpass	outperform, outdo, outclass
out-	out, outside	outline, outbuilding, outskirts
out-	proceeding from	outgrowth, outlook, output
over-	excessive	overwork, overproduce, oversell
over-	beyond	overlook, overlap, overtime
poly-	many	polygon, polymer, polytechnic
post-	after	postpone, postmortem, postgraduate
pre-	before	preview, prevent, preformed
quadr-	four	quadrangle, quadrilateral, quadrant
re-	back	revert, reject, retract
re-	again	reapply, refasten, reiterate
retro-	back, backwards	retrospect, retroactive, retrorocket

Prefix	Meaning	Examples
self-	by itself, by oneself	self-lubricating, self-employed, self-loading
self-	in itself, in oneself	self-evident, self-explanatory, self-assured
semi-	half	semicircular, semiconductor, semiannual
side-	to or on the side	sidetrack, sidelong, sidecar
sub-	under, below	submarine, subsoil, substrate
sub-	a lesser part of	subclass, subdivision, subroutine
super-	beyond, above	superstructure, supersonic, superheated
trans-	across	transmit, transaction, transect
tri-	three	triangle, tripod, trioxide
ultra-	extreme	ultrasound, ultraviolet, ultramicroscopic
un-	not	uncertain, unavoidable, unclear
un-	reverse	unfold, undo, unbend
under-	beneath, below	underground, underlie, undersurface
under-	inadequate	underdeveloped, underestimate, understaffed
uni-	one	uniform, unify, unique
up-	up	upgrade, upturn, upkeep
well-	well	well-timed, well-advised, well-defined

EXERCISE 24-3 Without consulting a dictionary, see if you can guess the meaning of the following words. Use the root and prefix lists to help you. After you have finished, you may consult a dictionary to check your answers.

detoxify

reflux

circumspect

interpose

subversion

implicate

centipede

isotherm

disclosure

collaborate

Other Words Nearby

Look at other words for hints as to the meaning of the unfamiliar word. Often the immediate context surrounding an unfamiliar word contains clues that enable you to guess the word's meaning. Let's look at an example. Do you know what the word *blowout* means? Probably not. If you come across it in a context like the following, however, you can probably make a good guess as to its meaning just by noting some of the other words near it:

> Everyone is familiar with newspaper accounts of the shoreline damage that oil causes. The damage occurs year by year on different coasts and sometimes results from *blowouts* at offshore drilling platforms as well as from ships.[6]

Even if we knew nothing about offshore oil drilling, we would suppose that the word *damage* in this context refers to accidents of some sort. Furthermore, these seem to be specific kinds of accidents: they involve discharges of oil, which cause "shoreline damage." From such clues you can infer that the word *blowout,* in this context, means either an explosion, a spill, a leak, or some other type of accident that releases oil into the sea. (*Explosion* comes closest to the exact meaning, as you might be able to guess if you know the word *blowup; a blowout* is an explosive release of oil out of the top of an oil well.)

Other Words in Parallel

Look for other words used in parallel with the unfamiliar word and see what common meanings they share. Often a new word may represent some item belonging to a list of items; if you know what all the other items are and what they all have in common, you may be able to guess the meaning—or at least narrow down the possible meanings—of the unidentified item. Or, you may come across a new word in some context that also contains a synonym or antonym of that word (two words are antonyms of each other if they are defined according to the same criteria but have opposite meanings). In such cases, writers typically put the related words (list items, synonyms, or antonyms) into parallel grammatical from—all nouns or all verbs or all adjectives, etc.—as discussed in Chapter 3. Thus, by looking for parallel grammatical form, you can often spot words related to the unfamiliar word you're trying to analyze, and these related words will provide clues for you. Here is an example, from a description of measuring instruments on a spacecraft:

> A solar flux radiometer measured the energy reaching each level of the atmosphere from the sun. Two other devices also tallied heat radiation, one of them possessing a window made of diamond.[7]

Do you know what the word *tallied* means? Even if you've never seen it before, you can guess what it means by noticing that it is in parallel to the word *measured*:

> A solar flux radiometer *measured* the energy. . . .
> Two other devices also *tallied* heat radiation. . . .

Both words are main verbs, both are in the past tense, and both have measuring instruments for subjects and measurable quantities for objects. Indeed, these two words mean essentially the same thing, i.e., they are synonyms. The connective word *also* supports this interpretation. So if you know what the word *measured* means, you know what *tallied* means.

Here is another example of parallel forms, this one being a two-item list from a discussion on plastic products:

> The insulated plastic coffee cup and the plastic egg carton are two products that often contain fluorocarbon gas.[8]

Do you know what a carton is? Even if you don't, you can guess what it is by noticing how the word *egg carton* parallels the word *coffee cup*. They are both noun compounds and the fact that they are both preceded by the same adjective (*plastic*) makes it especially easy to spot the parallelism. Just as a cup is a standard container for coffee, a carton is a standard container for eggs. So the words *cup* and *carton* have much in common, though they are not synonyms.

EXERCISE 24-4 In the passages below, try to guess the meaning of the italicized words by looking at related words in the immediate context. Be prepared to explain your reasoning.

A   Researchers at SPRL are currently collecting bottles of air at selected places to learn the actual amount of nitrous oxide entering the atmosphere. Such places include especially those *sites* that may be emanating N_2O at levels higher than used to occur in nature—fertilized fields at various seasons, *compost heaps,* animal feed lots, lawns, golf courses, and eutrophic lakes and ponds.[9]

B   The region around the Great Lakes contains some of the great population centers of the United States and Canada. To the *teeming* millions in this region the lakes hold promise of an inexhaustible source of water for municipal, industrial, agricultural, and recreational uses. But, as is well known, rapid *urban* and industrial development has *prematurely* exhausted the usefulness of some parts of the lakes, with the result that in those parts the waters are no longer *potable,* "fishable," or "swimable."[10]

Rhetorical Patterns and Visual Aids

Look at other aspects of the larger context, such as rhetorical patterns and visual aids. As discussed in Chapters 2 and 7, properly written prose is always structured according to certain patterns of development, or rhetorical patterns. These patterns are most noticeable in units of prose larger than the single sentence. Paragraph units, for example, are often structured in a general-to-particular pattern: the first sentence of the paragraph contains some sort of generalization, and the subsequent sentences contain details supporting this generalization. Some other rhetorical patterns common to scientific and technical English are classification, comparison-contrast, problem-solution, process description, and logical analysis. Some rhetorical devices commonly used to fill out these patterns are examples, analogies, details, conditional statements, definitions, lists, and questions.

Often you can guess the meaning of an unfamiliar word if you can see how it fits into a rhetorical pattern. For example, do you know what the word *drawback* means? If not, see if you can't make a good guess at its meaning in the following context:

> The *drawbacks* of the bubble chamber are mainly two. First, the chamber is dumb and slow. It cannot be programmed except to a limited degree to distinguish interesting events from uninteresting ones; the piston must be withdrawn and a picture of every event taken before traces of the event vanish. This can result in millions of films that must be culled—the second *drawback*—by hand. Technicians, or scanners, must look at every film, and upon finding worthy images, they must measure the angles and curves by hand. Once derived, of course, the data from the bubble chamber films are precise, but thousands of technician-hours must first be invested.[11]

If you recognize the general-to-particular rhetorical pattern of this paragraph, you'll notice how all of the sentences following the first one (that is, all of the particulars) seem to be pointing out problems associated with the bubble chamber. If the first sentence is a proper generalization, then, it must be the case that the word *drawback* means, more or less, "problems." And this is roughly what it does mean.

Connective words can be very useful in analyzing rhetorical patterns. In the above case, the connectives *first* and *second* are used effectively to bring out the particularly problematic features of the bubble chamber and thus tie the discussion to the opening generalization. (Other connectives and their uses are discussed in Chapter 16.)

Visual aids are very important rhetorical devices, and you should use them whenever you can to help guess the meanings of unfamiliar words. For example, even though you may know something about the problems of cleaning up oil spills in the ocean, chances are you don't know what the terms *fiber bundle, honeycomb screen, vortex, boom, squeegee,* and *hydrofoil* mean. Do you? If not, you'll still be able to read the following excerpt and guess what they mean, simply by referring to the visual aids where indicated:

The recent test series dealt with devices that pick up spilled oil directly, without fences or corrals. They are a bit like carpet sweepers in that they draw oil off the surface of the ocean as it moves beneath or above them. Almost all the devices were contrived to slow down the oil, make the slick thicken, and then suck it off the surface. One means of thickening the oil layer was *fiber bundles* of polypropylene arrayed in the surface layer of oil and water (Fig. 1). The idea here is to provide some viscous resistance in the top layer and make it move more slowly than the rest of the current; an oil slick would then be made thick enough, after passing through the fibers, to be cleanly pumped away. In another scheme, a succession of *honeycomb screens* of diminishing gauge was assembled for the same purpose (Fig. 2). A third system sought to concentrate

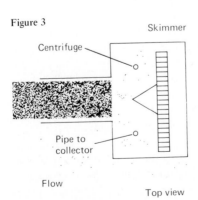

Figure 1

Bundle of polypropylene fibers

To collector

Water

Flow

Side view

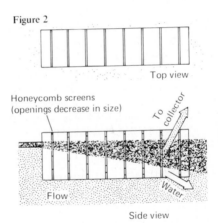

Figure 2

Top view

Honeycomb screens (openings decrease in size)

To collector

Flow

Water

Side view

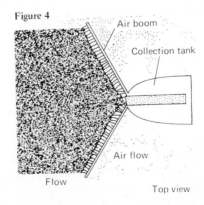

Figure 3

Skimmer

Centrifuge

Pipe to collector

Flow

Top view

Figure 4

Air boom

Collection tank

Air flow

Flow

Top view

Figure 5

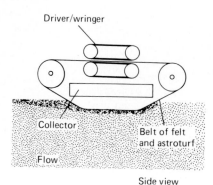

Figure 6

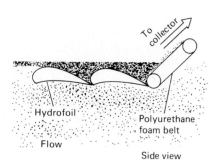

oil at the center of a *vortex,* where a suction pipe could then be applied to draw off the oil (Fig. 3). Another device used *booms* to extend a curtain of blowing air on the flowing oil slick (Fig. 4). Two systems used endless belts of oleophilic (oil-attracting) material. The belts of one of these systems would move at the same speed as the current, thereby reducing the slick's relative velocity to zero, and carry oil to a *squeegee* that would strip it from the belt for pumping away (Fig. 5). The other belt system used *hydrofoils* to slow and concentrate the slick while the belt served to elevate the oil from the water surface to a collector (Fig. 6).[12]

The visual aids accompanying this passage are virtually indispensable for understanding it. The author and illustrator seem to have known this, for they set up a form of parallelism between the italicized nouns and the visual aids, one series correlated with the other. Unfortunately, they were not completely consistent in carrying out this design. To be consistent, they should have used the same key terms in the visual aids as they used in the text. In four of the six figures, they did (Figures 1, 2, 4, and 6), but in the other two they didn't. In these latter two, you simply have to infer from the visual and written descriptions that *centrifuge* is a synonym for *vortex* and *driver/wringer* a synonym for *squeegee.*

PRINCIPLE 2
If you encounter an unfamiliar word that is not defined in the text, try to guess or at least narrow down its meaning by using contextual clues.

EXERCISE 24-5 In the passage below, try to guess the meaning of the words in italics by using word analysis and contextual clues. Be prepared to explain your reasoning.

OSHA: COSTLY CRUSADERS

On April 28, 1971, Congress passed the Occupational Safety and Health Act, which created an organization whose purpose was to document and cut down the number of industrial accidents. OSHA (Occupational Safety and Health Administration) was to be the *watchdog* and enforcer of safety laws. OSHA is authorized to set safety and health standards for nearly all non-government employers. This included (at the time of its formation) 5 million employers and 60 million workers; only employers covered by other federal safety programs or family-operated enterprises were *exempt*.

OSHA, therefore, is a very powerful and influential organization. It has the power to investigate a company on a worker complaint basis or on its own *incentive*. If OSHA finds something which it feels is a safety hazard, it can impose a fine of up to $1000 a day for the violation or it can seek court orders to close the facility down. In the event of any accidental death or disaster injuring five or more workers, OSHA is automatically entitled to investigate the scene and to impose fines or shut the facility down (with a court order).

Many Americans tend to think that occupational hazards are a thing of the past, like *sweatshops* and child labor. This is a very dangerous misconception. Workplaces now are about twice as dangerous as our nation's streets and highways. OSHA officials felt this misconception was due to a lack of industrial accident statistics. Until OSHA was formed, a relatively small percentage of industrial accidents were being reported. Early OSHA officials estimated that once accidents were reported and investigated the number would double.

Much of OSHA's early dealings were with four target occupations which were abnormally unsafe and had a high rate of accidents. They were: *longshoremen* (the workers were in constant danger of falling crates and unstable loads), meat packers (packers often cut themselves on the band saws and cutting tools), loggers (workers are easily crushed by logs and are often in situations where one slip means death or serious injury), and metal stamping plants (presses and stampers generate a continuous 115 dB, the threshold of pain being 120 dB). These industries were required to protect their workers from such *adverse* conditions. The employers felt that protective devices such as steel-toe boots, hard hats, *goggles* and ear protectors would provide adequate personal protection along with guards and safety devices on the machines. OSHA, however, felt that protective wear was not the answer and began to inquire about engineering the safety into the machines and into the workplaces. This line of thought threw OSHA and businesses into deeper conflict. Businesses had already started criticism toward the way OSHA went about its business. Businessmen felt that OSHA was investigating poorly and that they were getting *bogged* down by trying to enforce many small and relatively insignificant things while letting the major issues *slide*.

Take, for instance, OSHA's latest main objective: noise reduction in factories and in products. OSHA at first had set noise levels at 90 dB. Workers were given ear protectors which met those standards, but OSHA felt that the machines should operate quietly enough to allow workers to take off the ear *muffs*. However, the engineering involved to change the machines would cost billions of dollars. To make matters more *touchy*, OSHA is trying to decrease the noise level to 85 dB.[13]

24.3

CONSULTING AN AUTHORITY

Sometimes, no matter how many word-formation clues and contextual clues you use, you can't quite figure out what an unfamiliar word means. In such cases, you may have to consult an outside authority, that is, a knowledgeable person or a dictionary of some sort. Before taking the time required to do this, however, make sure you really *need* to know what the word means. Is it so important that you can't understand the passage without knowing the exact meaning of this word? Is the word likely to reappear later on? Can't you get by with just a vague idea of what it means or what it probably means? Remember, the time required to look up words in a dictionary or talk to an expert is time taken away from the reading process itself. There is a cost in both time and concentration whenever you consult an outside authority, and you must weigh the benefits against these costs.

When to Consult an Authority

Generally, you should consult an authority only for words that seem especially important in the particular passage you're reading. For example, if a word is a *topical* one, that is, if it represents the subject being discussed, we can assume it to be important. Topical words are often called key words in scientific and technical publications, and indeed topical words are key to both understanding and remembering the subject matter of discourse. For this reason, topical words are often used repeatedly, as are most important words in written prose. Thus, if a word is *used repeatedly,* it is probably an important one.

Sometimes, however, a word may be used only once and yet still be important enough to require exact knowledge. For instance, if it is part of an important *procedural* description, a description that you must know in step-by-step detail, then it is probably a word you should know the exact meaning of. Yet another circumstance signaling an important word is the use of examples or analogies to explain the meaning of important abstract concepts. Often an author will rely on the reader's knowledge of common words in trying to explain a difficult concept. These *explanatory* words, though appearing only once, briefly take on the importance associated with the concept being explained.

Whom to Consult

What sort of authority should you consult? In our view, the best authority is a knowledgeable person with whom you can talk directly, be it a supervisor, professor, fellow student, or someone else. Since the meaning of a word varies with the context in which it's used, there is a distinct advantage in being able to show your consultant the word in its actual context. That way, even if the consultant already knows the general meaning of the word, he or she can check it against the context and determine or verify its exact meaning. Also, when you deal with a consultant face to face, you can follow up his or her explanations with more questions if you like.

Dictionaries

A dictionary is also a useful authority, of course, provided it's one designed with your needs in mind. As a nonnative speaker of English, you probably need more information about words than is normally given in dictionaries designed for native speakers. For example, most nouns are normally used either as countable nouns or as uncountable nouns, and it is important to know the difference (Chapter 17). Native speakers learn very quickly whether a noun is normally countable or uncountable, but nonnative speakers often don't. Therefore a dictionary designed for nonnative speakers should indicate for each noun what its status is in terms of countability. Most dictionaries do not do this. Another feature you should look for in a dictionary is the use of sample sentences to supplement the definitions, as in the following example:

> *envision* (ĕn·vĭzh′ŭn), *v.t.* To have a mental picture of something beyond one's scope of vision. *It is difficult for most of us to envision the distant future.*

Sample sentences serve at least two important purposes. First, they make the meaning of a word more comprehensible. Definitions alone are often very abstract and thus harder to understand than concrete examples. Second, sample sentences illustrate how a word is typically used, not simply how it can be used. For instance, *envision* is typically used with a human subject (*us* in the above example) and a very inaccessible, possibly very abstract direct object. Although a single sample sentence cannot encompass all of the typical uses of a word, of course, it can at least be suggestive of them. Sample sentences are especially valuable in illustrating the use of idiomatic phrases and phrasal verbs. Among the dictionaries that satisfy these requirements are the *Longman Dictionary of Contemporary English,* the *Oxford Student's Dictionary of Current English,* and the *Oxford Advanced Learner's Dictionary of Current English.*

Even a well-designed dictionary, however, is likely to be deficient in at least one important respect: it will be out of date and incomplete in its treatment of scientific and technical words. The language of science and technology simply develops new terminology too quickly for dictionaries to keep up with it, and much of it is too specialized to be of interest to the general public. Thus, if you're interested in expanding your scientific and technical vocabulary, you might consider using a specialized dictionary if you can find one. Even a specialized dictionary, though, may be woefully incomplete and out of date. In particular, it may overlook many words from general everyday English which have highly specialized meanings when they are used in science and technology.

PRINCIPLE 3

If an unfamiliar word seems to be particularly important and you can't determine its meaning from contextual clues, consult an authority.

24.4

GETTING TO KNOW A WORD WELL

If you decide that a word is important enough to merit learning, try to learn it as well as you can. Don't be satisfied with simply learning what the word means; try to learn everything you can about it—its normal uses, its pronunciation, its related forms—so that you can use it productively in speaking and writing.

Usage

Note how the word is used. If it's a noun, is it countable or uncountable? What kinds of verbs or adjectives does it normally co-occur with? Does it occur in noun compounds and, if so, with what kinds of nouns? Is it restricted in the kinds of complements it can take?

If the word is a verb, is it a one-word verb or a two-word (phrasal) verb? Is it transitive or intransitive? Can it be followed by more than one object noun? Is it a state verb or an activity verb? What kinds of adverbs or nouns does it tend to co-occur with? What kinds of complements can it take? Does it occur in the passive form? If so, how often?

If the word is an adjective, what kinds of nouns or adverbs is it likely to co-occur with? Is it restricted in the kinds of complements it can take?

If the word is an adverb, what kinds of adjectives or verbs does it normally co-occur with?

Related Forms

Learn related forms. Most English words have one or more related forms, having the same root but differing in prefix or suffix and perhaps in grammatical function as well. For example, the word *adverse* has at least three closely related forms—*adversity, adversary,* and *adversely*—and a number of more distantly related ones—*converse, convert, inversion, subversive, perversity, animadversion,* and so on. The word *exempt* also has three closely related forms—*exempted, exemption, exemptible*—but only a few distantly related ones—*preempt, preemptive, peremptory,* and *peremptorily* are the only ones that come to mind. Notice how even with these two adjectives, *adverse* and *exempt,* there is considerable variation in the suffixes and prefixes one finds in their respective related words. Notice also that their closely related forms differ in grammatical function: *adverse* has two closely related noun forms (*adversity* and *adversary*) and a closely related adverbial form (*adversely*) but no closely related verb form at all, whereas *exempt* has a closely related noun form (*exemption*), verb form (*exempt, exempted*), and adjective form (*exemptible*) but no closely related adverb form.

Obviously, to learn the related forms of a word you cannot depend on a single pattern of relations. Each word has its own network of related forms, and you simply have to learn them one by one. The lists of roots, prefixes, and suffixes presented earlier in this chapter should help you at least to know what to look for in the way of possible related forms.

Pronunciation

Learn how the word and its related forms are pronounced. Although there are certain general principles of pronunciation in English, there are also many exceptions to these principles. Therefore, whenever you learn a new word, pay special attention to how it's pronounced. If you're consulting a knowledgeable person, ask him or her to pronounce the word for you, several times if necessary. If you're consulting a dictionary, make sure you know the phonetic symbols used to indicate proper pronunciation; a guide to these symbols can usually be found somewhere at the beginning or end of the dictionary. (For detailed instruction and practice in pronunciation, see Appendix C.)

> **PRINCIPLE 4**
> Try to get to know words well, especially important words. This means learning what they mean, how they are used, how they are pronounced, and what their related forms are.

EXERCISE 24-6   These final exercises are designed to test your ability to apply the principles described above. For each italicized word, first see if there isn't an informal definition of it somewhere in the passage. If not, try to guess the meaning of the word by using word-formation and contextual clues. For words that seem particularly important, make sure you know the exact meaning, even if you have to consult an outside authority. In general, try to learn as much as you can about each word within the time allotted.

A   **WATER POLLUTION: FROM RAW SEWAGE TO DETERGENT BUBBLES TO CARCINOGENIC CHEMICALS**
Since the early 1960's the American public has become increasingly *uneasy* about the quality of its water. This uneasiness has been directed toward both *"raw water"* (the water in waterways and in the ground) and processed drinking water.
    Public concern initially focused on massive pollution of our waterways by *untreated sewage* and other solid and liquid wastes from municipalities and industries. In the early 1960's the problem was dramatized by the *deterioration* of two great waterways, the Hudson River and Lake Erie. Oil slicks, dead fish, detergent bubbles, beer cans, and other *debris* had turned one of the most beautiful rivers in the world into "an open sewer." As for Lake Erie, about a fourth of the lake, encom-

passing 2600 square miles in the central basin, was all but *"dead."* The area was covered with a layer of *scum* and contained almost no oxygen and no fish. Across the nation, countless other once clear, cool, and lovely lakes and rivers also were rapidly deteriorating.

Subsequently *public complacence* was unsettled by a succession of scares about the presence of *insidious* chemicals in water. Detergent foam had emerged from household taps in some areas. Then various chemicals—DDT, mercury, and PCBs, among others—were discovered in unusually high concentrations in the *aquatic* food chain. These discoveries led to the issuance of fishing *bans* in Lake Erie, Lake St. Clair, the Hudson River, and other waterways. Mercury and other heavy metals were also detected in *trace quantities* in drinking water. Meanwhile various cancers were linked to drinking water in some localities. In 1974 asbestos-like fibers believed to be *carcinogenic* were found in the drinking water of Duluth, Minnesota. Suspected carcinogens were identified in New Orleans drinking water and in many water supplies throughout the country. Then the presence in drinking water of certain organic compounds believed to be carcinogenic *was linked to* the chlorination process.[14]

B   **HIT AND MISS: ELASTIC INTERACTIONS**

How big is a proton? Are the proton and the neutron, which weigh about the same, identical except for the neutron's lack of electrical charge? Are these particles hard, with sharp boundaries, like ball bearings, or soft, with *fuzzy* edges, like a powderpuff? And do they look the same at different energies? This last question is like asking whether one's hand looks the same to radar rays, under visible light, and under x-rays.

The proton and the neutron are the only strongly interacting particles found in natural, stable matter, and they form the nuclei of all atoms, together giving an atom almost its entire weight. Although physicists have now identified almost three hundred different elementary particle states that interact through the strong interaction, of these only the proton is permanently stable.

The proton and the neutron can be *mutually* studied by causing them to interact with one another and observing the results. Roughly speaking, when a neutron and a proton collide, either of two things can happen. If the collision is a *glancing* one, it is likely that the two particles will remain *intact* and scatter much like billiard balls. If, on the other hand, there is a *head-on* collision, the result can be the spectacular creation of as many as thirty new particles, mesons. These arise from a conversion of kinetic energy in the colliding particle into mass.

The central position of protons and neutrons in the subatomic world makes it *crucial* to understand in detail the structure, shape, and size of these particles. Many other experiments, if they are to work, depend upon good knowledge about protons and neutrons. And detailed knowledge about these important particles provides a *touchstone* for theories explaining the subatomic world; a *credible* theory must accord with the finely understood details of a few particles as well as provide *plausible* explanations of little-understood phenomena.

Jones and Longo are studying the glancing interactions. These are called elastic collisions, bounces, rather than crashes, or inelastic collisions, in which new particles result. Instruments are set up to determine how the beam and target particles *ricochet*. The experiment will learn about the surface of protons by observing the distribution pattern of neutrons and protons that have bounced off one another. The pattern will allow inferences about what happened when the neutron *grazed* the proton and hence about the proton's size and shape.

Jones *likens* the experiment to an exercise in analyzing the shape of an elephant by bouncing basketballs off it. With a large enough number of basketballs and a method of measuring where each basketball bounces away, a group of *blindfolded* scientists might eventually deduce the shape and size of the creature before them. If they could perform the same analysis again with tennis balls, they might discover details smaller than they had seen with basketballs. The benefits from a tennis-ball analysis, which might ensure that the blindfolded scientists did not mistake the elephant for a rhinoceros, correspond to the benefits of particle interactions at high energy-levels.[15]

REFERENCES

1   J. Wager, K. N. Trueblood, and C. M. Knobler, *Chem One*, McGraw-Hill, New York, 1976, p. 1.

2   B. Hiatt, "Planetary Exploration," *The Research News* **29**(10–12):23 (1978).

3   Hiatt, "Planetary Exploration," p. 26.

4   B. Hiatt, "Coal Technology," *The Research News* **24**(12):13 (1974).

5   J. Wei, "Membranes as Mediators in Amino Acid Transport," *The Research News* **31**(11/12):20 (1980).

6   B. Hiatt, "Oil on Troubled Waters," *The Research News* **27**(2):3 (1975).

7   Hiatt, "Planetary Exploration," p. 15.

8   B. Hiatt, "Ozone Zone," *The Research News* **28**(4/5):15 (1977).

9   Hiatt, "Ozone Zone," p. 19.

10   J. Wei, "Towards Cleaner Water," *The Research News* **29**(8/9):14 (1978).

11   B. Hiatt, "What's the Matter: Particle Physics at the University of Michigan," *The Research News* **26**(3/4):17 (1975).

12   Hiatt, "Oil on Troubled Water," pp. 12–13.

13   S. Hannah, "OSHA: Costly Crusaders," *Michigan Technic* **96**(4):16 (1978).

14   J. Wei, "Trace Contaminants and Advanced Technology in Wastewater and Water Treatment," *The Research News* **29**(8/9):5 (1978).

15   Hiatt, "What's The Matter," pp. 12–13.

ADDITIONAL READING

Practical
   P. Barr, J. Clegg, and C. Wallace, *Advanced Reading Skills,* Longman, London, 1981.
   T. F. Glazier, *The Least You Should Know about Vocabulary Building: Word Roots,* Holt, Rinehart and Winston, New York, 1981.
   I. K. Kankashian, *Contexts: A Strategy for Vocabulary Enrichment,* Regents, New York, 1981.
Theoretical
   D. F. Clarke and I.S.P. Nation, "Guessing the Meanings of Words from Context: Strategy and Techniques," *System* **8**:211–220 (1980).
   J. M. Ulijn, "Conceptual and Syntactic Strategies in Reading a Foreign Language," in E. Hopkins and R. Grotjahn (eds.), *Studies in Language Teaching and Language Acquisition,* Brockmeyer, Bochum, Germany, 1981.
   C. F. Van Parreren and M. C. Schouten-Van Perreren, "Contextual Guessing: A Trainable Reader Strategy," *System* **9**:235–241 (1981).

APPENDIXES

A

PUNCTUATION

A.1

SENTENCE PUNCTUATION

Punctuation marks serve as useful "road signs" to a reader. They help the reader see the grammatical structure of a sentence and thus more easily grasp its meaning. Punctuation can also help the reader see the rhetorical relationships between sentence elements.

The most commonly used—and *misused*—punctuation marks in scientific and technical writing are:

The comma	,
The semicolon	;
The colon	:
Parentheses	( )
Dashes	— —

These punctuation marks should be studied and used effectively by *every* technical writer. Unfortunately, they often are not. Many inexperienced writers believe—mistakenly—that punctuation marks are supposed to indicate the sound pattern of a sentence, that is, the pauses one would make if reading the sentence aloud. It is true that vocal pauses and punctuation marks often coincide, but this is not always the case, and even where they do coincide, a vocal pause by itself is hardly a reliable guide as to which punctuation mark to use.

In short, punctuation marks do not indicate sound patterning (at least not directly); instead, they indicate *grammatical structure, meaning,* and *rhetorical relationships.*

Independent Clauses

Three marks of punctuation are used to show a close relationship between independent clauses:

1 Use a *comma* if the two independent clauses are coordinate to each other, and the second clause begins with a coordinating conjunction (*and, but, or, so,* etc.):

The pressure in a gas or in a liquid is different at points that differ in height, and properties that depend on the pressure are also different.

Computer-linked scanning devices have been developed for bubble chamber films, but they cannot function without human helpers.

2   Use a *semicolon* if the two independent clauses are coordinate to each other, but there is no coordinating conjunction:

The structure of a simple liquid such as argon or methane is determined primarily by the repulsive forces between molecules; the attractive forces act only as a "glue" that holds the fluid together.

There are many geometries that are different from Euclidean; however, only the elliptic and hyperbolic geometries are called *non*-Euclidean.

3   Use a *colon* if the second clause elaborates on or explains the first:

All chemical equations should be balanced: they should show the same number of atoms of a given kind and the same net charge on each side.

The concept of randomness can best be explained by the use of examples: the sequence of heads and tails in tossing a perfect coin is said to be random.

EXAMPLE

Technology never exists in a social vacuum: it is embodied in products, processes, and people. By the same token, technology does not "transfer," or circulate, in a social void; it circulates throughout diverse institutional channels and mechanisms. . . . Thus social values cannot be separated either from technology itself, or from its mode of transfer. When these values, which are usually determined by the "parent" culture, conflict with those of developing countries, the technology transferred can deter—rather than promote—those countries' social objectives

    The most significant sources of technology transfer are consulting firms and the transnational corporations (TNCs) for manufacturing, extraction, and service. TNCs disseminate product and process technologies. But

STATEMENT FOLLOWED BY EXPLANATION

STATEMENT FOLLOWED BY CONTRASTING STATEMENT

"decisional technologies" exist as well, and these are vital: } CLOSELY RELATED
they comprise the know-how required to diagnose   } COORDINATE CLAUSES
complex problems and formulate solutions. As Argentine
physicist Jorge Sabato puts it: "The ability to conduct a   } STATEMENT FOLLOWED
feasibility study with its own means is the touchstone   } BY EXPLANATION
revealing when a country has conquered technological
autonomy."[1]

Introductory Subordinate Elements

Use a comma to separate introductory subordinate elements from the rest
of the sentence (and thus make it easier for the reader to locate the sentence
subject):

> Since most current plantings are transplants from the greenhouse and
> are several months old, they can tolerate stronger herbicides.

> However, the supply of rubber has so far been able to keep up with
> the slow increase in demand.

Parenthetical Information

Three different marks can be used to indicate parenthetical (nonessential)
information:

1   Use *commas* in most cases:

> The so-called British system of weights and measures, which is
> now on its way out in Britain, has its roots in ancient history.

> Galileo originally intended to be a doctor, on his father's advice
> that there was much more money in healing ailments than in
> proving theorems.

2   Use *parentheses* for references, abbreviations, definitions, qualifica-
     tions, and other subsidiary information that may be of interest only to
     certain readers:

> It has long been known that the vertebrate electroretinogram (ERG)
> in response to flashes of light consists of various components that
> have been ascribed to electrical sources located mainly in the
> receptor layer and in the inner nuclear layer of the retina.

> The nucleus of each atom is composed of protons and neutrons
> (except for the isotope of hydrogen of atomic weight 1, whose
> nucleus consists of a single proton).

3   Use *dashes* for emphasis (unless writing for conservative readers):

> Mixtures can be categorized as either homogeneous—uniform in properties throughout the sample—or heterogeneous.

> Many of the experimental methods of modern chemistry were developed originally by men and women regarded as physicists—for example, the methods of spectroscopy and those of structure determination by the diffraction of x-rays.

EXAMPLE

At the small end of the physical world, there is a realm in which it is difficult, *if not impossible,* to measure or manufacture anything. There, a stray cosmic ray can destroy a wire only a few atoms thick. There, the Heisenberg uncertainty principle prevents simultaneous measurement of both the velocity and position of a particle. There, the very act of examining a structure—*with the beam of an electron microscope, for example*—will degrade or destroy it.

    It is a realm too fascinating—*and promising*—to resist. When engineers explore it, they find a wonderland of faster switches, more efficient circuits, higher reliability, and lower costs. As scientists work on scales of atomic diameters and angstroms (*an angstrom equals a ten-billionth, 10^{-10}, of a meter*), they are discovering new and unexplored phenomena that increase their understanding of the physical world.[2] [italics added]

Marginal annotations:
- PARENTHETICAL INFORMATION
- PARENTHETICAL INFORMATION, EMPHASIZED
- PARENTHETICAL INFORMATION, EMPHASIZED
- PARENTHETICAL SUBSIDIARY INFORMATION

Lists

Unformatted lists—those not separated from the text—are set off either by a colon or by dashes or parentheses:

1   Use a *colon* if the list ends the sentence:

> The new technology is most widely known for its microelectronics: pocket calculators, microcomputers, and other products that have revolutionized information processing and communication.

> To produce 100,000 Btu's, one needs the following quantities of various fuels: $2\frac{1}{2}$ lb of natural gas, or 5 lb of petroleum, or 8 lb of coal, or 14 lb of wood.

2   Use *dashes* or *parentheses* if the list does not end the sentence:

> Every aspect of the peptide neurotransmitters—how they are made, what they do, where and how they act, how they are disposed

of, and even whether they are true transmitters or serve some other function—is being studied with great intensity.

The charge of an electron has been shown to occur in integral multiples (1, 2, 3, 4, etc.) of its basic value; therefore magnetic charges should also occur in integral multiples of some basic value.

To separate the items of a list:

1   Use *commas* in most cases:

The three parallel lines are as follows: $y = y_1$, $y = \frac{1}{2}(y_1 + y_2)$, and $y = y_2$.

There are many words in our language that indicate sets: *school* of fish, *swarm* of bees, *herd* of cattle, *squadron* of planes, etc.

2   Use *semicolons* if the items are long or are punctuated internally:

We have assembly plants in the following locations: Cedar Rapids, Iowa; Totowa, New Jersey; Edmonds, Washington; and Castle Rock, Colorado.

An efficient project cost system must accomplish three important objectives: (1) check actual and predicted costs of ongoing projects against the estimated cost; (2) obtain production rates for use in estimating of new work and create historical files; and (3) forecast the project final cost.

In formatted, displayed lists, that is, those set off from the body of the text, punctuation is optional:

The positions available are

Recovery plant supervisor
Management analyst
Administrative aide
Clerk-typist

The positions available are the following:

Recovery plant supervisor,
Management analyst,
Administrative aide, and
Clerk-typist.

A.2
WORD AND PHRASE PUNCTUATION

At the word and phrase level, the most frequently misused punctuation marks are:

The apostrophe   '

The hyphen   -

Quotation marks   " "

Underlining   _____

These are most often misused with regard to the functions discussed below.

Possession

1   Add an *apostrophe* and an *s* to all nouns except plurals ending in *s*:

Roberta's watch	the company's policy
my assistant's desk	the people's concern
the children's toys	the men's room

[Exception: Personal pronouns (*it, they, you,* etc.) never take the apostrophe to show possession. The correct possessive forms of these words are *its, their, your, our, my, her,* and *his* when they are used as adjectives and *its, theirs, yours, ours, mine, hers,* and *his* when they are used as pronouns.]

In cases of joint possession, attach *'s* only to the last noun: *Dun and Bradstreet's projections, John and Mary's sailboat.* When referring to combined cases of individual possession, on the other hand, attach *'s* to each of the nouns: *John's and Mary's birthdates.*

2   Add just an *apostrophe* (no *s*) to plural nouns ending in *s*:

the workers' pension plan	our parents' friends
the Joneses' house	the neighbors' garden

Occasionally you may want to show possession with a singular noun which ends in *s* or with an *s* sound. In such cases, follow rule 1 above unless the result would be hard to pronounce:

Bass's new hiking boots   but   Sears' new hiking boots

Alice's restaurant   but   Moses' restaurant

Noun Compounds

Nonspecialist readers may have difficulty interpreting long noun compounds. There are two simple ways to alleviate such problems:

1   Use an occasional *preposition.* Scientific and technical English abounds with terms made up of long strings of nouns: terms like *diffusion convection mass transfer resistance* and *4Pb0 · PbSO$_4$ lead acid battery paste formation.* Although such noun compounds may cause no trouble for specialist readers, they can be very confusing to nonspecialists, especially if there are not enough clues in the immediate context to serve as a guide to interpretation. Therefore, avoid using such terms out of context (for example, in titles or abstracts); instead of *3B pump maintenance recall rate,* try *rate of recalls for maintenance of the 3B pump* or *maintenance recall rate for the 3B pump.* It may take a little more space, but it enables the nonspecialist reader to grasp the meaning of the term much more easily and quickly.

2   Use an occasional *hyphen.* Another way of carving up long compound nouns into more digestible pieces is to pair off closely related words by means of hyphens:

> Diffusion-convection mass-transfer resistance
>
> High-frequency semiconductor layers
>
> Automatic score-keeping feature
>
> Dow coal-liquefaction process
>
> High-energy laser-beam weapons
>
> Manual materials-handling tasks

This grouping of related words helps prevent momentary misinterpretations on the part of a reader who is not familiar with such terms, especially when adjectives are involved. Notice, for instance, how *manual materials handling tasks* without the hyphen could momentarily be misconstrued as tasks involving the handling of manual materials. By inserting a hyphen between *materials* and *handling,* the writer alerts the reader to the fact that the adjective *manual* modifies *materials-handling* or *tasks,* not *materials.*

End-of-Line Hyphenation

To divide a word at the end of a line, use a *hyphen.* If you have started to write a word at the end of a line but find you don't have room to complete it, you can hyphenate it and carry a part over to the beginning of the next line. Bear in mind, though, that this device exacts a slight toll on the reader: it briefly interrupts the smooth flow of words and forces the reader to guess at the meaning of the hyphenated word before seeing the second part of it. So, do not overuse this technique and, when you do use it, try to make it as easy as possible for the reader to guess the meaning of the full word after seeing only part of it. The following rules are designed for that purpose.

1   *Never divide a syllable.* Syllables are the building blocks of words. You can often guess the meaning of an unfamiliar word just by putting together the "meanings" of the syllables it is made up of. If you see the syllable *chem-* leading off a word, you can pretty well guess what the rest of the word might be, but if you see only half the syllable (*ch-*), you have no idea what follows: *chart, chapter, chiropractic, chlor-ine, choices, chest,* to mention but a few. So be sure to maintain the integrity of syllables when you divide a word. If you don't have time to check the dictionary (all good dictionaries show the syllable divisions of words), you can use these two rules of thumb: (1) all syllables have at least one vowel sound; (2) all syllables can be pronounced easily and, if necessary, can be shouted easily to someone at a distance. *Chem-* satisfies these conditions, but *ch-* doesn't.

INCORRECT	CORRECT
ma-rket	mar-ket
repo-rt	re-port
thr-ough	through
proje-ct	pro-ject
dupl-ic-ate	du-pli-cate
persp-e-ctive	per-spec-tive
phl-ebi-ti-s	phle-bi-tis

2   *If there is a natural break in a word, try to divide it there.* Many words divide naturally into two parts: *work-shop, self-disciplined, turn-key, break-water, anti-toxin, electro-magnetic, thermo-meter,* etc. By divid-ing the word at the end of the line according to its natural parts, you make it easier for the reader to recognize the two parts. Notice how funny it looks, for example, to see a word divided like this: *sol-id-state.* It's much more natural and easier to read divided as *solid-state.*

Titles of Written Works
<u>Underline</u> (or *italicize* if you have access to a typewriter with an italics font) the names of books, journals, magazines, monographs, and other publica-

tions that are bound separately:

> J. Galbraith, <u>Designing Complex Organizations</u>, Addison-Wesley Publishing Company, Menlo Park, Calif., 1973.

> D. O'Brien et al., "Responsibility for Inspection," *Journal of the American Concrete Institute*, Vol. 69, No. 6, June 1972, pp. 320–333.

Put *quotation marks* around the names of articles, reports, memos, and other pieces of writing that are either not published or not bound separately:

> J. D. Borcherding, "An Exploratory Study of Attitudes That Affect Human Resources in Building and Industrial Construction," Department of Civil Engineering, Stanford University, Stanford, Calif., 1972.

> W. Larkin and D. Burns, "Sentence Comprehension and Memory for Embedded Structures," *Memory and Cognition*, 5 (1977), pp. 17–22.

> M. J. Zakkak, "Cost Control Simulation," thesis presented to the University of Texas, Austin, TX, 1976, in partial fulfillment of the requirements for the degree of Doctor of Philosophy.

Note: Long formal technical reports are often cited with the descriptive name in quotation marks but the report number underlined or in italics:

> "PCS, Project Control System/360," <u>GH 20-0376-3</u>, International Business Machines Corp., 1971.

> S. H. Wearne, "Contractual Responsibilities for the Design of Engineering Plants," *Report No. TMR5*, School of Technological Management, University of Bradford, England, 1975.

A.3

CAPITALIZATION

When referring to something by its official name, *capitalize* all words in the name other than articles, short prepositions, and conjunctions:

Personal names:	Phyllis Gomez, Lawrence Van Vlack
Personal titles:	Project Officer, Professor of Mechanical Engineering, M.D.
Places:	115 South Cache Street Jackson, Wyoming United States of America

Organizations:	Dupont Chemical Company, Arizona Girls Leadership Club, Brooklyn College
Days, months, eras:	Monday, Memorial Day, Christmas, October, the Twentieth Century, the Eighties
Programs:	the Professionals-in-Training Program, the Stanford University Degree Program in Physics
Natural languages:	French, Spanish, Chinese
Computer languages:	BASIC or Basic, FORTRAN or Fortran, ALGOL, Pascal
Model names:	the IBM Selectric, the Ford Mustang

REFERENCES

1   Denis Goulet, "The Dynamics of International Technology Flow," *Technology Review*, May 1978, p. 32.

2   William J. Cromie, "Room at the Bottom," *Mosaic*, May-June 1981, p. 25.

B

INFORMAL CONVERSATIONAL EXPRESSIONS

A great deal of scientific and technical communication is carried out by means of ordinary conversation—in the office, in the classroom, in the laboratory, in hallways, at lunch, over the telephone, at conferences, etc. In fact, it's probably safe to say that *most* scientific and technical communication takes place in this way. And this doesn't even take into consideration the amount of ordinary social communication that occurs through conversation—social communication that helps establish good relations between coworkers and thus helps make technical communication easier. For these reasons, it's well worth anyone's time to develop effective conversational skills.

One prominent feature of informal conversational language is the use of idiomatic expressions. Sometimes the meaning of such expressions cannot be determined by putting together the meanings of their component parts. *To kick the bucket,* for example, does not have anything to do with kicking or buckets (at least not in the idiomatic sense); rather, it means "to die." Because conversational expressions often have such restricted and unpredictable meanings, they have to be learned one by one, which makes it difficult to learn a large number of them. Nonetheless, you should try to learn as many as you can.

The following list is designed to help you get started. It contains 230 of the most commonly used conversational expressions in technical and scientific settings in the United States (though they are characteristic of American English, most are also used in British English). Each entry contains (1) the expression itself, (2) one or more synonyms for that expression, and (3) an example of how the expression is used.

INFORMAL CONVERSATIONAL EXPRESSIONS

Expression	Synonym(s)	Example
about-face	reversal	After first rejecting our offer, she did an about-face and accepted it.
about to	ready to, all set to	We were about to start the meeting when Jack called to say he'd be late.
above all	especially, most importantly	Above all, be sure to test all the samples using the same procedure.

Expression	Synonym(s)	Example
account for	explain	We are unable to account for the extra chlorine in the water supply.
all along	the whole time, from the beginning	I felt all along that we were on the wrong track.
all in all	all things considered, in general	All in all, despite problems with the weather, we had a successful trip.
all set	ready	Everything is all set for the visitors' arrival.
allow for	take into consideration	Unfortunately, our calculations failed to allow for possible errors.
amount to	total, add up to	Our expenses will probably amount to more than $100,000.
appeal to	be attractive to	This new model should appeal to potential buyers.
arrive at	reach, make	The committee expects to arrive at a decision sometime next week.
ask about	inquire	I asked the project manager about hiring a new technician, but she said there were no funds for it in the budget.
ask for	request	We'll have to ask the manager for more supplies.
(not) at all	of any kind, in any way	They don't have any software specialists at all.
(be) at fault	wrong, deserving of blame	I'm afraid I'm at fault: I forgot to reset the stack switch.
(be) aware	know about	Are you aware of David's medical problems?
back to square one	at the starting point again	We're not making any progress with this approach. Let's go back to square one and try again.
back and forth	first in one direction, then in the opposite direction	A pendulum swings back and forth.
back up	(1) move backward	Back the truck up to the loading dock.

Expression	Synonym(s)	Example
	(2) support, reinforce	Be sure to back up your recommendations with good arguments.
beat around the bush	avoid stating the main point	He didn't want to embarrass them, so instead of saying what was on his mind he kept beating around the bush.
blow up	detonate, explode	If too much pressure builds up in the sterilizer, it might blow up.
bottom line	main point	The bottom line is that they're not going to support us after our current grant expires.
(be) bound to	certain to	If we send out enough grant proposals to other agencies, we're bound to get some support somewhere.
brand-new	absolutely new	Barbara traded in her old car for a brand-new one.
break down	(1) stop functioning	The new machine has already broken down.
	(2) analyze, take apart	We shall begin by breaking the process down into 15 steps.
breakthrough	a major development or advance	IBM is about to make a new breakthrough in computer technology.
bring up	introduce a topic of discussion	I hate to bring up this topic at this time, but I think we really need to talk about it.
brush up on	renew one's knowledge of	I'll have to brush up on my French before going to Paris next year.
build on/upon	to use as a foundation	I'd like to get a job where I can build on my university training.
burn the midnight oil	work late at night, work overtime	To get my report finished on time, I'll have to burn the midnight oil.
burn up	destroy by fire or heat	Unfortunately those files were all burned up in that fire we had last month.
by and large	for the most part, in general	We've had some problems on this project, but by and large it's been quite a success.

Expression	Synonym(s)	Example
by the way	incidentally, while we're on the subject	By the way—since you've raised the subject—why did you decide to use the 9200?
by word of mouth	by unofficial communication	Many job openings are advertised by word of mouth, not by posted advertisements.
call on	visit	May I call on you in your office sometime next week?
call (up)	telephone	I'll call you (up) as soon as I arrive. By the way, what's your number?
(be) capable of	able to	Ms. Gray's latest report shows that she is capable of handling tough assignments.
(be) careful to	take measures to, be sure to	You must be careful to protect our trade secrets from our competitors.
carry out	accomplish, complete	You will be expected to carry out your duties without extra help.
carry the ball	be in charge, have primary responsibility	We've done our share; let's let Jones carry the ball from now on.
catch up on	do what was postponed earlier	I hope to catch up on my reading during my vacation.
catch up with	overtake, draw even with a competitor after being behind	Although our competitors have been working on this process for a long time, I think we can catch up with them soon.
change one's mind	revise one's opinion about something	At first I thought their plan was good, but now I've changed my mind.
change over to	convert to	We shall change all our operations over to the SI system in January.
check over	inspect, examine	Check the equipment over one more time before you start.
check up on	investigate	We'd better check up on that company's credit rating before signing any contract with them.
check with	ask for advice or approval	It sounds like a good idea to me, but you should check with Dr. Lee before you go ahead with it.

Expression	Synonym(s)	Example
come across	discover, find	I thought I had gotten all the bugs out of the program, but then I came across some new ones.
come along	progress, develop	How are your hydroponic tomatoes coming along?
come in handy	be useful	This new tool should come in handy.
come out	be published, be issued	The new OSHA regulations should come out sometime next week.
come to terms	reach an agreement	After three days of negotiations, the company and the union finally came to terms.
come to the point	state the main point	I wish you'd come to the point instead of beating around the bush so much.
come up with	develop, find	We should be able to come up with a solution to this problem by tomorrow.
consist of	be composed of	Brass consists of copper and zinc.
count on	depend on	You can count on that company to deliver the software on time.
cut down on	reduce	We can make a profit only if we cut down on expenses.
cut off	stop, terminate	Our supplier has threatened to cut off further shipments unless we pay our bill.
cut out	remove	The administration intends to cut all unnecessary items out of the budget.
day in and day out	regularly	Sue comes to work day in and day out, even when she's not feeling well.
day off	holiday	I'd like to take a day off next week so that I can visit my sick aunt.
deal with	confront, face	I'm too tired to deal with that problem right now; let's take it up tomorrow.
a dime a dozen	plentiful, commonplace	Skilled carpenters used to be a dime a dozen; now it's hard to find one anywhere.
depend on	rely on, count on	Linda's a good researcher; I think we can depend on her to design the experiment right.

Expression	Synonym(s)	Example
(be) devoted to	be concerned with	Most of the staff meeting will be devoted to discussing the annual report.
do over	do again, redo	Your homework has too many mistakes in it; you'll just have to do it over.
down-to-earth	practical, realistic	We're spending too much time philosophizing about this; let's be a little bit more down-to-earth, OK?
drop off	fall off, decline	We expect their production of 360's to drop off as soon as they start manufacturing the 420x.
dwell on	be overly concerned with	Why dwell on the problems of the past? We've got enough things to worry about right now!
end up with	have as a result	If you hire that technician, you'll end up with serious trouble in the lab.
enlarge upon	explain in more detail	I see that this idea is new to you, so let me enlarge upon it for a minute.
faced with	confronted with	If this product doesn't sell, our firm will soon be faced with bankruptcy.
fall behind	proceed too slowly, lag behind	Production will fall behind if the supplies don't arrive soon.
fall off	decrease, drop off	The present rate of inflation should fall off by several percentage points in the next year.
feed into	flow into	The paper feeds into the machine, where it is printed on both sides.
few and far between	rare	Skilled carpenters used to be found all over the place; now they're few and far between.
figure on	plan on, expect to	Sam knew that going into business for himself would be difficult, but he didn't figure on going bankrupt.
figure out	solve	No matter how hard I try, I can't seem to figure out this homework problem.
fill in	insert material in	To complete this report form, all you have to do is fill in the blanks.

Expression	Synonym(s)	Example
fill out	complete	Please fill out this application form before you leave.
find out	discover	We shall find out the truth one way or another.
finish up	complete	We can't finish up this job without more concrete.
focus (one's attention) on	concentrate on	In this report I shall focus on the steps we took to improve our analytical procedure.
follow up on	pursue, develop	Perez has good ideas, but he seldom follows up on them.
for good	forever, permanently	I think Shirley has left the academic world for good; she's got everything going for her in industry.
for the time being	for the present	The lab routine seems to be working fine for the time being. If problems come up later, we can always change it.
from A to Z	completely, from start to finish	Just to make sure everybody knows the procedure, let's go through it from A to Z.
from time to time	occasionally, once in a while	From time to time my husband has to go to the clinic for therapy.
get down to business	talk business after polite formalities	They spent ten minutes talking about mutual acquaintances before getting down to business.
get going	start	I'm tired of sitting around waiting for everybody to show up; I wish we could get going.
get in touch with	contact by phone or mail	Is there any way we can get in touch with you while you're on your trip?
get off	mail	I'll ask my secretary to get this material off to you as soon as possible.
get rid of	dispose of	We don't need this stuff any more. Will you please get rid of it.
give someone a buzz/ring	telephone	I'll give you a buzz at home tonight, OK?

Expression	Synonym(s)	Example
gloss over	discuss superficially and thus hide certain weaknesses	Your report would be okay, Johnson, except that it glosses over some potentially very serious problems.
go ahead	proceed	When I heard it was on sale for 25% off, I decided to go ahead and buy it.
go by	be called by, use as a name	Your last name is hard for me to pronounce. Is there anything else you go by?
go-getter	an energetic, ambitious person	John's a real go-getter, isn't he? That's his third promotion in less than a year!
go halves/ 50-50	split the costs	I know you invited me out, Tim, but let's go halves on the bill, OK? It wouldn't be fair for you to pay it all yourself.
go over	review, re-examine	Let's go over this procedure one more time to be sure that everyone understands it.
go over someone's head	bring a complaint or proposal directly to your boss's supervisor without first getting your boss's permission	Sometimes the best way to get something done is by going over your boss's head. In some companies, though, it's the fastest way to get fired!
hand in	turn in, submit	Do you think you can hand in your report by 5:00?
hands-on experience	direct, personal experience	Kay should be just the person we need: she has a lot of hands-on experience with x-ray machines.
hang up (the phone)	put the receiver back on the hook	Excuse me, Sue, but I've got to hang up; my supervisor wants to talk to me.
have a lot on the ball	have many good qualities	Carol has a lot on the ball; I think she'll go far in her career.
have a word with	talk with	Max, could I have a word with you when you get a minute?
hinge on	depend on	The success of this new surgical technique hinges on how skilled the surgeon is.
hit the ceiling	become angry	Mr. Diaz will really hit the ceiling when he hears about this!

Expression	Synonym(s)	Example
hold off (on)	postpone	Mr. Davis says he wants to hold off on the decision until the monthly review is completed.
hold up	delay	These repairs shouldn't hold up production for more than a few hours.
hold water	have any validity	I'm sorry, but your argument just doesn't hold water.
how come	why	How come you haven't complained about this before?
impose on/upon	inconvenience, be a burden to	I don't want to impose on you, but I need a letter of recommendation. Would you mind writing one for me?
improve on something	make something even better	I know it's a good product, Owens, but isn't there some way we can improve on it?
in a nutshell	in summary, in short	I'll just skip the details for the time being and tell you basically what happened. To put it in a nutshell, all our amoeba samples were contaminated.
in case (of)	in the event of	In case of an accident, call Dr. Wilson.
in charge of	responsible for	I'm putting you in charge of this part of this experiment.
in-house	within the company	Rather than using an outside agency, wouldn't it be more economical to develop our own in-house training program?
in the same boat	faced with the same problem	The models we're competing against all failed the collision test, too. In fact, it looks like we're all in the same boat on this.
(be) in the way	be an obstruction, a hindrance	I wish we could move this table somewhere else; it's just in the way here.
in time	within the required time	In order to finish my data-processing in time, I'll have to work night and day.
inquire about	ask about	The project manager just called to inquire about our progress.

Expression	Synonym(s)	Example
interfere with	hinder	Prof. Asada doesn't believe in interfering with the creative work of research scientists.
jot down	quickly write down	Let me just jot down your phone number so that I can get in touch with you.
jump to the conclusion that	make a quick judgment without adequate justification	Mr. Young looked at my first two printouts and jumped to the conclusion that I was arguing for Method A. In reality though, I think Method B is better.
keep abreast of	keep up with, stay current with	We should be keeping abreast of the latest developments in our field.
keep down	maintain at a low level, suppress	I'd like to keep our expenses down if at all possible.
keep in mind	bear in mind, not forget or overlook	Keep in mind that all the samples must be assayed before we can move on to the next stage of the experiment.
keep in touch	stay in communication (with)	We've enjoyed doing business with you. Please keep in touch.
keep on	continue	The one thing that drives me crazy is that Joe keeps on smoking in the lab even though we asked him not to.
keep out	prevent from entering	Be sure to keep all strangers out of the building.
keep pace with	be competitive with, grow at the same rate with	Without our own computer we cannot keep pace with our competitors.
keep track of	know the location or status of	It's hard for me to keep track of all my papers; maybe I need a filing system.
know-how	practical knowledge	Susan has the know-how it takes to get the job done right.
know the ropes	know the customary procedures, be experienced	It'll take a while before you know the ropes around here.
lag behind	move more slowly than, fall behind	The design group is lagging behind schedule.

Expression	Synonym(s)	Example
lay off	temporarily dismiss from employment	If the national economy keeps getting worse, many companies will have to lay off some of their employees.
lean away from	avoid	We try to lean away from dependence on any one source of funding.
lean toward	favor	Although I lean toward option A, the department head seems to favor option B.
leave out	omit	Unfortunately, we left out some important details in our writeup of the experiment.
leave someone holding the bag	avoid responsibilities, thus forcing someone else to fulfill them	I think the construction firm wants to withdraw from its commitment and leave us holding the bag.
look for	seek	We aren't looking for perfection, but we do expect you to turn in reports on time.
look forward to	await with eagerness	I look forward to hearing from you at your earliest convenience.
look over	inspect, examine	I'd appreciate it if you would look these blueprints over before we start laying the foundation.
look up	search for and find	You can look her phone number up in the directory.
(it) looks like	(it) appears that, (it) seems to be the case that	It looks like the company is going to hire some new people.
make a mountain out of a molehill	exaggerate the real importance of something	Are you sure those cracks are really that dangerous, Mike? Are you sure you're not just making a mountain out of a molehill?
make certain	be sure	Make certain to turn out all the lights when you leave.
make do	accomplish one's purpose	We don't have new equipment, but we can make do with what we've got.
make sure	be certain	Please make sure that all the test tubes are thoroughly cleaned.

Expression	Synonym(s)	Example
narrow down	reduce the number of	Given these data, I think we can narrow down the alternatives to Plan A and Plan D, right?
now and then	occasionally	Now and then I get the urge to take a long walk by myself.
old hand	someone with long experience in doing something	Fortunately my boss is an old hand at solving personnel problems like this.
on the whole	in general, for the most part	We had some bad weather, and the car needed repairs. Still, on the whole, it was a good trip.
on time	at the appointed time	Ms. Valdez always likes us to be on time for meetings.
once in a while	occasionally	We should have a day off once in a while, don't you think?
out-of-date	old-fashioned, obsolete	Too many of our plants are out-of-date; we should be building more new ones.
out of order	not functioning	Don't bother trying to get the TV to work; it's out of order.
out of the question	impossible, not even to be considered	Your proposed solution, I'm afraid, is out of the question; it's just far too expensive.
overstaffed	have too many employees	We're so overstaffed I'm afraid we'll have to lay some people off.
pay attention to	listen to, look at	Harry complains that his kids never pay attention to him when he's talking.
phase out	eliminate by stages	Management has decided to phase out our current accounting system and eventually replace it with a better one.
pick up	acquire, get	This project has been a nerve-wracking experience—I'll bet I've picked up a few gray hairs working on it.
plan on	expect to	You should plan on taking at least three weeks to complete this part of the project.

Expression	Synonym(s)	Example
point of view	perspective, opinion	According to Bob's point of view, the rabbits make better test animals than the rhesus monkeys.
point out	direct someone's attention to, emphasize	When you show our new equipment to the inspection team, be sure to point out its safety features.
pull together	cooperate	Our company is going through tough times right now, but if we all pull together we can get by.
put back	return	After you've finished using the tape recorder, please put it back in the closet.
put off	postpone	I'd like to put off my decision until tomorrow.
put our heads together	consult, plan together	If we put our heads together, I'm sure we can find a solution to this problem.
put out	extinguish	Would you please put out your cigarette?
put two and two together	observe what's obvious, make an obvious inference	All you have to do is put two and two together and you can see that Robertson and Williams don't like each other.
qualify for	fulfill necessary conditions for	Because of your weak background in microbiology, you do not qualify for this position.
quite a few	a fairly large number of	I thought we'd only have one or two rejects in the entire sample, but actually we ended up with quite a few.
red tape	bureaucratic forms, official regulations	There's so much red tape involved, it may take us months to get our request approved.
reinvent the wheel	do outdated research	We want to do state-of-the-art research, not reinvent the wheel.
resort to	have recourse to (for lack of any other options)	Since we can't afford to buy a dialysis machine, I guess we'll have to resort to renting one.
right away/ now	immediately	Let's buy it right away before prices go up.

Expression	Synonym(s)	Example
run across/into	encounter, meet accidentally	While I was in Singapore on a business trip, I ran across an old friend of mine.
run around in circles	get nowhere, make no progress	I've been working on this design problem all day with nothing to show for it. I just seem to be running around in circles.
run low on	have only a small quantity of	We seem to be running low on supplies; let's put in an order for more.
run out of	use up, exhaust	Let's get the project finished before we run out of funds.
second-rate	of mediocre quality	Ellen got her degree at a second-rate school but has been a first-rate employee for us.
see eye to eye	agree	We just don't see eye to eye on this issue: she wants to keep using the same method and I want to try a new one.
set up	establish, create	Our company has just set up a new training program in intercultural communication for employees going overseas.
shoot for	aim for, have as a goal	Lou's been putting in a lot of overtime lately; he must be shooting for a promotion.
shoot the breeze/bull	talk casually	During lunch hour I like to shoot the breeze with my colleagues.
shorthanded	lacking sufficient help, understaffed	As shorthanded as we are, it will probably take us three weeks or more to fill your order.
shut down	stop, close	We don't want to shut down the assembly line if we can help it.
six of one, a half dozen the other	there's no real difference between them	Why bother voting for either candidate? They're just six of one, a half dozen the other.
sleep on it	think about it for a day or so	Your proposal sounds good, but I'd like to sleep on it a while before making a decision.

Expression	Synonym(s)	Example
slow down/up	go slower, reduce the speed of	These machines are running too fast. We should slow them down.
so far	up to now	So far we haven't had any problems with cost overruns.
sooner or later	eventually	This stirrer may still be working okay, but don't forget, it's 20 years old. Sooner or later it's going to break down.
speak up	speak louder	We can't hear you in the back row. Could you speak up, please?
spin one's wheels	make no progress	I can't solve this homework problem; no matter what I try, I just keep spinning my wheels.
split hairs	make a big issue over minor details	There's no sense splitting hairs over such a petty matter.
stand a chance	have a chance	Do you think we stand a chance of having our paper accepted?
stand for	represent	In this equation S_l stands for the longitudinal tensile strength and S_r for the radial tensile strength.
stand out	be prominent	The color will stand out more if you put it against a white background.
start out with	begin with	Ms. Chen started out with three employees in 1974; today she has 75.
stick with	stay with, keep	I'd like to buy a new car, but for now I think I'll stick with the one I've got.
straighten out	put in order	We'll have to call in the supervisor to straighten this mess out.
switch on/off	activate/deactivate	Would you mind switching on that light over there?
take apart	disassemble	In order to see what's wrong with this motor, we'll have to take it completely apart.
take into account	consider, be aware of	In designing this machine tool, we must be sure to take into account all the design criteria.
take turns	alternate with	Alan, why don't you and Betty take turns checking the pressure readings?

Expression	Synonym(s)	Example
talk shop	talk about business matters	My wife doesn't like it if I talk shop at parties.
team up with	work with	I'd like you to team up with Ed Kurolski on this project since he's been working in a related area.
tear down	demolish	The company intends to tear the old plant down to make room for a new one.
think of	remember, consider	Have you thought of all the possible consequences of going ahead with your plan?
think over	reflect on, meditate over	Mr. Peters, I'd like to think over your offer for a few days before making a decision.
through the grapevine	by rumor, by hearsay	We've heard through the grapevine that Frank's going to be our next section chief.
time off	free time, time away from the job	My daughter's in the hospital—I need some time off so I can go visit her.
turn down	deny, reject	I'm sorry to say your proposal has been turned down.
turn in	submit, hand in	Ms. Miller wants us to turn in our trip reports as soon as possible.
turn into	be converted into	Wine will turn into vinegar if you leave it uncorked for long.
turn on/off	activate/deactivate, switch on/off	Do you mind if I turn the air conditioner off?
turn out	happen, end up	We thought our experimental design was flawless, but it turns out that we failed to control an important variable.
understaffed	not have enough employees	Since we're so understaffed at the moment, most of us have been working overtime.
under the weather	sick, not very well	I'm feeling a little under the weather, Ms. Haddad. Could I take the afternoon off?
up-to-date	current	I wish I were more up-to-date about current developments, but I don't have much time for reading these days.

Expression	Synonym(s)	Example
use up	exhaust, use completely	We've used up all of our travel funds for the year; any trips from now on will have to be paid for out of our own pockets.
used to	accustomed to	I'm not used to eating with chopsticks, but I'm willing to give them a try.
wear out	become useless	The fan belt in my VW is supposed to last 30,000 miles before it wears out.
wind up	complete, conclude	If we can wind up the first series of trial runs by next week, we'll have enough time to finish the other two on schedule.
wipe out	destroy, eliminate	If this new software doesn't sell, we'll be wiped out.
work out	(1) resolve	She and I argue a lot, but if we could just sit down and talk things over I think we could resolve our differences.
	(2) develop	How are things working out at your new job?
would rather	prefer to	I would rather read *The New York Times* once a week than get the local paper every day.
wrestle with	struggle with, ponder, worry about	We've been wrestling with this problem for a long time.

ADDITIONAL READING

(American idioms unless noted otherwise)

L. A. Berman, and L. Kirstein, *Idiom Workbook,* IML, Silver Spring, Md., 1979.

R. J. Dixson, *Essential Idioms in English,* Regents, New York, 1971.

R. E. Feare, *Practice with Idioms,* Oxford University Press, New York, 1980.

L. Goldman, *Getting Along with Idioms: Basic English Expressions and Two-Word Verbs,* Minerva Books, New York, 1981.

Longman Dictionary of English Idioms, Longman, New York, 1979 (mainly British English, advanced).

J. Seidl, and W. McMordie, *English Idioms and How to Use Them,* 4th ed., Oxford University Press, New York, 1978 (features British idioms).

H. C. Whitford, and R. J. Dixson, *Handbook of American Idioms and Idiomatic Usage,* Regents, New York, 1973.

S. Wiener, *A Handy Book of Commonly Used American Idioms,* Regents, New York, 1958.

C

PRONUNCIATION

If you plan to use English for spoken communication, whether for oral presentations or for conversation, it is important that you pronounce the sounds correctly. This includes not only the individual sound segments (*p, b, s,* etc.) but also the stress patterns of words and the intonation patterns of phrases and sentences.

The best way to work on your pronunciation is to practice with a native speaker of English. If you can't find such a person, a good tape-recording of a native speaker will suffice. (Cassette tapes keyed to this appendix are available; contact Professor Thomas Huckin, Department of Humanities, University of Michigan, Ann Arbor MI 48109, USA.) Using a live native speaker as a model is best, however, because he or she can both serve as a good model and tell you how well you are imitating.

In choosing a model, of course, you should consider the fact that American English and British English differ significantly in their pronunciation. Indeed, the differences can give rise to serious misunderstanding. Whether you should learn American English or British English is a choice that only you can make, based on such factors as your previous training in English, your current training, and your career plans.

C.1
ON USING THIS APPENDIX
The contents of this appendix are based on "standard American English" pronunciation (i.e., the pronunciation used by national television newscasters in America). This fact should be kept in mind, especially with regard to the material on vowels, word stress, and sentence intonation, since the pronunciation of these sounds varies noticeably from one English dialect to another, even within the United States. (The consonants and consonant clusters, by contrast, are pronounced pretty uniformly across dialects.) Thus, if you want to use the material to develop a British English pronunciation, you should first have it checked over and appropriately modified by a native speaker of British English.

Diagnosis
The best way to use the material in this appendix is to first have your pronunciation checked by a native speaker of English, to see what sounds are giving you trouble. This can best be done by having the native speaker sit down with you and listen while you read aloud a passage such as the one printed below. (If you provide your listener with a photocopy of the passage,

it will be easier for him or her to make notes of the sounds that give you trouble.) Read over the passage ahead of time, so that you are familiar with it. Then, while you are reading it out loud, have your listener pay attention to your vowels and consonants, your word stress, and your sentence intonation. (The following checklist is designed to help your listener do this.)

Checklist for Listener of Reading Passage

While the student reads the passage aloud to you, listen to how he or she pronounces the various sound segments (vowels, consonants, consonant clusters, etc.). Make a note of those that are not pronounced correctly. On a second reading, try to note whether or not the student is putting stress on syllables that require stress and whether or not the student is applying proper intonation to the sentence as a whole. The following checklist gives more detailed suggestions of things to listen for.

1   *Sound segments.* Does the student pronounce the vowels correctly? The consonants? The consonant clusters (like the *skw* sound in *squeeze*)?

2   *Word stress.* Words containing more than one syllable typically have one syllable stressed or emphasized more than the others, as, for example, the first syllable of *criticized*. The stress should not only be located on the right syllable but should also be prominent enough to contrast sharply with the other syllables in the word, i.e., *CRITicized*. In general, does the student apply stress correctly?

3   *Sentence intonation.* Sentences should be uttered with fluctuating pitch, not in a monotone. In particular, the most important words of a sentence should be pronounced with relatively high pitch and the least important words with relatively low pitch. Questions requiring a "yes" or "no" answer should be pronounced with distinctly rising intonation at the end, whereas other questions as well as all statements should be pronounced with falling intonation at the end. In general, does the student apply correct intonation to sentences?

DIAGNOSTIC READING PASSAGE

Engineers are sometimes criticized for designing devices that do not use the full value of the heat they consume. Do engineers deserve to be so criticized? If not, what can they say in their defense? One response that engineers can make to the charge is that the Carnot principle puts a theoretical limit on any attempt to squeeze more than a certain maximum of work out of a unit of heat.

Actually, engineers deserve to be congratulated in having created, in the modern steam-cycle electrical generating station, an engine that approaches the Carnot cycle in the effectiveness of its use of heat. The efficiency of the steam cycle, which is what powers the generators, is achieved when hot, high-pressure steam that enters the turbine chamber is directed ultimately to a condenser whose temperature is low (about 27°C, or 80°F) and whose pressure is reduced to about one tenth that of the atmosphere. Great care has been taken to maximize the temperature and pressure differences between the boiler, or internal reservoir of heat, and the condenser, or external reservoir to which heat is rejected.[1]

Practice

Once your listener has identified your pronunciation problems, you can begin working to solve them. This requires steady practice. Spend at least several minutes every day practicing each of the sounds that give you trouble. Then, from time to time, try to have your pronunciation checked again by your native-speaker listener. Do this until the listener is satisfied that your pronunciation, though perhaps not perfect, will at least not cause communication problems for other listeners.

For practice material, we suggest using the drill material included in the remainder of this appendix. It is divided into three sections: (1) sound segments, (2) word stress, and (3) sentence intonation.

C.2

SOUND SEGMENTS

This section provides practice material for the vowels and consonants of the language, including consonants that often occur together as "consonant clusters." We have not attempted to account for all of the sound segments of English, but only for those that traditionally prove troublesome for nonnative speakers.

Each practice list is headed by a phonetic symbol drawn from common dictionary usage. This is purely a convenience, necessitated by the fact that any one particular English sound can often be spelled in two or more different ways.

Vowels

There are 14 vowel sounds in English that form the core of the vowel system and so should be mastered. Most of these are not "pure" vowels; rather, they are spoken with the tongue moving slightly, causing the vowel quality to change somewhat as the vowel is being articulated. In some cases, a so-called single vowel is actually more of a diphthong, or two vowel sounds spoken in quick succession.

The best way to master these vowels is to try to imitate a native speaker or someone whose pronunciation is pretty much like that of a native speaker. The following lists can be used for this purpose. If you can find a good speaker to serve as a model, ask her or him to pronounce the words listed under each vowel sound that gives you trouble. Repeat each word in turn, trying to imitate the speaker exactly. If you can't do it immediately, don't give up! Often, if you keep working at it, you can perfect a sound through gradual approximation.

VOWELS

/a/	/æ/	/ā/
rod	add	rate
módel	ángle	decáy
operátion	reáction	operátion
hot	graph	equátion
próblem	stándard	scale
beyónd	mathemátics	paint
fáther	mechánical	weigh

/ɔ/	/e/	/ē/
lawn	set	speed
call	véctor	wheel
haul	accélerate	équal
ought	énergy	yield
saw	éffort	mean
áwkward	prevént	adiabátic
more	compléx	free

/i/	/ī/	/o/
línear	size	cold
ímpulse	light	coat
equilíbrium	inclíned	mótion
scientífic	idéal	Carnót
índex	gýroscope	flow
withín	pólarize	télescope
bit	height	compónent

/au/	/oy/	/U/
now	noise	pull
abóut	destróy	push
down	emplóyer	book
loud	oil	refér
pówer	soil	úrgent
aróund	avóid	could
south	toy	good

/ū/	/ʌ/
rule	rúbber
redúce	ímpulse
joule	up

/ū/	/ʌ/
tool	indúction
Bernóulli	cóuple
humídity	nóthing
screw	númber

Sometimes the difference between two vowel sounds is quite small, and yet this difference can be the distinguishing factor between two entirely different words. For example, the words *age* and *edge* differ in their pronunciation only in how the vowel is pronounced: in *age* the vowel is pronounced with the tongue elevated to midheight in front and the jaw and tongue muscles somewhat tensed, whereas in *edge* the tongue is slightly lower in front and the muscles are more relaxed. Even though these movements differ only slightly, it is important that you make them properly if you want to be easily understood.

The following exercises are designed to help you articulate the slight differences between vowel sounds that can distinguish different words. These are all common words. The words in each pair differ only in how the vowel is pronounced. Practice one pair at a time, going down the list until you can feel a clear difference when making the two vowel sounds. Try to have a native speaker listen to you while you recite.

SOME VOWEL CONTRASTS

/ē/	/i/	/ā/	/e/
feel	fill	age	edge
seeks	six	sale	cell
each	itch	weight	wet
decéased	desíst	sprayed	spread
leap	lip	láter	létter
wheel	will	attáined	atténd
reach	rich	rake	wreck

/o/	/ɔ/	/Ur/	/ɔr/
so	saw	first	forced
flow	flaw	were	wore
cold	called	burn	born
coat	caught	shirt	short
hole	haul	confírm	confórm
coke	caulk	stirred	stored
		fur	4
		expert	export

Consonants

English consonants in general are produced by momentarily restricting the stream of sound as it flows through the mouth. Each consonant is distinguished basically by three factors: (1) where the restriction occurs, (2) whether the restriction is complete or not, and (3) whether or not the vocal cords are vibrating. For example, the *p* sound is produced by completely restricting the sound flow at the lips and stopping the vocal cords from vibrating. The *b* sound is produced in the same way, except that the vocal cords continue to vibrate. We can characterize this one difference as a difference of "voicing": the *p* sound in English is "unvoiced" whereas the *b* sound is "voiced."

The 17 most troublesome English consonants for nonnative speakers are *b*, *ch*, *d*, *f*, *h*, *j*, *l*, *p*, *r*, *s*, "soft" *th*, "hard" *th*, *v*, *w*, *y*, and *z*. They can be roughly described as follows:

b	Lips completely together, vocal cords vibrating. Example: *base*
ch	Body of tongue touching roof of mouth, then released; vocal cords not vibrating. Example: *check*
d	Tip of tongue touching roof of mouth just behind teeth, then released; vocal cords vibrating. Example: *detail*
f	Upper teeth touching lower lip; vocal cords not vibrating. Example: *fan*
h	Mouth slightly open, air flowing forcefully past vocal cords but vocal cords not vibrating. Example: *heat*
j	Like *ch* (body of tongue touching roof of mouth, then released) but vocal cords are vibrating. Example: *jet*
l	Tip of tongue touching roof of mouth behind teeth, sound stream flowing around sides of tongue; vocal cords vibrating. Example: *law*
p	Like *b* (lips completely together, then released) but vocal cords are not vibrating. Example: *push*
r	Lips curled out, extended forward, and held closely together, then quickly retracted; vocal cords vibrating. Example: *ratio*
s	Body of tongue almost touching roof of mouth behind teeth, allowing air to pass with hissing sound; vocal cords not vibrating. Example: *simple*
sh	Tongue almost touching roof of mouth but allowing air to pass continuously; vocal cords not vibrating. Example: *shut*
"soft" *th*	Tip of tongue held lightly between teeth, allowing air to pass; vocal cords not vibrating. Example: *thermal*

"hard" th	Like soft *th* but with vocal cords vibrating. Example: *this*
v	Like *f* (upper teeth against lower lip) but vocal cords are vibrating. Example: *value*
w	Lips briefly extended forward in the form of an *O*, then quickly retracted; vocal cords vibrating. Example: *wave*
y	Body of tongue tensed, held close to roof of mouth but not touching it, thus allowing air to pass; vocal cords vibrating. Example: *yield*
z	Like *s*, but with vocal cords vibrating. Example: *zero*

These are not complete or exact descriptions, of course, but only rough sketches; better descriptions, including diagrams of the articulatory movements involved, can be found in the references listed at the end of this appendix.

No description, however, no matter how exact it may be, will by itself enable you to achieve perfect pronunciation. The best way to work on your English pronunciation, as we mentioned earlier, is to try to imitate the pronunciation of a native speaker or a nonnative speaker whose pronunciation is very good. The following lists of words are best used this way. If you can find a suitable speaker of English to serve as a model, ask her or him to pronounce the words listed under each consonant sound that gives you trouble. Repeat each word in turn, trying to imitate the speaker exactly.

CONSONANTS

/b/	/ch/	/d/
base	check	detail
bus	reach	addítion
bend	touch	redúce
rúbber	charge	évidence
lab	catch	dock
absórb	arch	door
cúbic	chápter	áttitude

/f/	/h/	/j/
fan	heat	jet
fact	hot	join
fail	humídity	lógic
relíef	dehýdrate	gýroscope
laugh	unhóok	edge
síphon	héavy	énergy
44	hope	judge

/l/

law
línear
rélative
eléctron
parallél
lével
lówer

/p/

push
pówer
tip
pump
mápping
prepáre
point

/r/

rátio
resístance
rócket
rígid
súmmary
prefér
wire

/s/

símple
sécond
scíence
force
subtráct
66
cénter of mass

/sh/

shut
sheet
percússion
coeffícient
shéllfish
flash
relátion

soft /th/

thérmal
théorem
length
width
math
thin
earth

hard /th/

this
that
there
ráther
lathe
those
óther

/v/

válue
vápor
éven
évery
solve
conservátion
vívid

/w/

wave
work
awárd
rewíre
pówer
dwélling
twist

/y/

yield
yard
únit
útilize
únion
cálculate
ángular

/z/

zéro
zone
váporize
ózone
phase
éasy
zígzag

As with some of the vowels, the differences between certain consonants can be quite small. For example, as we noted before, the sounds *b* and *p* are articulated in exactly the same manner but for one difference: *b* is voiced and *p* isn't. The two words *bull* and *pull* are pronounced differently only insofar as one begins with the vocal cords vibrating and the other doesn't.

The following lists represent some of the most troublesome consonant contrasts for nonnative speakers. In all of the pairs listed (except for hard *th* and soft *th*), the two words are pronounced exactly alike except for the particular consonant being contrasted. Practice with one pair at a time; try to make the difference in consonants as clear as you can.

SOME CONSONANT CONTRASTS

/l/	/r/		/f/	/v/
lock	rock		fan	van
lamp	ramp		few	view
law	raw		life	live
flame	frame		fee	V
glow	grow		proof	prove
colléct	corréct		face	vase
tool	tour		first	versed
lével	léver		half	have
false	force			
rúbble	rúbber			

soft /th/	hard /th/	/s/	/sh/	/z/
thésis	these	sip	ship	zip
north	northern	sue	shoe	zoo
bath	bathe	see	she	Z (zee)
thin	this	said	shed	Z (zed)
éther	éither	face	fácial	phase
width	withering	class	clash	
wrath	rather	loose		lose
thérapy	their	hats	hatch	
		fúrnace	fúrnish	

/p/	/b/		/j/	/y/
pan	ban		jet	yet
P	B		joke	yoke, yolk
push	bush		jarred	yard
pump	bump		jot	yacht
rip	rib		jeer	year
cámper	cámber			
plaque	black			
pole	bowl			

Consonant Clusters

Many English words contain strings of consonants, or consonant clusters. For example, the word *scrapes* begins with three consecutive consonant sounds (s, k, r) and ends with two more (p, s); if we spelled the word according to the way it's pronounced, it would look like this: *skrāps*.

There are more than 40 different kinds of consonant clusters in English, of which 30 commonly cause trouble for nonnative speakers: *bl, br, byu, dr, fl, fr, fyu, gl, gr, dl, kr, ks, kt, kw, kyu, pl, pr, ps, sk, skr, sl, sp, spl, spr, st, str, sw, thr, tr,* and *ts.* The following lists are provided to give you practice with them. If you have trouble pronouncing a particular consonant cluster, here's a trick you can try. Pretend there's a space between the consonants and then work on them separately, gradually reducing the space so that it disappears. For example, let's say you're having trouble with the *bl* cluster at the beginning of a word like *blue.* Divide the word into two parts with a space between the *b* and the *l,* like this: *b   lue.* Pronounce each part separately, pausing briefly in between. When you feel you have command of each consonant separately, gradually begin shortening the pause between them until, in effect, there is no pause:

b   lue

b   lue

b lue

b lue

b lue

b lue

blue

There, you've got it!

CONSONANT CLUSTERS

/bl/	/br/	/byu/
blue	bridge	beáuty
bleach	break	bureáucracy
bláckboard	cálibrate	bútane
blíster	bróken	abúse
blast	álgebra	debut
block	broad	imbúe
blówtorch	bright	bútyl

/dr/

drive
draw
draft
address
drill
drum
drý cell

/fl/

flow
flúid
flat
éffluent
flux
defláte
flight

/fr/

free
frame
fríction
refrígerate
refráction
frózen
infraréd

/fyu/

few
fúsion
fúture
refúel
sulfúric
refúse
diffúsion

/gl/

glass
glue
gleam
glow
glaze
conglomerátion
Éngland

/gr/

grow
grádual
ingrédient
mílligram
grádient
grew
green

/kl/

clear
clamp
clutch
núclear
decláre
cýclotron
conclúde

/kr/

crack
cross
mícroscope
incréase
crítical
crúde
crane

/ks/

lacks
makes
áxiom
óptics
x́-ray
ínflux
compléx

/kt/

efféct
cóntact
reáctor
véctor
duct
predíct
liked

/kw/

quálity
quart
equátion
báckward
quótient
equípment
requést

/kyu/

cube
perpendícular
cúrious
mércury
Curie
cúcumber
excúse

/pl/

plan
plane
plot
unplúg

/pr/

príntout
proof
prime
propórtions

/ps/

ellípse
keeps
hopes
pumps

/pl/

supplý
expláin
súrplus

/pr/

compréssion
recíprocal
prevént

/ps/

sýnapse
tapes
loops

/sk/

skill
scálar
schédule
disk
task
Pascál
skýscraper

/skr/

screw
scratch
describe
discréte
súbscript
scrape
scrawl

/sl/

slant
slow
slip
slot
slope
slate
slice

/sp/

spécial
space
spárk
spíral
respónd
speak
expéct

/spl/

splash
splice
split
expláin
explóre
explóde
displáy

/spr/

spring
spray
spread
sprócket
dispróve
exprópriate
expréss

/st/

steel
stándards
státic
elástic
cónstant
cost
resísts

/str/

strong
strúcture
strain
stróntium
extréme
restríct
abstract

/sw/

swing
sweep
switch
persúade
unswépt
Swéden
swell

/thr/

three
through
throw
thróttle
thrust
unthréaded
thréatened

/tr/

try
true
tráffic
tráining
trigonómetry
Detróit
contról

/ts/

únits
convérts
its
creátes
sátellites
debts
prétzel

C.3

WORD STRESS

In spoken English the different vowel sounds of a word are often pronounced with different degrees of emphasis, or "stress." The word *vector,* for example, has two vowel sounds, the first of which is given much greater stress than the second: in pitch, in length, and in loudness. That is, a stressed vowel is normally pronounced at a higher pitch and is normally longer and louder than an unstressed vowel. Conversely, an unstressed vowel is pronounced with lower pitch and is shorter and weaker; indeed, an unstressed vowel often even loses its identity as a particular type of vowel. For example, the *o* vowel in *vector,* being unstressed, is pronounced not as an *o* but rather as a neutral, unidentifiable, "weak" vowel. (We shall hereafter use the symbol ə for this type of vowel.) Putting all these facts together, the word *vector* should be pronounced like this:

$$V \quad E \quad C$$
$$t \quad ə \quad r$$

Try pronouncing it yourself. Make sure you produce a sharp difference between the two vowels, even if you have to exaggerate. Nonnative speakers are often reluctant to make the difference as sharp and clear as it should be, so force yourself if necessary.

There are actually not two but three degrees of stress in English. The example given above involved only two vowels, and so two degrees of stress were sufficient to describe it. Longer words, however, often invoke a third, intermediate degree, which we shall call "partial stress." Thus, the three degrees of word stress in English are full stress, partial stress, and no stress. Let us take the word *vectorial* to illustrate the difference. This word is pronounced with four vowel sounds, the second of which receives full stress: *vecTORial.* The first and third vowels receive partial stress, and the fourth vowel receives no stress at all. Putting these facts together, we can depict the correct pronunciation of *vectorial* roughly as follows:

$$T \quad O \quad R_i$$
$$vec \quad\quad ə\ l$$

Try it yourself.

These two examples illustrate several important points about the English stress system. First of all, there is normally only one fully stressed vowel in a word. Secondly, the location of this fully stressed vowel can vary from one word to the next, even when two words are related (as are *vector* and *vectorial,* for example). Thirdly, the spelling system used for written English is not a reliable guide for pronunciation. Written vowels are not always pronounced (e.g., the underscored *e* in *refin*e*d*). Those vowels which are pronounced but which are unstressed often are not given the pronunciation suggested by the spelling (e.g., the *o* in *vector* is not pronounced like an *o*). Two words with the same spelling may have different stress patterns (e.g., the word *contrast*: when used as a noun it is pronounced *CONtrast;* when used as a verb, however, it is pronounced *conTRAST*).

EXERCISE C-1   The following is a list of words commonly used in science and technology. For each one, indicate by means of an accent mark (´) which vowel you think should be fully stressed. Do not use the dictionary while doing the exercise, though you may use it later to check your answers. The first two have already been done for you.

párticle	arithmetic	dynamic
mátter	resistance	automobile
energy	calculate	minimum
cyclotron	machine	potential
biology	technical	pressure
reverse	process	variable
physics	turbine	

We mentioned above that even when two words are obviously related, they may not have full stress on the same vowel. This is particularly striking in the case of certain two-vowel words that are spelled exactly alike, such as the example we gave above regarding the words *conTRAST* (a verb) and *CONtrast* (a noun). In such cases, the verb form has full stress on the second vowel and the nonverb form (usually a noun) has it on the first vowel. Some other examples of this type are as follows:

VERB	NONVERB
conflíct	cónflict
decréase	décrease
extráct	éxtract
impórt	ímport
inclíne	íncline
incréase	íncrease
objéct	óbject
perféct	pérfect
permít	pérmit
presént	présent

progréss   prógress
recórd   récord
rejéct   réject

A more common occurrence concerns two or more words having the same stem but different suffixes, so that they are related but of different length. In such cases, the longer word often has its fully stressed vowel to the right of where it is in the shorter word. For example, consider our old friends *vector* and *vectorial*. These words are obviously related, both having the same stem. However, the shorter form has full stress on the first vowel (*véctor*) whereas the longer form has it on the second vowel (*vectórial*). In other words, the location of the full stress has shifted one step to the right for the longer word.

Although this kind of stress shift is a common phenomenon in English, there are many exceptions to the rule. That is, there are many cases where one member of a related word-pair is longer than the other and yet has the same stressed vowel, e.g., *resíst* and *resístance*. In fact, there are even a few cases where the longer word has its stressed vowel to the *left* of where the stressed vowel is in the shorter word.

EXERCISE C-2 Below are some sets of related words. Examine each and see if you can correctly mark the location of the fully stressed vowel. (The first six words are already marked for you.) As in the previous exercise, do not use your dictionary until *after* you've completed the exercise.

périod	dénse
periódic	dénsity
periodícity	densitómeter
instinct	supplement
instinctive	supplementary
argument	social
argumentative	socialist
argumentation	socialistic
distribute	electron
distribution	electronic
electric	plastic
electricity	plasticity

sequence	molecule
sequential	molecular
develop	alternate
development	alternative
developmental	alternation
insulate	component
insulation	componential
microscope	probable
microscopic	probabilistic
separate	economy
separable	economic
separation	economize
separability	economically
category	concept
categoric	conceptual
categorize	conceptually
categorically	conceptualize
categorization	conceptualization

Although all English words have a standard stress pattern (as indicated in any good dictionary), there are cases where this pattern should be violated for purposes of pointing out a contrast to some other word nearby. Take, for example, the word *internal*. Normally, this word is pronounced with the second vowel receiving full stress: *intérnal*. However, if you recall the last sentence of the diagnostic reading passage, you'll remember that the writer had deliberately set up a contrast between the words *internal* and *external*.

> Great care has been taken to maximize the temperature and pressure differences between the boiler, or *internal* reservoir of heat, and the condenser, or *external* reservoir to which heat is rejected.[1]

In such a case, since the difference between these two words lies in the first vowel sound of each (*in-* versus *ex-*), it is appropriate, indeed required, to stress these two vowels instead of the normal second vowels. In other words, in order to point out the contrast between these two words in a context such as this one, you should pronounce them: *ínternal* and *éxternal.*

C.4
NOUN COMPOUNDS

When two nouns are strung together and used as a single word, or noun compound, full stress is usually applied only to some vowel in the first noun, not the second. For example, consider the nouns *gravity* and *meter*. When these words are pronounced independently, they are stressed like this: *grávity* and *méter*. When they are joined together, however, only the first word is stressed: *grávity meter*. Some other examples are *círcuit breaker, vácuum tube, spárk plug, aír conditioner, Bernóulli effect, pháse reaction, ignítion system, héat loss, óil pump,* and *fíre extinguisher*. An example from the diagnostic reading passage is *génerating station.*

This stress pattern is usually found only when the two words are in fact nouns. Do not expect to find it for adjective plus noun combinations, even when such a combination may be commonly used as a "single" word. For example, the term *chemical reaction* is commonly used as a single word, just as *phase reaction* is. Since the word *chemical* is an adjective, not a noun (note the adjectival suffix *-ical*), the combination is accordingly pronounced with stress on both words: *chémical reáction*. Some other adjective plus noun examples are *hydraúlic préss, inclíned pláne, mechánical shóvel, kinétic énergy,* and *Bóyle's láw.*

EXERCISE C-3 The following are two-word combinations used as single words. Some are noun compounds, others are adjective plus noun combinations. Indicate by means of one or more accent marks what the stress pattern should be. (Be sure to put the accent on the correct vowel when there is more than one vowel in the word!) Try pronouncing each combination or compound while you are doing the exercise.

water softener	electric motor
molecular weight	fracture mechanics
edge dislocation	network molecules
energy gap	linear density
photographic film	heat pump
thermal conductivity	vacuum cleaner
radiation detector	nuclear reactor

shock absorber	particle accelerator
polynomial equation	ionic bond
valence electrons	x-ray diffraction
periodic table	Bragg's law
civil engineering	computer specialist
crystal-field theory	plane-polarized light

C.5

SENTENCE INTONATION

Correct pronunciation of English includes more than just sound segments and word stress; it also includes the intonation of sentences. English sentences are typically spoken with up-and-down pitch variations. These variations in pitch have important communicative functions, and so listeners are trained to listen for them. In particular, the most important words of a sentence are normally pronounced with relatively high pitch and the least important words are pronounced with relatively low pitch. Also, the difference between types of sentences is marked by a distinctive pitch pattern at the end: questions requiring a "yes" or "no" answer are pronounced with rising intonation, whereas all statements (and all other types of questions) are pronounced with abruptly falling intonation at the end.

To simplify the following discussion, we shall assume that English sentence intonation is constructed from two basic tones: high and low. These terms should be interpreted in a relative sense, not in any absolute one. That is, every native speaker of English makes a distinction between these two tones, though the exact pitch of each tone differs from speaker to speaker. In fact, even for the same speaker either of these tones can be pronounced at somewhat varying levels of pitch.

Signaling Important Words

One of the principal functions of sentence intonation is to draw the listener's attention to the most important words of the sentence. What do we mean by "important words"? These are usually any of the following:

1   Nouns, verbs, adjectives, or adverbs that represent *new information,* that is, information that has not been mentioned before or even suggested.

2   Any word or word prefix that represents an *opposition or contrast* of some sort.

3   Any word that the speaker feels deserves *special emphasis.*

We can illustrate the first two of these categories with examples drawn from the diagnostic reading passage. The passage begins like this:

> Engineers are sometimes criticized for designing devices that do not use the full value of the heat they consume. Do engineers deserve to be so criticized? If not, what can they say in their defense?

In the first sentence, all of the nouns, verbs, adjectives, and adverbs, of course, represent new information: this is where the concepts conveyed by these words are first introduced. The speaker can signal the fact that this is new information—and thus important information—by pronouncing each of these words (i.e., the italicized ones) with high pitch:

> *Engineers* are *sometimes criticized* for *designing devices* that do *not use* the *full value* of the *heat* they *consume*.

(Conversely, each of the other words in the sentence should be pronounced with low pitch.)

In the second sentence, only one word (*deserve*) represents new information:

> Do engineers *deserve* to be so criticized?

The remaining words of this sentence either represent given information (*engineers, so criticized*) or are auxiliaries (*do, to be*).

In the third sentence, the word *not* is an important word because it represents an opposition or contrast:

> If *not*, . . .

Therefore, it should be pronounced with high pitch. Other words that should also be given high pitch in this sentence (because they represent new information) are *what, say,* and *defense*.

Signaling the Sentence Type

The other principal function of sentence intonation is to alert the listener to what type of sentence it is. In particular, sentence intonation is used to distinguish yes or no questions (i.e., those questions requesting a "yes" or "no" answer) from all other sentence types: yes or no questions are pronounced with rising intonation (low → high) and all other sentence types (statements, commands, exclamations, other types of questions, etc.) are pronounced with abruptly falling intonation (high → low).

Unlike the signaling of important words, which can occur anywhere in the sentence, the signaling of sentence type is normally done at or near the end of the sentence. Specifically, *the signaling of sentence type begins with the last important word in the sentence.* For example, consider the second sentence in the reading passage:

>Do engineers *deserve* to be so criticized?

This is a yes or no question, and so it requires rising intonation. The last important word in the sentence is *deserve* (which happens to be the *only* important word in this case), and so the rising intonation pattern should begin on this word (to be more exact, on the *fully stressed vowel* of this word):

>Do engineers de ^{s}erve to be so critized?

In other words, the sentence is pronounced with low pitch until the second vowel of the word *deserve*, where it rises quickly to high pitch. This high pitch is then maintained to the end of the sentence.

The third sentence, however, is not a yes or no question; accordingly, it requires falling intonation. The last important word in this sentence is *defense,* which, like *deserve,* is normally stressed on the second vowel. Thus, the falling intonation pattern should begin on this second vowel. That is, the speaker will begin to pronounce this vowel sound with high pitch but will quickly let the pitch drop to a low level at the end of the vowel:

>. . . de^{fe}\nse.

If we combine this end-of-sentence signaling with the signaling associated with important words, we end up with the following intonation contour for this sentence:

>If ^{n}o$_{t,}$ what   can they ^{say} in their de^{fe}\nse?

In other words, the speaker will pronounce all four of the important words with high pitch but will make sure that the pitch on the last one drops abruptly from high to low. (The pitch on *not* should also drop somewhat because it marks the end of a clause.)

Signaling the Items of a List

A third important use of intonation occurs when a speaker is reciting a list. In such cases, each item of the list (except the last) should be pronounced with a *sustained* high pitch. For example, if we have a sentence like this:

>The three states of matter are solids, liquids, and gases.

we note that it contains an unformatted list of three items: *solids, liquids, gases.* The first two items should each be pronounced with sustained high pitch (i.e., high pitch not only on the first, stressed vowel but also on the second, unstressed vowel); this makes it immediately clear to the listener that a list is being presented. The third item, on the other hand, receives high pitch only on the first vowel, not on the second. The resulting intonation pattern will be roughly as follows:

$$\text{solids,} \quad \text{liquids,} \qquad\qquad \text{gas}^{\text{e}}_{\text{s.}}$$
$$\text{. . .} \qquad\qquad\qquad \text{and}$$

Notice how the words *solids* and *liquids,* in contrast to *gases,* are pronounced at a sustained high pitch level even through their unstressed second vowels.

A special kind of list is one that asks the listener to make a choice from among two or more items. Here, too, each item of the list except the last should be pronounced with sustained high pitch. Consider, for example, this sentence:

Is the test *destructive* or *nondestructive?*

Normally the word *destructive* is pronounced with high pitch only on the stressed second vowel:

$$\text{de}\ ^{\text{struc}}\ \text{tive}$$

In the sentence above, however, it is part of a two-item list and so should be pronounced differently, with sustained high pitch carrying through the last, unstressed vowel:

$$\text{. . . de}\ ^{\text{structive}}\ \text{. . .}$$

This way, the listener is alerted right away to the fact that he or she is being asked to make a choice between two possible answers. To make matters complete, we can describe the intonation contour for the full sentence as being approximately like this:

$$\text{Is the test de}\ ^{\text{structive}}\ \text{or}\ ^{\text{non}}\ \text{destructive?}$$

NOTE: If you listen to a native speaker's sentence intonation, you will find that it actually contains considerably more variation of pitch than we are depicting here. Intonation can be used in many different ways to express many different—often subtle—shades of meaning. Unfortunately, it takes years of constant interaction with native speakers to master such subtleties.

What we are emphasizing here are the more basic aspects of English inton-ation, those which form the foundation of the intonation system and which, happily, can be readily mastered by applying the principles described above.

EXERCISE C-4   Convert each of the following sentences into a form that indicates the intonation pattern it should have. You may assume that the important words of the sentence are those appearing in italics. (The first sentence has already been done for you.)

A   *No silicon analog* of *graphite* is *known.*

Answer: No sil $_i$ $_{con}$ ^{an} $_{a}$log of ^{graph} $_i$ $_{te\ is}$ ^{kno} ^{w} $_{n.}$

B   The *range* of *speeds observed* for *chemical reactions* is *enormous.*

C   Is there *good evidence* that *forces exist* between *nonpolar molecules*?

D   *What* is the *weight* of an *aluminum* atom?

E   *Every applied scientist* and *engineer—mechanical, civil, electrical,* or *other*—is *vitally concerned* with the *materials available* to *him* or *her.*

F   *Which* is a *simpler concept: speed?* or *velocity?*[2]

EXERCISE C-5   Read over the following passage and underline what you think are the most important words. Then, using these underlined words as aids, convert each sentence into a form indicating the intonation pattern it should have. When you have finished, check your result by reading it aloud to your teacher or native-speaker listener.

> How big is a proton? Are the proton and the neutron, which weigh about the same, identical except for the neutron's lack of electrical charge? Are these particles hard, with sharp boundaries, or soft, with fuzzy edges? The central position of protons and neutrons in the subatomic world makes it crucial to understand in detail the structure, shape, and size of these particles. Detailed knowledge about these important particles provides a testing ground for the-ories explaining the subatomic world.[3]

REFERENCES

1   B. Hiatt, "Heat into Work: The Second Law of Thermodynamics," *The Research News* **28**(11/12):5 (1977).

2   L. H. Van Vlack, *Elements of Materials Science and Engineering,* 3d ed., Addison-Wesley, Reading, Mass., 1977, p. 3.

3   B. Hiatt, "What's the Matter: Particle Physics at the University of Michigan," *The Research News* **26**(3/4):12 (1975).

ADDITIONAL READING

A. Baker, *Ship or Sheep? An Intermediate Pronunciation Course,* Cambridge University Press, New York, 1977.

M. E. Clarey and R. T. Dixson, *Pronunciation Exercises in English,* Regents, New York, 1963 (tapes available).

J. Morley, *Improving Spoken English,* University of Michigan Press, Ann Arbor, 1979 (tapes available).

D.L.F. Nilsen and A. P. Nilsen, *Pronunciation Contrasts in English,* Regents, New York, 1973.

E. C. Trager, *PD's in Depth,* ELS, Portland, Ore., 1982 (tapes available).

INDEX

a/an, 353–365

Abbreviations, 410

Ability, expressing degrees of, 437–438

Abstract:
description and examples of, 180–184, 334–335
descriptive, 181
informative, 181–182
and summary, compared, 182

Accented syllables, 525

Acronyms, 410, 417

Action verbs, 446

actually, 344
vs. *currently* or *presently,* 347

Adjective clauses (*see* Relative clauses)

Adjectives:
demonstrative, 335–337, 379–381
using for cohesion, 418–420
normally used with *the,* 365, 367–368
ordinals, 368
others, 368
superlatives, 368

Admissions applications, 257–274

Adverbial phrases, using for emphasis, 329–332

Adverbs, conjunctive, 341–352
using for emphasis, 329–332

after all vs. *finally,* 347

all ready vs. *already,* 455

although, 329, 346–347

Ambiguity, avoiding, 336–337

American English pronunciation, 513

American Society for Engineering Education, 5–8, 10–11

Analogies, 476
using, 293–297

Anaphoric reference, 418–420

Anaphoric use of the definite article, 372–373

Antonyms, 470

Apostrophe, the, 354, 490

Application, letters of, 257–274

Applying for admission or employment, 257–274

Arguments:
basic strategies for, 64–67
of character and credentials, 64–67
criteria in, 75, 77–84, 202
effectiveness of, 59–61, 64–65, 71–73, 75–80
of emotion, 64–67
expectations about, 59–61
of fact, 67–69, 70–73, 203
of logic and reason, 64–67
making claims in, 59–61, 67–75
organization of, 67–73, 75–77
pattern for, 75–77
of policy, 69–73, 203
providing proof, 59–61, 67–80
relation between arguments of fact and policy, 71–73
types of, 67–73

Aristotle, 64

Article:
definite *(the),* 365, 367–387
used for cohesion, 418–419
indefinite *(a/an),* 353–365

Articles, journal (*see* Journal articles)

as, 347

as follows, 420–421

Associated nominals, 424–426

Audience analysis:
for oral presentations, 47–55, 164–166, 168, 170
procedure for, 52–55
for written documents, 47–55

Audiences:
    analysis of, (*see* Audience
        analysis)
    characteristics of, 50–55
    complaints from, about technical
        writing, 9–10, 53
    designing communication for, 47
    goals of, 54–55
    managers as, 51–55
    nature of, 50–55
    needs of, 47, 54–55
    values of, 54–55
Authoritative sources, consulting for
    unfamiliar words, 476–479
Auxiliary verbs:
    modal, 429–440
    omitting in relative clauses, 396

Background knowledge, 290, 295
Bar charts and bar graphs,
    132–134, 172
Barton, Lisa, 119
because, 329, 346
Becker, Alton L., 75, 88
besides, 342
Blocked letter format, 245–246
Briefings, oral (*see* Oral
    presentations)
British Ecological Society, 334–335
British English pronunciation, 513
Business letters, 237–275
    basic formats, 240–247
    examples of, 367–369
by contrast, 345

can, 429–438
Capital letters, using, 357–358,
    493–494
Career planning, 258–259
Cataphoric use of the definite arti-
    cle, 374
Claims in arguments, 59–61, 67–75

Clarity:
    in sentences, 312–340
    in visual aids, 142–146
Clauses:
    independent, 485–487
    relative (*see* Relative clauses)
    subordinate, 325–326, 329–332,
        349–350
Coefficient, Remrak, 296
Cognate forms of words, 478–479
Coherence:
    in paragraphs, 17
    using noun compounds to help
        establish, 417–418
Cohesion, 407–428
Colon, the, 349–352, 485–488
Combining sentences, 322–328
Comma, the, 349–351, 353, 477,
    481, 485–487
Comma splices, 322
Commands, signaling in speech,
    531–532
Communication:
    audiences for, 47–55
    complaints about, 10
    job-related tasks in, 9–10, 167
    routes of, 53–54
    stages of, 53–54
    uses of, 53–54
    (*See also* Technical
        communication)
comparing to (with), 348
Complaint, letters of, 249–252
    responding to, 252–254
complement vs. *compliment,* 455
Complements, verb, 449
Compound nouns (*see* Noun
    compounds)
Concatenated nouns (*see* Noun
    compounds)
Conciseness, 333–339
Conclusions:
    in journal articles, 279–284
    in long informal reports, 193–195
    in oral presentations, 169, 173

conclusively vs. *in conclusion,* 348

Conditional statements, making, 435–436

Conjunctions:
  coordinating, 349–350, 485–486
  subordinating, 341–342, 346–350
    using for emphasis, 329–332

Conjunctive adverbs, 341–352
  using for emphasis, 329–332

Connectives, 341–352, 472
  using for emphasis, 329–332

consequently, 342

Consonant clusters, pronouncing, 522–524

Consonants, pronouncing, 518–521

Consultants, using, 476

Contextual clues, using while reading, 459–482

Continuation pages, 246

Contracts, 336

Contrast, indicating in speech, 530–531

Conversational expressions, common, list of, 495–511

Conversational "rules," 321–322

conversely, 329

Coordinating conjunctions, 349–350, 485–486

Copy line in letters, 247

could, 429–438

Countability, 353–365, 477
  of nouns, 477

Countable nouns, 354–364, 381
  informal usage of, 363

Counters, 359–360, 365

Credentials and character in arguments, 64–67

Credibility, professional, 333–335, 341, 453

Criteria:
  in arguments, 75, 77–84, 202
  formal, 79

Criteria *(Cont.):*
  general, 78
  informal, 79
  special, 78

Dangling modifiers, 322

Dashes, 485, 488

Data presentation in visual aids, 125–128, 130–148

Davis, Richard M., 6, 9–10

Day, Robert, 276, 280

Definiteness [*see* Article, definite *(the)*]

Definitions:
  contextual, 459–461
  dictionary, 477
  formal, 356–357
  using, 296–297

Demonstrative adjectives, 335–337, 379–381
  using for cohesion, 418–420

Description using visual aids, 123–125

despite, 329

Details:
  emphasizing, 321–322
  organizing, 180–185

Dictionary, the:
  consulting, 459–461, 476–479, 492
  selecting, 477
  using, 342, 356, 454

Direct object of sentence, 304, 306–307, 310–312

Discussions, structuring of, 184–185

Dodge, Richard W., 115

Duplicate pronouns, avoiding, 393–394

each, using, 447–448

Editing for emphasis, 321–340

Editing sentences, 303–319
  flow chart for, 313
  procedure for, 312–317
effect vs. *affect,* 455
Emotion in arguments, 64–67
Emphasis:
  creating through repetition,
    408–409
  editing for, 321–340
  indicating in speech, 530–531
  in relative clauses, 397–398,
    400–401
Employment applications, 257–274
Enclosure line in letters, 247
End-of-line hyphenation, 491–492
Equative sentences, 310–311
Equative shift, 310–311
Equipment for oral presentations,
  174
Errors:
  checking for, 341–345
  sequencing, 453
  in spelling, 453–457
  substantive, 453
even so, 343
even though, 346
Event verbs, 446
Examples, using, 293–297
Exclamations, signaling in speech,
  531–532
Expectations in arguments, 59–61
Expressions, common conversa-
  tional, list of, 495–511
Extraposition of relative clauses,
  391

Fact, arguments of, 67–73
Feasibility reports, 199–211
  criteria for, 202–205
  foreword to, 203
  summary of, 205
  title page of, 200–210
  (*See also* Technical reports)
Festinger, Leon, 87

Field testing, 299–300
Final reports, 211
finally vs. *after all,* 347
Flow chart for article usage,
  379–381
Foggy language, avoiding, 333–339
foreword vs. *forward,* 455
Forewords, 87–104
  assumptions in short form,
    98–99
  choosing full or short form,
    101–102
  general form for, 87–97,
    101–102
  introducing a problem: full form,
    88–91
    short form, 98–102
  vs. introduction to the discussion,
    191–192
  outline of full form, 97
  outline of short form, 99
  samples of, 89, 98, 100, 105,
    107
  short form for, 98–102
  of technical reports, 87–104,
    179, 194
Formal report, 198–199
  title page for, 199–201
  (*See also* Technical reports)
Formats, basic letter, 241–247
  blocked, 245–246
  semiblocked, 244
  unblocked, 242–243
Formatting, 155–162
French, 347
Full-form noun phrase repetition,
  408–409
furthermore, 329, 342

Generalizations, making, 441–442
generally, 346
Generic reference, 356, 364, 372
Gestures, physical, 322
Gilbert, Ed, 5–6

"Given" information, 289–290, 304–305, 377
  vs. "new" information, 310–312
Gobbledygook, 8, 327–328
Grammatical functions of words, determining, 462–464

Headings, using key words in, 293
"Heavy" noun phrases, 306–309
Hierarchically structured text, 293, 297–298
Highlighting important points, 125
however, 329, 342
Hyphen, the, 354–356, 490–492
Hypotheses:
  devising alternative, 283
  testing, 283
Hypothetical statements, making, 435–436

Idiomatic expressions, common, list of, 495–511
if, 346, 435–436
Illustrations, using, 293–297
Imagery, 290
Implied uniqueness, 386–387
  using *the* for, 378
Important words, signaling in speech, 530–531
in addition, 342
in comparison, 345
in conclusion, 344
  vs. *after all,* 347
  vs. *conclusively,* 348
in fact, 344
in general, 329, 344
in other words, 343
in particular, 344
  vs. *particularly,* 348
in practice, 345
in theory, 345
indeed, 344
Indefinite article *(a/an),* 353–365

Independent clauses, 485–487
Indirect object of sentence, 304, 306–307, 311–312
  shift of, 311–312
Inference, strong, 282
Inflection, verbs with irregular forms of, 449–451
Informal usage of countable nouns, 363
Information:
  "given," 289–290, 304–305, 377
    "new" information vs., 310–312
  ordering, 303–319
    using noun compounds to facilitate, 417
  parenthetical, 351–352, 487–488
  selecting, 287–301
information as an uncountable noun, 364
ing complements, verbs taking, 449
ing-form relative clauses, 399, 446
Initials line in letters, 247
instead, 345
instead of, 347
Instructions, writing lists of, 326–327, 336
Intensifiers, 322
  using for emphasis, 329–332
Intonation, sentence, 322, 397–398, 530–534
Introduction vs. foreword, 191–192
Introductions:
  assumptions in short form, 98–99
  general or full form for, 87–97
    outline of, 97
  guide for choosing full or short form, 101–102
  introducing a problem: full form, 88–91
    short form, 98–102
  of letters, 103
  in oral presentations, 169, 171–173

Introductions *(Cont.)*:
  samples of, 89–90, 100–101,
    105
  of scientific or technical articles,
    103–104
  short form for, 98–102
    outline of, 99
Introductory clauses, processing of,
  308–309
Introductory subordinate elements,
  487
Irregular verbs, 448–451
it, 347
it vs. *this,* 419–420
Italics, using, 492–493
its, 490
its vs. *it's,* 455

Jargon, 290, 297
Job application letter, 257–274
Journal articles, 275–285
  Discussion section, 280–284
  Introduction, 276–278
  Materials and Methods section,
    278–279
  Results section, 279–282

Key words, 291–294, 322
  learning the meaning of, 476
Kimel, William R., 7–8

Latin, 465
lead vs. *led,* 344, 456
Letterhead stationery, 246
Letters, 237–275
  basic formats, 241–247
  continuation pages, 246
  copy line, 247
  enclosure line, 247
  initials line, 247

Letters *(Cont.)*:
  letterhead stationery, 246
  types of: application for
    employment or admission,
    257–274
  complaint, 249–252
    responding to, 252–254
  request, 252–255
    responding to, 254–256
  transmittal, 247–250
"Light" noun phrases vs. "heavy"
  noun phrases, 306–309
Line drawings, 137–138
Line graphs, 132–133
Listing instructions, 326–327, 336
Lists:
  intonation used with, 532–533
  in paragraphs, 27, 298
  punctuating: formatted, 489
    unformatted, 488–489
  types of, 36
    fully formatted, 36
    partially formatted, 36
    unformatted, 36–37
Logic and reason in arguments,
  64–67
Long informal report, 189–198
  conclusion of, 193–195
  introduction to the discussion,
    191–192
  recommendations in, 193–195
  (See also Technical reports)

MacIntosh, Fred H., survey by,
  9–10, 333
Main points:
  emphasizing, 321–322
  stating, 179
Managers:
  characteristics of, 51–52
  complaints about technical writing
    from, 9–10
  faulty assumptions about, 51

Managers *(Cont.):*
  importance of, 51–53
  nature of, 51–55
  variety of roles, 51–52
Manuals, writing, 378
Manuscript, field testing the, 299
Mathes, J. C., 87
may, 429–438
Memo, technical, 186–189
Memory:
  long-term, 290
  short-term, 307–309
Middendorf, William H., 4
might, 429–438
Miller, George, 307
Mintzberg, Henry, 51–52
Misspelling, 453–457
Modal verbs, 429–440
Modified noun phrases, using *the*
    with, 374
Modifiers, 323
Monsees, Melford E., 7–8
moreover, 342
must, 429–438

Names, using noun compounds as,
    416–417
nevertheless, 343
"New" information, 289–290,
    304–305
  indicating in speech, 530–531
  putting "given" information be-
    fore, 304–305
Nonspecialist readers, 179, 218,
    293–297, 483
  using noun compounds with,
    293, 412
Noun compounds, 411–414, 491
  construction of, 412
  interpretation of, 413–414
  pronunciation of, 529
  use of, 416–418
"Noun-heaviness," 335
  avoiding, 338

Noun phrases:
  excessive use of full, 336–337
  "heavy," 306–309
  "light," 306–309
  need to repeat full, 337
  optimal ordering of, 304–309
Nouns:
  countable, 354–364, 381
  normally used with *the,* 368–369
  "two-way," 361–362
  uncountable, 353–354, 359–364,
    380
Number agreement between sub-
    ject and verb, 447–448

Object:
  direct, 304, 306–307, 311–312
  indirect, 304, 306–307, 311–312
  of preposition, 304, 306–307,
    310
"Objectivity," 329
Obligation, expressing degrees of,
    429–433
of course, 343
of phrases, using *the* with nouns
    followed by, 374
Omitting *that* or *which* in relative
    clauses, 402
on the other hand, 329, 343, 348
  vs. *on the contrary,* 348
only, 330–331
Operating instructions, writing, 336
Optimization, stylistic, 324
Oral presentations, 167–178
  audience analysis for, 47–55,
    164–166, 168, 170
  the closing summary in, 169, 173
  delivery of, 175–176
  determining your purpose, 168
  equipment for, 174
  formal, 168–176
  informal, 176–177
  the introduction in, 169,
    171–173

Oral presentations *(Cont.):*
  nervousness in, 175–176
  outlining in, 168, 171
  patterns of organization in, 168,
    171
  practice for, 173–175
  preparation for, 167–173
  questions in, 174–175
  reading from a manuscript in,
    175
  repeating important points in,
    174
  style in, 175
  supporting information in, 168,
    170–171
  transitions in, 174
  visual aids in, 169, 171
Ordinal adjectives:
  used with *a/an,* 385
  used with *the,* 368
Organization:
  of arguments, 67–73, 75–77
  in oral presentations, 168, 171
  in paragraphs, 23–33, 298
Outlines of arguments, 75–77
Outlining in oral presentations, 168,
    171

Paragraphs, 17–34
  adequate content, 17
  coherence, 17
  continuity between, 20
  definition, 17
  development, 23–33
  ordering: by cause/effect, 24–25
    by chronology, 23–24
    by comparison/contrast, 26–27
    by general to particular, 28
    by listing, 27, 298
    by other patterns, 28–29
  parallelism in, 40–44
  patterns of organization in,
    23–33, 298
  predictive nature of, 19

Paragraphs *(Cont.):*
  topic statements, 17–21, 298
  unity, 17
Parallelism, 35–44
  formatting, 36–39
  in lists, 36–39
  misleading, 40, 324
  in paragraphs and larger units,
    40–44
  used for interpreting unfamiliar
    words, 470–471
Paraphrases, using, 295–297
Parentheses, 485, 487–489
Parenthetical information, 487–488
Part-whole relationships, using for
    cohesion, 424–426
Participial relative clauses, 399, 446
particularly vs. *in particular,* 348
Passive-active alternation, 310
Passive sentences, 310, 431, 448
Past tense, 442
Personal data sheets, 257–274
Personal goals, determining your,
    258–259
Phonetic symbols, 479
Photographs, 136
Phrase punctuation, 490–493
Phrases, adverbial, using for em-
    phasis, 329–332
Pie diagrams, 135
Pike, Kenneth L., 75, 88
Pitch *(see* Intonation, sentence)
Place names, using *the* with, 369
plain vs. *plane,* 456
Platt, John, 282–283
Policy, arguments of, 69–73, 203
Popper, Karl, 283
Possession, using punctuation to
    indicate, 490
Preciseness, need for, 336–337
Prefixes, word, 465
  list of, 467–469
Preposition(s), 491
  object of, 304, 306–307, 310
Present perfect tense, 443

Present tense, 441–442
Presentational phrases, 420–421
Previous mention, using *the* to indicate, 372–373
principal vs. *principle,* 456
Probability, expressing degrees of, 433–436
Problem statements, 88–102
Procedural descriptions, 476
Professional credibility, 333–335, 341, 453
Progress reports, 211
Progressive *(-ing)* form of verbs, using the, 445–447
Pronouns, 335–337
    duplicate, avoiding, 393–394
    relative, 389, 391–402
    using for cohesion, 418–420
Pronunciation, 479, 513–535
    British English vs. American English, 513
    diagnosis of, 513–515
    of noun compounds, 529
    sentence intonation, 322, 397–398, 530–534
    of sound segments, 515–524
        consonant clusters, 522–524
        consonants, 518–521
        vowels, 515–517
    word stress, 525–529
Proof in arguments, 59–61, 67–80
Proofreading, 453–457
Proper nouns derived from common nouns, 369
Proposals, 213–235
    editing of, 225–226
    formal: abstract for, 216, 230–231
        appendixes for, 220
        background section for, 218
        budget for, 220–222, 234
        description of proposed research for, 218–219, 232–233

Proposals, formal *(Cont.):*
        description of relevant institutional resources for, 219
        introduction for, 216, 218, 231–232
        list of references for, 219
        personnel, 219–220, 234–235
        short, 215, 230–235
        title page for, 216–217
    informal, 215, 220–225
    letter of transmittal for, 247–250
    review of, 226–229
Punctuation, 485–494
    sentence, 485–489
    word and phrase, 490–493
Purpose, stating your, 292

Questions in oral presentations, 174–175
Quotation marks, 490
    using with titles, 493

Raney, Russel R., 3
rather, 345
rather than, 347
Readability, 287–340
Readability "formulas," 289, 322
Readers:
    characteristics of, 50–55
    designing communication for, 47
    expectations of, 59–61
    needs of, 47, 50–55
    nonspecialist, 179, 218, 293–297, 412, 483
    specialist, 180–185, 218–219, 275–276, 297, 332, 424–426
Reading, selective, 293
Reading strategies, 459–482
Recommendations, 193–195
Reduced relatives, 396–402
Redundancy, unnecessary, 323

Reference:
  nonunique, 353–358
  repeated (see Repeated
    reference)
  unique, 353, 367–381, 386–387
References, punctuating, 492–493
Related word forms, learning, 478
Relative clauses, 323, 389–405
  nested, 389
  participial, 399, 446
  placement of, 390–391
  reduction of, 396–402
  used for cohesion, 410–411
  using relative pronouns with,
    391–402
  using the with nouns followed by,
    374
Relative pronouns, 389, 391–402
Remrak coefficient, 296
Repeated reference, 336–337,
    407–428
  by associated nominals, 424–426
  full-form, 408–409
  short-form, 409–411
  by synonymous terms, 422–424
Repetition:
  for emphasis, 408–409
  excessive, 324
Reports (see Technical reports)
Request, letters of, 252–255
  responding to, 254–256
Résumés, 257–274
Rhetoric: Discovery and Change,
    75
Rhetorical patterns, using to
    interpret unfamiliar words,
    472
Rhythm, sentence, 397–398
Roots, word, 465–467
"Rules" of conversation, 321–322

Sayles, Leonard R., 52
Schiff, Peter M., 11

Selective reading, 293
Self-assessment, 258–259
Semantics of noun compounds,
    413–414
Semiblocked letter format, 244
Semicolon, the, 485–486, 489
Sentence combining, 322–328
Sentence editing, 303–319
  flow chart for, 313
  procedure for, 312–317
Sentence intonation, 322,
    397–398, 530–534
Sentence length, 322–323
Sentence punctuation, 485–489
Sentence subjects:
  putting topical information in,
    304–306
  using key words in, 293
Sentence type, signaling in speech,
    531–532
Sequencing errors, proofreading
    for, 453
Shared knowledge, using the when
    referring to, 377–378
Short-form relative clauses,
    396–402, 416
Short-form repetition, 409–411, 416
Short informal report:
  formatting, 188–189
  heading, 186–189
  (See also Technical report)
Short sentences, 322–323
should, 429–438
Signal words, 341–352
  using for emphasis, 329–332
since, 329, 342, 346
Souther, James W., 115
Spanish, 347
Specialist readers, 180–185,
    218–219, 275–276, 297,
    332, 424–426
specifically, 329, 344
Spelling errors, proofreading for,
    453–457

Standard terminology, 297
Statements, signaling in speech, 531–532
stationary vs. *stationery*, 456
Stative verbs, 446
Stems, word, 465–467
Stevenson, Dwight W., 87
Stress, word, 525–529
"Strong inference," 282
Style, 321–340
    in oral presentations, 175
Stylistic rules for relative pronouns, 394–402
Subheadings, using keywords in, 293
Subject of sentence (*see* Sentence subjects)
Subject-verb number agreement, 447–448
Subordinate elements, introductory, 487
Subordinating conjunctions, 341–342, 346–350
    using for emphasis, 329–332
Subordination, 325–326
Substantive errors, proofreading for, 453
Suffixes, lists of word, 462–463
Summaries, 179
    and abstracts, compared, 182
    construction of, 104
    designing for managers, 104–106, 115–116
    designing for particular audiences, 106, 110
    examples of, 105, 107, 194
    in oral presentations, 169, 173
    stressing information in, 104
    in technical reports, 104–110
Swales, John, 277–278
Syllable divisions, 492
Synonymous terms, using for cohesion, 422–424
Synonyms, 470

Tables as visual aids, 135–136
Technical communication:
    complaints about, 10
    importance of, 3–7
    inadequate training for, 7–9
    needed skills for, 9–10
    time required for, 5–6
Technical discussions, structuring of, 184–185
Technical memo, 186–189
Technical reports, 179–211
    abstract, 180–184
    details in, 180–185
    discussion, 184–185
    foreword, 87–104, 179, 194
    main points in, 179
    nonspecialist readers of, 179
    readers of, 47–57, 179
    specialist readers of, 180–185
    (*See also* Feasibility reports; Formal report; Long informal report; Short informal report)
Technical terms, avoiding unnecessary, 335
Tenses, verb, 441–447, 449–451
    present perfect, 443
    simple past, 442
    simple present, 441–442
the, 367–387, 418–419
their vs. *there* vs. *they're,* 456
therefore, 329, 342
this:
    vs. *it,* 419–420
    and *these,* 335–337, 379–381, 418–420
though, 346
thus, 342
Title page, 199–201
Titles:
    personal, 493
    punctuating, 492–493
Topic, defining the, 292
Topic sentences (*see* Topic statements)

Topic statements, 17–21, 293, 298, 325
"Topical" information, 305–306
Topical words, 322
  learning the meaning of, 476
Trade-offs, stylistic, 324
Transitional words, 341–352
Transitions in oral presentations, 174
Transmittal, letters of, 247–250
Truthfulness in visual aids, 146–148
Turk, Christopher, 334–335
"Two-way" nouns, 361–362

Unblocked letter format, 242–243
Uncountable nouns, 353–354, 359–364, 380
Underlining, 490, 492–493
Unique reference, 353, 367–381, 386–387
Unity in paragraphs, 17
Unmarked complements, verbs taking, 449
Unnecessary words, 333–339
Usage, word, 478
Users, field testing with, 299–300

Verb tense (see Tenses, verb)
Verbs, 429–451
  with irregular forms of inflection, list of, 449–451
  main, 441–451
  modal, 429–440
  not used in passive constructions, list of, 448
  not used in the progressive form, list of, 446
  omitting in relative clauses, 396
  of state, 446
  taking -ing complements, list of, 449
Visual aids, 119–163
  advantages of, 119

Visual aids (Cont.):
  bar charts/graphs, 132–134, 172
  conventions of visual perception, 129–130
  designing: for clarity, 142–146
    for most visual effect, 120–123
    for relevance, 137–142
    for truthfulness, 146–148
  effectiveness of, 120–123
  formatting, 155–162
  independence from the text, 149
  integration with the text, 148–149
  line drawings, 137–138
  line graphs, 132–133
  in oral presentations, 169, 171
  photographs, 136
  pie diagrams, 135
  selection of, 128–137
  strengths and weaknesses of types of, 130, 132–137
  tables, 135–136
  types of, 130–137
  using, 293–297
    to interpret unfamiliar words, 472–474
  when to use, 123–128
Vocabulary building, 459–482
Vowels, pronouncing, 515–517

weather vs. whether, 456
wh questions, signaling in speech, 531–532
which vs. that, 395
while, 346
Whiz deletion, 396
Whole-part relationships, using for cohesion, 424–426
who's vs. whose, 456
will, 429, 438–439
Word prefixes, 465
  list of, 467–469
Word processor, using a, 454
Word punctuation, 490–493

Word stems, list of, 466–467
Word stress, 525–529
Word usage, learning correct, 478
Wordiness, avoiding, 333–339
Words:
    analyzing, 462–469
    interpreting unfamiliar, 459–482

would, 429–438

Yes/no questions, signaling in
    speech, 531–532
Young, Richard E., 75, 87–88
your vs. *you're,* 456